Flood Risk Assessment and Management

Flood Risk Assessment and Management

How to Specify Hydrological Loads, Their Consequences and Uncertainties

Andreas H. Schumann
Editor

Ruhr-University Bochum, Germany

Springer

Editor
Prof. Dr. Andreas H. Schumann
Ruhr-University Bochum
Chair of Hydrology and Water
Management
Bochum, Germany
andreas.schumann@rub.de

ISBN 978-90-481-9916-7 e-ISBN 978-90-481-9917-4
DOI 10.1007/978-90-481-9917-4
Springer Dordrecht Heidelberg London New York

Cover illustration: bridge destroyed during the flood 2002 at the Mulde River in Germany. Photographer: Prof. A. Schumann.

Printed on acid-free paper

Springer is part of Springer Science+Business Media (www.springer.com)

Preface

This book was planned with two intentions. The first one was to close the gap between the holistic view on flood risk, which was established in the last decade by risk oriented planning, focused on socio-economic consequences of floods, and a flood hydrology, which was still based on a safety oriented approach inherited from structural flood protection in the past. In safety oriented planning it was sufficient to specify a single flood event which was assessed as the limit of flood safety. If this event was exceeded the system was at risk. This remaining risk was not considered any further. Nowadays it is widely accepted that the consideration of remaining risks is an essential component of Integrated Flood Management. Integrated Flood Management, as proposed e.g. by the WMO and the Global Water Partnership, demands risk management. Risk management calls for identification and assessment of risk. Risk has to be assessed and eliminated or at least minimised if it is unacceptable. In this process we are faced with many uncertainties, which are mainly hydrological uncertainties. These uncertainties have to be specified and considered with regard to multiple failure modes and the complex relationships between hydrologic loads and social vulnerabilities.

The second intention was the propagation of new instruments of flood risk management, which were developed within the framework of the National Research Program "Risk Management of Extreme Flood Events" (RIMAX), funded by the German Federal Ministry of Education and Research. With regard to this program it was not feasible to present all projects here as the number of these projects was large. Thirty eight projects were supported between the year 2005 and the year 2010. Instead of entire projects a selection of tools and ideas will be presented here which were developed and applied in some of these projects. These components are on the one hand essential for flood risk estimation and management and on the other hand at the cutting-edge in this field of research. This selection from RIMAX-projects has to be not comprehensive as another publication about RIMAX-result is under preparation by the RIMAX-project steering group. Several innovative solutions, which were provided by RIMAX-projects, were not integrated in this book. On the other hand some aspects of flood risk estimation, which the RIMAX-programme did not entail, but should be discussed in their relevance for flood risk estimations (e.g. regionalization of flood probabilities) were integrated here.

The resulting book is not a report about research projects or a collection of papers which were presented at a scientific conference. It is also definitely not a textbook; yet some general aspects are included in the introduction of the different chapters. The readership of this book is expected to consist of hydrologists and water managers which are interested in recent developments of hydrological tools and methodologies for flood risk estimation, assessment and management.

The editor wishes to thank not only the authors, with whom he co-operated for 2 years, but also the Chief-engineer of the Institute of Hydrology and Water Management, Dr. Markus Pahlow, and the secretaries Mrs. Smolka and Mrs. Mueller who supported him with the compilation of chapters.

Bochum, Germany Andreas Schumann

Contents

Contributors

Johannes Cullmann Federal Institute of Hydrology, Koblenz, Germany, cullmann@bafg.de

Jörg Dietrich Institute of Hydrology, Water Resources Management and Environmental Engineering, Ruhr-University Bochum, Bochum, Germany, joerg.dietrich@rub.de

Uwe Haberlandt Institute for Water Resources Management, Hydrology and Agricultural Hydraulic Engineering, Leibniz Universität Hannover, Hannover, Germany, haberlandt@iww.uni-hannover.de

Yeshewatesfa Hundecha GFZ German Research Centre for Geosciences, Potsdam, Germany, hundey@gfz-potsdam.de

Bastian Klein Institute of Hydrology, Water Resources Management and Environmental Engineering, Ruhr-University Bochum, Bochum, Germany, Klein@bafg.de

Heidi Kreibich Helmholtz Centre Potsdam, German Research Centre for Geosciences, Potsdam, Germany, kreib@gfz-potsdam.de

Ralf Merz Institute for Hydraulic and Water Resources Engineering, Vienna University of Technology, Vienna, Austria, merz@hydro.tuwien.ac.at

Bruno Merz Helmholtz Centre Potsdam, German Research Centre for Geosciences, Potsdam, Germany, bmerz@gfz-potsdam.de

Mark Musall Karlsruhe Institute of Technology, Karlsruhe, Germany, Musall@kit.edu

Franz Nestmann Karlsruhe Institute of Technology, Karlsruhe, Germany, Nestmann@iwg.uka.de

David Nijssen Institute of Hydrology, Water Resources Management and Environmental Engineering, Ruhr-University Bochum, Bochum, Germany, david.nijssen@rub.de

Peter Oberle Karlsruhe Institute of Technology, Karlsruhe, Germany, Peter.Oberle@kit.edu

Markus Pahlow Institute of Hydrology, Water Resources Management and Environmental Engineering, Ruhr-University Bochum, Bochum, Germany, markus.pahlow@rub.de

Gerd H. Schmitz Institute of Hydrology and Meteorology, University of Technology, Dresden, Germany, muich@rcs.urz.tu-dresden.de

Andreas H. Schumann Ruhr-University Bochum, Chair of Hydrology and Water Management, Bochum, Germany, andreas.schumann@rub.de

Dirk Schüttemeyer Meteorological Institute, University Bonn, Bonn, Germany, schuettemeyer@googlemail.com

Clemens Simmer Meteorological Institute, University Bonn, Bonn, Germany, csimmer@uni-bonn.de

Thomas Sommer Dresden Groundwater Research Centre, Dresden, Germany, tsommer@dgfz.de

Annegret Thieken alpS – Centre for Natural Hazards and Risk Management and University of Innsbruck, Innsbruck, Austria, Thieken@alps-gmbh.com

Yan Wang Institute of Hydrology, Water Resources Management and Environmental Engineering, Ruhr-University Bochum, Bochum, Germany, yan.wang@rub.de

Chapter 1
Introduction – Hydrological Aspects of Risk Management

Andreas H. Schumann

Abstract Flood risk is widely under discussion. It is understood as a result of interactions between mankind and nature. In this chapter the definitions of risk, risk estimation and risk assessment are discussed. The main emphasis is given to hydrological aspects of risk and challenges for flood hydrology. It is shown that the problem of hazard specification cannot be solved without consideration of the needs of risk management. It is shown that the flood probability alone is not sufficient if multiple failure modes have to be considered in planning. At the other side operational flood risk management depends strongly on hydrological forecasts. Based on this discussion the different chapters are summarized to give an impression about the overall content of this book and its relevance for flood risk estimation and management.

Contents

1.1 Determinants of Flood Risk

Flood risk management became an item on the political agenda of the European Union (EU) due to severe floods in Europe around the turn of the century. In the year 2007 the EU issued the European Flood Directive (European Commission, 2007).

A.H. Schumann (✉)
Ruhr-University Bochum, Chair of Hydrology and Water Management, Bochum, Germany
e-mail: andreas.schumann@rub.de

A.H. Schumann (ed.), *Flood Risk Assessment and Management*,
DOI 10.1007/978-90-481-9917-4_1,

It demands the assessment and management of flood risk with the aim to reduce adverse consequences of flooding. In the Directive flood risk is defined as "the combination of the probabilities of a flood event and of the potential adverse consequences for human health, the environment, cultural heritage and economic activity associated with a flood event". This combination of probabilities and consequences interprets risk as an expected value of consequences. However, there is a need to apply a more holistic view on flood risks. First of all, consequences are not unambiguously related with flood probabilities. Harm, losses, damages, perturbations or stress result from hazards only if vulnerability to theses impacts exists. Hence consequences are uncertain but not probabilistic. The vulnerability of a society depends strongly on socio-economic conditions. Socio-economic categories which are determining the extent of adverse consequences of flooding are considered in the concept of "vulnerability". There are many different definitions of the term "vulnerability". A very short one was given by the U.N.: "Degree of loss (from 0 to 100%) resulting from a potentially damaging phenomenon" (U.N. 1992). A more common definition was formulated by Kasperson et al. (2010) "the degree to which a system or unit (such as a human group or a place) is likely to experience harm due to exposure to perturbations or stresses". These authors discuss three dimensions of vulnerability: 1. the exposure to stresses or perturbations, 2. the sensitivity to the stress or perturbation including to anticipate and cope with the stress and 3. the resilience, the ability to recover from the stress, to buffer themselves against and to adapt to future stresses and perturbations. For a social system, e.g. a community, vulnerability can be defined by "A set of conditions and processes resulting from physical, social, economical and environmental factors, which increase the susceptibility of a community to the impact of hazards" (U.N. ISDR 2002). With regard to floods Merz et al. (2010) characterised the vulnerability *V* by a combination of exposure *E*, susceptibility *S* and response capacity *RC*. The exposure *E* is specified by social, economic, ecologic and cultural consequences that may be affected by floods. The susceptibility *S* is the degree to which the system is damaged by floods. The response capacity *RC* describes the ability to respond to and to recover from a flood.

1.2 Importance of Detailed Flood Characterisations in Risk Estimations

Exposure, susceptibility and response capacity are categories which depend on different flood characteristics in a nonlinear way. The loss susceptibility *S* will be extremely high if it is related to a probable maximum flood, but low for a "normal" flood, e.g. with a return period of 10 years. Also, the response capacity depends on the severity of the flood event that is being considered. There are attempts of more differentiated views on flood risk. Merz et al. (2009) argue that "Low probability/high damage" events are more important from the societal point of view than it is considered by the expected annual damage. They suggest a penalizing of events with disastrous consequences by integrating risk aversion into decision making. In this argumentation high damages are connected with floods which have a low probability. This one-to-one assignment is doubtful. Normally, extreme large floods will result in large damages. However, the severity of socio-economic consequences depends on multiple flood characteristics, which can not be specified by a single probability. To give an example for this thesis: damages in agriculturally used floodplains will differ with seasons. Another significant factor which influences consequences is the spatial extent of a flood event, which determines not only the degree of a disaster, but has also impacts on the response capacity. A flash flood is confined to the direct neighbourhood of a river and often to small watersheds only. Floods in alluvial plains of large river basins with widely extended inundated areas could paralyse national economies.

Usually flood events are specified only by the probabilities of their flood peaks. These probabilities are derived from statistical analyses of yearly discharge maxima or discharges above a threshold. Other flood characteristics are considered as conditioned on the peak. The statistical relationships between multiple flood characteristics can not be handled by the probability of the peak. If these probabilistic interdependencies between flood characteristics, e.g. between the shape of the hydrograph, its volume and its peak, are disregarded, the risk estimation for a flood event of a certain dimension will not be reliable. The estimation of these interrelationships is difficult as the database is often insufficient for statistical analyses of multivariate characteristics. Here we are confronted with a number of epistemic uncertainties. Uncertainties can be differentiated into three different types: (1) the epistemic uncertainty, which is the uncertainty attributable to incomplete knowledge about a phenomenon that affects our ability to characterise it, (2) the aleatoric uncertainty, inherent in a nondeterministic (stochastic, random) phenomenon and (3) the surprisal uncertainty which results from totally unexpected factors, e.g. from human behaviour (Table 1.1). The probabilistic methodologies which are widely applied in flood hydrology are an attempt to describe the second type, but the outcome will be affected strongly by the first type. As the epistemic uncertainty may be reduced with time if more data are collected and more research is completed, scientific activities should be focussed on it. Aleatoric uncertainty can not be reduced by further studies, as it expresses the inherent variability of a phenomenon. With regard to risk communication the problem exists that the aleatoric uncertainty has to be specified with a large impact of epistemic uncertainty. Even if the epistemic component could be reduced, uncertainties in general and epistemic uncertainties in particular will be an integral part of flood risk management. This becomes evident also in flood control. Uncertainties are one reason why river training, construction of dykes or flood storages and other technical measures with the aim to reduce the probability of flooding never ensure complete safety to floods (the other main reason is the aleatoric uncertainty and the need to limit flood control on a reasonable goal).

If epistemic uncertainties are neglected then the illusion of flood safety or of perfect flood control arises. The remaining risk endangers flood affected regions and their inhabitants if epistemic and aleatoric uncertainties are disregarded. Technical flood protection structures may increase this risk. This “levee effect” of flood protection structures was described by Tobin as follows: “the structure may generate a false sense of security to the extent that floodplain inhabitants perceive that all

Table 1.1 Types of uncertainties

Type of uncertainty	Epistemic uncertainty	Aleatoric uncertainty	Surprisal uncertainty
	Incomplete knowledge about a phenomenon that affects our ability to describe it, can result from ignorance, from scarce data and from unknown biases	Nondeterministic (stochastic, random) phenomenon, usually modelled by probability distributions	Covers matters which are unexpected Mostly arising from human factors
Ways to handle	Can be reduced with the acquisition of more information and by research	Can not be reduced by further studies, but can be quantified with standard probability theory	Minimize the chance of error and the unexpected occurrence Ensure that the system does not fail when the unexpected occurs
Type of knowledge	Ignored unknowns Known unknowns	Known unknowns	Unknown unknowns

flooding has been eliminated" (Tobin, 1995). Possible effects of this false sense of security are reduced preparedness and a significant economic development of flood protected areas. As a result the risk will be increased. Even if these effects are difficult to specify in terms of monetary values, they become evident in public discussions after floods which mirror the unfulfilled expectations of flood control by technical measures.

The adverse consequences of floods depend on the resistance of affected structures or systems. In safety engineering these interactions between load and resistance are considered. Here the probability of a failure is defined by superimposing the probability density functions (pdfs) of the load and of the bearing capacity of a structure. With regard to the different failure modes of flood protection systems it becomes necessary to specify the hazard in greater detail by its strength, intensity or extension. A detailed characterisation of typical failure modes is essential for a relevance ranking of multiple flood characteristics. To give some examples for this thesis: A dyke e.g. is endangered by overtopping but also by seepage. With regard to these two different failure modes the water level of the flood peak is not sufficient to specify the total risk of a failure. Other examples are: The function of flood storages depends on the volume but also on the shape of the hydrograph, or the stableness of a building during a flood depends on the hydrodynamic load and the buoyancy. More holistic hydrological analyses are obviously essential to characterize the hazard with regard to these different failure modes. An integration of such information in flood risk assessments provides options to identify the weakest points in flood protection systems and the impacts of possible failures. To give a practical example of this approach: The author was asked to analyse the flood risk in a river basin which was affected by land subsidence caused by mining activities. Some areas are drained by pumping stations. The estimation of flood risk was based on a detailed modelling of the flood processes and an analysis of possible damages within the inundated areas. The question was raised how the pumping stations should be considered within the risk estimation process. If the pumping stations would be turned off during a flood, the resulting inundations would be higher. The operating company precluded a failure of the artificial draining system, e.g. caused by an areal power blackout, completely. They argued that in such cases an emergency power supply would be established locally. Nevertheless, the flood simulations were done twice: with operating pumping stations and without pumping. The inundated areas of both scenarios were compared. Based on this comparison it was possible to identify the neuralgic points of the pumping system. The operating company improved the planning of the emergency power supply system based on the most critical locations within the basin. In this risk assessment, the duration of the floods and their volume were specified with probabilities which were derived from statistical analyses of rainfall events with different durations.

1.3 Hydrological Information for Flood Risk Management

Hydrological information plays the major role in flood risk management (Plate, 2009). Flood risk management can be subdivided into two different parts. In a narrow sense flood risk management describes the process of managing an existing flood risk situation (Plate, 2002). In a wider sense it includes the planning of activities which will reduce flood risk. This planning can be directed to an optimisation or improvement of existing systems with the aim of being prepared for a flood and to minimise its adverse impacts or to set up a new or revised system. A system has not necessarily to be a constructional system. The installation of an early warning system based on meteorological and hydrological forecasts, a system of flood insurances or a legal system specifying building codes could also reduce flood risk at different levels. In all cases the performance of such measures has to be assessed within the process of risk analysis. Risk analysis provides the base for decisions on maintaining, improvement or abandonment of existing flood control systems. It should include the estimation and evaluation of residual risks for the case that existing flood protection systems fail. In all cases where the residual risk has to be reduced or an increase of risk caused by changes seems to be unbearable, the planning of further measures has to be started again. In this planning process the causes and consequences of potential disasters including the remaining risks have to be estimated. These analyses have to be repeated for each of the planning alternatives to mitigate not only adverse consequences of flooding, but also the risk that planned systems have a lower performance than expected.

The relationship between hazard, consequences and planning decisions was specified by Plate (2002) with the following definition of risk *RI*:

$$RI(\boldsymbol{D}) = \int_{0}^{\infty} K\,(x|\boldsymbol{D}) f_x\,(x|\boldsymbol{D})\,dx \tag{1.1}$$

where $K(x|\boldsymbol{D})$ describes the consequence function, depending on x, the magnitude of the event causing the load (e.g. water level) and $\boldsymbol{D}$ is the vector of decisions that influence the consequences K of event x. The function $f_X(x|\boldsymbol{D})$ is the pdf of the occurrence of x. Flood control measures can modify this pdf. Other decisions have impact on the consequences, indirectly by reducing hydrological loads or directly by reducing their adverse consequences. Thus the function K is conditional of x and $\boldsymbol{D}$. In his definition Plate considers two types of decisions, decisions which influence the consequences of flooding and decisions which modify the pdf of floods. With regard to the argumentation given under point Section 1.2, the magnitude of the event x is also a vector, as different flood characteristics have to be considered. A single pdf of x is insufficient to describe the multivariate probabilistic structure of flood events, which in turn determines the consequences of flooding.

We have to differentiate between "risk analysis", which is an examination of the complex item "risk", its elements (hazard and vulnerability) and their relationsships and "risk assessment", which determines the importance, size, or value of risk. Risk assessments have to be based on risk analyses. Plate (2002) differentiates between three parts in flood risk assessments for the planning stage (Fig. 1.1):

- Flood risk analysis, which consists of hazard determination, vulnerability analysis and risk determination,
- Disaster mitigation, which can be achieved by technical and non-technical measures and
- Preparedness, which involves planning for disaster relief and early warning and evacuation.

As discussed above, hydrological input is essential for risk analyses. Hydrology has to characterize floods as part of the hazard determination and it is a precondition

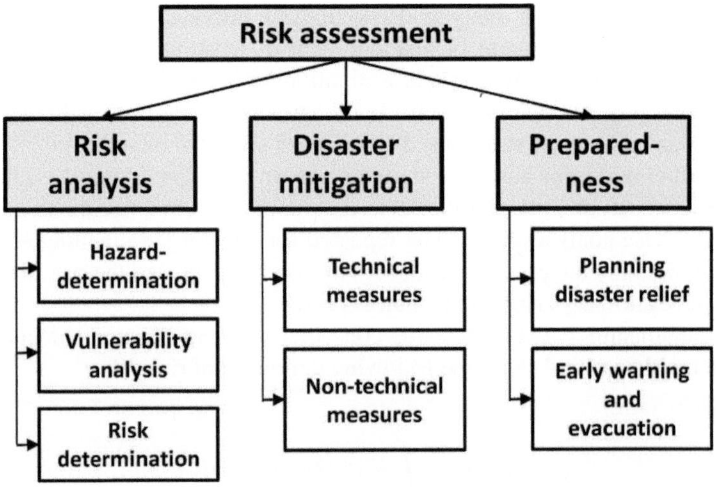

Fig. 1.1 Project planning as part of risk management from Plate (2002)

for assessments of vulnerabilities. Hydrological information is needed to specify measures of disaster mitigation by their efficiencies and remaining risks. Preparedness can be ensured only if the degree of a potential disaster is known, which depends on the hazard, and if the options for early warnings, which depend on hydrological and meteorological forecasts, can be assessed in a realistic way.

Hydrological information will be never complete. Flood risk changes in time (Merz et al., 2010). Many of these changes are highly uncertain and can not be assumed in advance. Any improvement of a flood protection system requires monitoring of hydrological and socio-economic changes and a reassessment of risks depending on the newest information. But also the options to provide the necessary information increase. New technologies and methods are under way. Some approaches which were developed in the last years or which are still in the stage of development are presented here.

1.4 The Content of This Book

In this book multiple facets of applications of flood hydrology in risk management are discussed. Special emphasis is given to uncertainties in operational flood management and in the planning of technical flood protection measures. The two stages of flood risk management, the operational stage and the planning stage, subdivides this book in two parts.

In the first part hydrological aspects of operational flood management are presented. Often early warnings and damage prevention (e.g. evacuation) depends on hydrological forecasts. These forecasts are associated with different levels of uncertainty. These uncertainties depend strongly on the forecasting lead time and the time of reaction of the hydrological system (watershed, river basin or river reach) of interest. For fast reacting systems a forcing of forecast models by observed data from different sources could be insufficient. In such cases the forecast horizon is limited by the time it takes the flood producing precipitation to reach the river profile of interest. With regard to operational flood protection measures this time span could be too short. The lead time can be extended by quantitative precipitation forecasts based on numerical weather prediction (NWP) models. High-resolution NWP can be coupled directly with flood forecasting systems. However, the quality of forecasted precipitation is often not sufficient for flood forecasts in fast-reacting basins where the precipitation fields can vary significantly with time and space. Precipitation forecasts in hilly regions are characterised by a wide range of uncertainties. For flood risk management the current uncertainties in NWP, as well as the handling of these uncertainties have to be taken into account. In Chapter 2 a brief introduction to the generation of weather forecasts with particular focus on the accuracy of rainfall prediction is given. Special emphasis is placed on meteorological ensemble forecasts which can be used to mirror uncertainties. Precipitation forecasts are provided for areas, but if flood forecasts are based on rainfall measurements a spatial estimation of rainfall is needed to provide the input into a hydrological model. In Chapter 3 conventional and geostatistical methods are presented for the spatial interpolation of the point measurements to raster cells and areas. It is focussed on the application of stationary and non-stationary geostatistical methods. The latter type of methods becomes increasingly important as additional information e.g. from weather radar can be considered for the estimation of mean areal rainfall. In this chapter also methods for conditional spatial simulation of precipitation are discussed. Those simulation approaches preserve the high spatial variability of rainfall and can be used for uncertainty assessments.

The benefits resulting from flood forecasts depend on their uncertainties. In Chapter 4 tools are presented which can be used to characterise these uncertainties and to reduce them in nowcasting by assimilation of observed data. It is focused on ensemble methods. Ensembles can be combined with data assimilation to produce "best guess" forecasts based on the Bayes' theorem. It starts with the well-known Ensemble Kalman Filter which is widely applied to update state variables of hydrological models. Other ensemble forecasts which are discussed here are based on meteorological ensembles forecasts and parameter ensembles. This forecasting part ends with an overview of currently available neural network designs with the purpose of a timely warning for operational flood risk management, which is given in Chapter 5. As neural network models are very effective with regard to their computational requirements they provide new options for operational scenario analysis and ensemble forecasts.

The second part of this book is dedicated to hydrological methods which are essential for the planning stage of flood risk management. There are two ways to specify the hydrological hazard: a deterministic one which is based on a transfer of precipitation into runoff by deterministic models and the probabilistic characterisation of flood risk with methods of mathematical statistics. The most common characterisation of a design flood is the flood peak and its probability. Despite many methodological developments since the pioneering work of the mathematician Gumbel, the general problem of flood statistics consists in their high epistemic uncertainty. Derived from a rather limited database, the resulting assessments about flood probabilities are highly variable in time. This is caused mainly by the stochastic character of flood inducing processes. This temporal variability of statistical results is very problematic for long-term planning. The occurrence of a single extreme event may modify the results of flood statistical analysis significantly. Another problem consists in the limited spatial information value. Flood regionalisation methods can be applied to reduce the uncertainties of flood estimations from local data in gauged catchments by pooling flood data within a region. These methods can be used also to provide flood statistical information for ungauged catchments, where no local streamflow data are available. Chapter 6 summarizes the most important methods and recent findings from the literature with a focus on the ungauged catchment case. Regionalisation methods, which are based on a transfer of flood information from hydrological similar catchments to catchments, where flood statistics have to be estimated, are discussed together with multiple regressions between flood statistics and catchment attributes and geostatistical methods, which use spatial proximity as a measure of hydrological similarity.

Precipitation series are essential for deterministic simulations of design floods. These series are often to short. The database for flood estimations can be improved by a coupling of stochastic rainfall generators with deterministic models. Chapter 7 is dedicated to stochastic rainfall synthesis focusing on methods for generation of short time step precipitation as required for flood studies. Here different characteristics of rainfall as stochastic process are discussed. Alternating renewal models, time series models, point process models are described as well as disaggregation and resampling approaches. The applicability of daily and hourly space-time precipitation models is demonstrated with two case studies.

Compared with observed data, generated flood series cover a wider range of possible flood situations. As described under point Section 1.2 such a broad characterisation of possible flood situations is essential to estimate the risk of failures. Also in the planning process of technical retention facilities it is necessary to use simultaneously multiple flood characteristics (peak, volume and shape). Nevertheless, a probabilistic specification of such events is needed. This can be done by multivariate statistical methods which are described in Chapter 8. The proposed Copula method can be applied to derive multivariate distributions from multivariate frequency analyses of correlated random variables. It has the main advantage that the marginal distributions of these random variables can be different.

After the characterisation of the hydrological hazard the consequences which would result from them has to be specified. For this purpose the next three chapters are dedicated to risk estimations and consideration of risks in flood management planning. In Chapter 9 the options of hydraulic modelling are presented. Two important aspects are discussed: the high spatial resolution of these models, which is essential to provide flood risk information at the local scale and the need for efficient algorithms which ensure short computational time requirements and an operational use of these models in nowcasting. Both aspects are contradictory. It is shown how the focus of modelling approaches can be shifted between flood risk planning and operational management to assess flood height, inundated areas and flood velocities in a problem-oriented way.

A special flood risk for urban areas results from sewer system and rising groundwater levels. Chapter 10 describes these problems of interactions between groundwater aquifers and sewer systems. A coupled modelling approach which considers these interdependencies is presented. It combines individual modules considering different model geometries, time synchronization and data exchange. The coupled model was applied for the City of Dresden (Germany). It describes the impact of floods on groundwater and can be used for a mapping of subsurface flood hazard in flood endangered urban regions. As described before, risk combines hazard and consequences.

Risks depend on hazard and susceptibility to hazards. Vulnerability has to be considered in risk management as an integral part of the analysis of consequences. Chapter 11 gives an overview about the assessment of direct economic losses as consequences of flooding. The basic concepts of damage assessments are introduced and the factors that influence flood damage are discussed. A damage model which can be used to estimate flood losses of private households is presented. It is based on extensive surveys of flood damages in Germany.

The book closes with Chapter 12 which presents a planning approach for technical flood retention facilities where the uncertainty of the hydrological hazard is considered. Copulas were applied for a multivariate statistical description of flood scenarios. With regard to the known unknowns the multivariate statistical characteristics of flood scenarios were handled by imprecise probabilities. These imprecise probabilities are specified by Fuzzy Numbers and integrated in a Multi Criteria Decision Making framework, which was developed for flood retention planning in a river basin.

In total this book is an attempt to integrate multiple facets of flood risk, where priority is given to hydrological methodologies, which were developed recently or which are still under development. It has the intention to show new options in such a way that it could be useful for practitioners and scientists.

References

European Commission (2007) Directive 2007/60/EC of the European Parliament and of the council of 23 October 2007 on the assessment and management of flood risks, available at: http://eurlex.europa.eu/LexUriServ/LexUriServ.do?uri=OJ:L:2007:288:0027:0034:EN:PDF. Last Accessed on May, 2010

Kasperson JX, Kasperson RE, Turner BL (2010) Vulnerability of coupled human-ecological systems. In: Rosa EA, Dieckmann A, Dietz T, Jaeger CC (eds) Human footprints on the global environment MIT Press, Cambridge

Merz B, Elmer F, Thieken AH (2009) Significance of "high probability/low damage" versus "low probability/high damage" flood events. Nat Hazards Earth Syst Sci 9:1033–1046

Merz B, Hall J, Disse M, Schumann A (2010) Fluvial flood risk management in a changing world. Nat Hazards Earth Syst Sci 10:509–527

Plate EJ (2002) Flood risk and flood management. J Hydrol 267:2–11

Plate EJ (2009) Classification of hydrological models for flood management. Hydrol Earth Syst Sci 13(10):1939–1951

Tobin GA (1995) The levee love affair: a stormy relationship. Water Resour Bull 31:359–367

United Nations, Department of Humanitarian Affairs (1992) Internationally agreed glossary of basic terms related to disaster management (DNA/93/36). United Nations, Geneva

United Nations, International Strategy for Disaster Reduction (ISDR) (2002) Living with risk: a global review of disaster reduction initiatives (preliminary version). UN ISDR, Geneva, July

Chapter 2
Uncertainties in Weather Forecast – Reasons and Handling

Dirk Schüttemeyer and Clemens Simmer

Abstract The generation of precipitation forecasts by means of numerical weather prediction (NWP) models is increasingly becoming an important input for hydrological models. Over the past decades the quality and spatial resolution of meteorological numerical models has been drastically improved, which makes it now possible to incorporate high-resolution NWP output directly into flood forecasting systems. The quality of forecasted precipitation, however, is still close to insufficient because rainfall constitutes merely the very end of a complex of interlinked process chains acting at a broad range of spatial and temporal scales. Consequently the precipitation fields can vary significantly with time and space and inherit wide ranges of uncertainties. For the purpose of flood risk management it is of particular interest to investigate both the potential and implications of the related variations and uncertainties. For this endeavour the general background and current uncertainties in NWP as well as the handling of the uncertainties has to be taken into account. This chapter gives a brief introduction into the generation of weather forecasts with a particular focus on the accuracy of rainfall prediction. It includes in this context the relatively new field of ensemble forecasting and discusses ways to link numerical NWP with radar-based precipitation nowcasting.

Contents

D. Schüttemeyer (✉)
Meteorological Institute, University of Bonn, Bonn, Germany
e-mail: schuettemeyer@googlemail.com

A.H. Schumann (ed.), *Flood Risk Assessment and Management*,
DOI 10.1007/978-90-481-9917-4_2,

2.1 Introduction

Flood risk management at various lead times is taking advantage from the improvement in quality and spatial resolution of meteorological numerical models used for quantitative precipitation forecasts (QPF) and numerical weather prediction (NWP) in general. The forecast quality necessary for a successful prediction of flood events to support civil protection actions, however, still poses a considerable challenge for operational weather forecast systems. The challenge results from the fact, that weather development and precipitation in particular results from the intrinsically nonlinear characteristics of fluid dynamics on a sphere coupled with complex thermodynamic processes e.g. phase changes of water, surface heat and water exchange at the surface, interaction with radiation from the sun and the atmosphere itself and so forth (compare e.g. Dirmeyer et al., 2009). The resulting precipitation fields – the by far most important and decisive input for hydrological models – constitute merely the very end of this complex of interlinked process chains acting at a broad range of spatial and temporal scales. Consequently the precipitation fields can vary significantly with time and space and inherit wide ranges of uncertainties.

The prediction of precipitation is closely connected to weather prediction in general. The quality of weather prediction especially in the medium range scale (several days) has been improving since the beginning of applying numerical methods roughly by 1 day per decade. This means, that the quality of the current 2-day forecast will roughly improve up to the quality of the current 1-day forecast within a decade. This statement also holds for the medium range forecast of precipitation (Bougeault, 2005). Considerable progress has also been achieved over the last decade in numerical weather prediction (NWP) on shorter and smaller scales e.g. by improved cloud microphysics modelling and the exploitation of the increasing available computing power for higher spatial model resolutions. The latter is also important for data assimilation strategies to further enhance the prediction capabilities of current NWP models. Especially for high resolution hydrological forecasts in fast responding catchments different combinations of Nowcasting and very short range NWP could be more suitable and are under development. The major aim is to produce reliable forecasts with high resolution in space and time. Most important for these applications are Precipitation Radar (PR) data, which are nowadays available from many national Radar networks[1] almost immediately after the measurement process with horizontal resolutions of kilometres and temporal resolution of minutes. The German Weather Service (DWD) for example operates one of the most advanced Radar networks in the world. The network is currently upgraded to polarisation diversity, which will allow – besides improving quantitative precipitation estimation (QPE) – to distinguish between different hydrometeors. Together with the anticipated improvements in the treatment of cloud and precipitation processes in weather forecast models, this will enlarge the

[1] The OPERA initiative of the national European weather services strives to unify the different formats and procedures in order to allow for European radar data composites.

potential of Radar information for data assimilation into forecast models and for improving precipitation nowcasting (usually up to a few hours) based on PR data alone.

The nonlinearity of the climate and weather system dictates, however, that prediction skill drops with increasing time starting from the initial condition. Thus, there remains an upper temporal bound to deterministic predictability of the atmosphere (Lorenz, 1963) currently believed to be between 10 and 15 days. Furthermore, the chaotic nature of weather and the non-Gaussianity of prediction errors on the short and very short scales – when convective events tend to dominate the system state – further limit deterministic predictability and require new concepts. It must be accepted that a complete description of the weather prediction problem should be formulated in terms of the time evolution of an appropriate probability density function (PDF) in the phase space of the atmosphere (see e.g. Molteni et al., 1996). In such a probabilistic framework the potential user of a weather forecast obtains information on the likelihood of a range of weather states and developments. Ensemble prediction methods are one (if not the only) method to obtain such information because it is computationally unfeasible to forecast the complete evolution of the probability density function in the future (Ehrendorfer, 1994a, b).

For ensemble forecasting a limited number of forecasts are generated by integrating a numerical model of the weather system forward in time starting with a limited set of distinct and plausible initial conditions (Leith, 1974). Figure 2.1 depicts such a probabilistic forecast system from the start until its final verification, where the different atmospheric states at the beginning (denoted by grey dots) are illustrated for an ensemble forecast with disturbed initial conditions. By means of uncertainty analysis during the forecast time the best members of the ensemble are

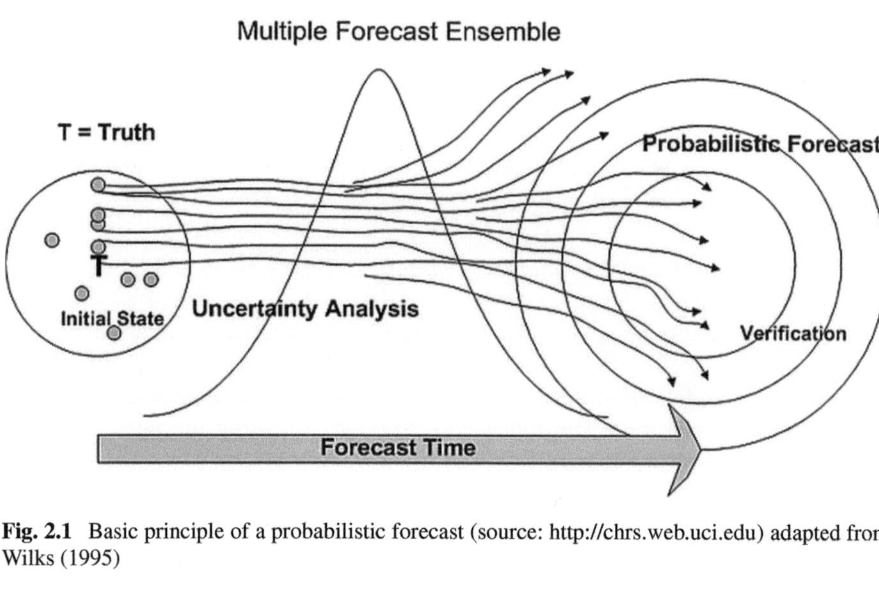

Fig. 2.1 Basic principle of a probabilistic forecast (source: http://chrs.web.uci.edu) adapted from Wilks (1995)

selected and thus enhance model system performance. The individual integrations can be done by the same model (*single model ensemble*), by different versions of the model in order to accommodate effects of model uncertainties, or by a set of different models (*multi model ensemble*). An example of this process is given by Stensrud et al. (2000), who tested different sets of ensemble conditions separately, with both perturbed initial conditions and different model physical parameterisations. These methods are already used by the leading weather prediction centres like the European Centre for Medium-range Weather Forecast (ECMWF), the National Centers for Environmental Prediction (NCEP), the Meteorological Service of Canada (MSC), and also by the German Weather Service (DWD) in the near future.

One of the most crucial problems in ensemble prediction is the construction of an ensemble, which takes due account of the numerous sources of uncertainty in the forecast. In addition, the allocation of the demanding computational resources is an important consideration for the implementation of operational ensemble forecasting systems.

2.2 Background and Current Uncertainties in Weather Forecast

During the last decades numerical weather prediction models have been developed, which explicitly model many of the important precipitation related processes, leading to an improved simulation of the nonlinear atmospheric evolution. In these numerical prediction systems, the atmospheric state at any time step is given by the predicted values of the model variables within a three-dimensional domain given the a priori information on the initial state of the weather system denoted as *analysis*. The latter is typically generated by an optimal merging of observations with a prior forecast (=background) through a statistical process. For every NWP model a number of coupled partial differential equations for the temporal development of the 3D-fields of the state variables (usually pressure, temperature, humidity, cloud water, precipitation water, and wind as a minimum) have to be integrated. These integrations have to be performed

- by taking into account the *boundary conditions* at the ground and the upper part of the atmosphere
- by *parameterizing the non-resolved processes* like turbulence, radiation, convection, cloud microphysics, exchange processes at the surface,...
- by *nesting* models of different resolution (thus creating additional boundary-value problems) – or – *variable grid-resolutions* (creating the need for scale-adaptable parameterisations).

Different examples for so-called meso-Gamma scale models are the 5th generation Penn State University/NCAR mesoscale model MM5 (Grell et al., 1994), its

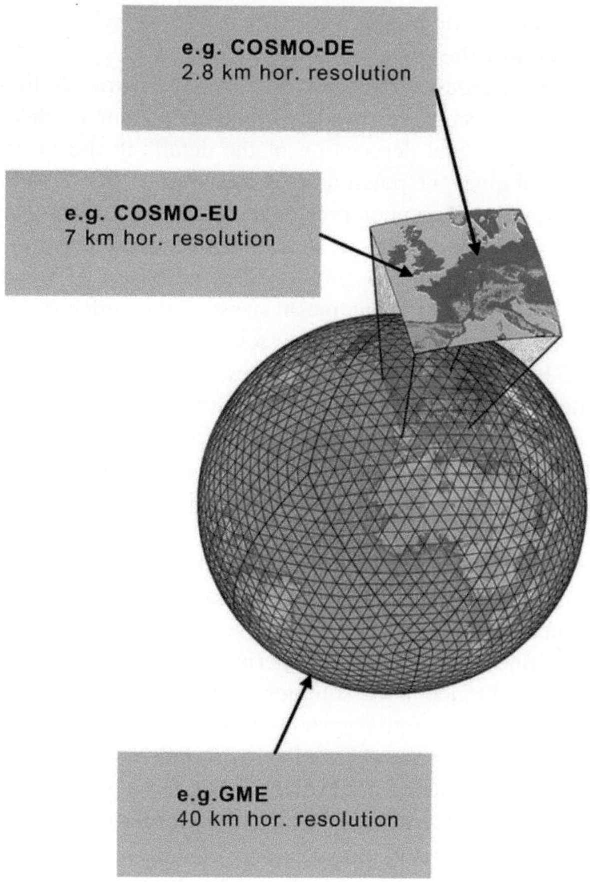

Fig. 2.2 Typical model setup for NWP using the COnsortium for Small-scale MOdeling (COSMO) model system as an example (source: DWD)

successor Weather Research & Forecasting (WRF), and the COnsortium for Small-scale MOdeling (COSMO) model. These models are non-hydrostatic limited area atmospheric weather prediction models, meaning that vertical velocity – and thus convection – is predicted at least partially directly, and that lateral boundary values for all predicted state variables need to be imported from a larger scale – usually global – model. Figure 2.2 gives an overview of a typical setup for such a numerical model system, utilizing the operational setup for the COSMO-model, centred over Europe and Germany. The operational version of COSMO currently runs with a horizontal resolution of $2.8 \times 2.8\ \text{km}^2$ ranging over 421 grid cells in longitude and 461 grid cells in latitude at its finest resolution. The atmosphere is vertically resolved into 50 terrain following layers. In general, the finest grid might reach from $1 \times 1\ \text{km}^2$ to $5 \times 5\ \text{km}^2$ grid size for the horizontal domain and 20–60 vertical layers

For the nesting procedure or interpolation of boundary conditions from a driving host model (with lateral as well as top boundary conditions) the EU domain is taken from the same model with 7 km resolution which is again nested in a global model (currently the so-called GME).

An increasing forecast domain size demands higher computational resources given the same resolved spatial scale. Finer model resolution usually leads to a better physical description of the details in the future state of the atmosphere; at least it gives the potential to do so.

Enhanced computer resources led in the past to a particular focus on the improved representation in the models of convectively driven precipitating systems. A large body of literature on numerical simulations of convective storms suggest – as a rule of thumb – that mesh sizes of the order of 1 km will suffice to simulate deep moist convection (e.g. review by Wilhelmson and Wicker, 2001). A study of Bryan et al. (2003) advocates, however, that in order to realistically represent deep convection, mesh sizes of the order of 100 m are necessary; they concede, however, that simulations on a 1 km grid are already able to reproduce the basic convective circulation itself, even if several aspects of it (e.g. precipitation amount, system phase speed) are modelled incorrectly. Most numerical studies on severe convection have been performed for the USA, so the applicability of the results in other regions, particularly mountainous areas, still remains to be shown. A first attempt to gain a deeper understanding of convective systems was made in the context of the COPS (Convective and Orographically-induced Precipitation Study, Wulfmeyer et al., 2008) experiment that took place in southwestern Germany and eastern France in the summer of 2007. During COPS the pre-convective conditions and deep convective systems were observed during most of their stages of development with ground-based and airborne instruments. Based on surface, in-situ, and remote sensing data, 4D data sets of key meteorological variables were acquired in order to understand convection initiation processes and hence improve mesoscale model forecasts of convective precipitation by e.g. advanced mesoscale data assimilation projects (Zus et al., 2008). Figure 2.3 shows the basic ideas of COPS, which might also explain the general background for improving model performance for NWP.

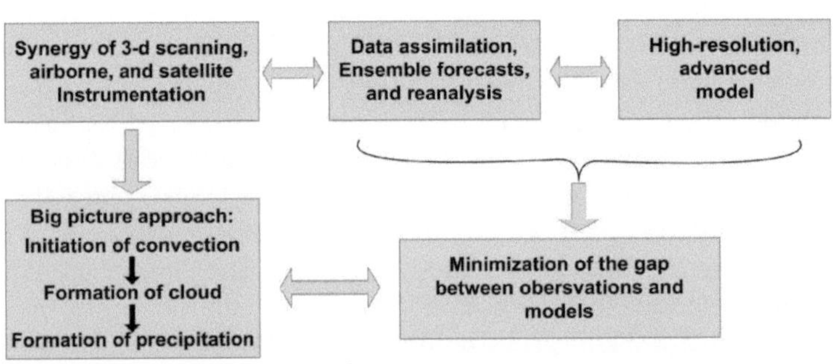

Fig. 2.3 Basic ideas of the Convective and Orographically-induced Precipitation Study (COPS) study that took place in summer 2007 for 3 months over the Black forest in Southern Germany (source: Wulfmeyer et al. (2008))

2.3 Data Assimilation Strategies

Besides model quality the generation of initial conditions for the higher resolved prediction systems becomes more demanding. As more weather satellite platforms and denser weather radar coverage become available, advanced cloud and precipitation assimilation algorithms are often advocated as the solution to the problem of mispredicted rainfall (Macpherson, 2001; Marécal and Mahfouf, 2002; Fillion and Mahfouf, 2003; Hou et al., 2001). The general idea of data assimilation is described in Fig. 2.4 where an optimal merging of observations with a prior forecast (=background) through a statistical process (described below) forms the basis to adjust the different model runs. For any data assimilation scheme the precipitation location and its structure as well as the correct movement are key topics to be addressed.

There are two general approaches to data assimilation: variational and non-variational. Variational assimilation methods effectively deal with the solution of linear estimation problems thereby incorporating the model dynamics (see e.g., Courtier, 1997) and aim to minimize a cost function that depends on the error covariance matrix of both the background state and the observations (Bouttier and Courtier, 1999). In this framework the so-called analysis state is the most likely one, given the observations and their statistical errors (Kalnay, 2003). This framework is, however, under-determined and an a priori estimate (background) of the model is needed.

Typical examples for variational implementations are Optimal Interpolation (OI), where local measurements are directly taking into account or three-dimensional variational (3DVAR) data assimilation, where a first guess, background error statistics, and observations are combined. With these three sources of information, cost-function minimization is performed in order to produce an "optimal" analysis.

In recent years four-dimensional variational assimilation (4DVAR) techniques became increasingly popular, also due to the fact that computational resources

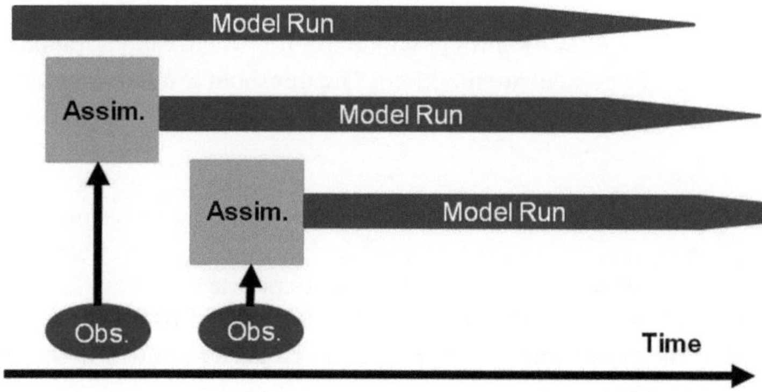

Fig. 2.4 Basic principle of data assimilation, where observations together with a prior forecast are utilized to adjust the model run (adapted from F. Ament, private communication)

became larger. 4DVAR is based on the same information as 3DVAR; but 4DVAR uses this information over a certain time period – the assimilation window – (e.g. 3, 6 or 12 h) in a model-consistent way. 4DVAR produces the initial condition, which leads to a free model run closest to the observation during the whole assimilation period. It has demonstrated its potential for mesoscale and storm-scale forecasting using Doppler radar winds and reflectivity data (Sun and Crook, 1997). In addition Zupanski et al. (2002) demonstrated for a Great Plains tornado outbreak that the NCEP 4DVAR system is well able to analyse precursor features for tornado activity, including wind shear, humidity, high and low level jet streaks, and convective available potential energy, although they also admit that model errors are a critical limitation. However, often even sophisticated and development-intensive data assimilation algorithms like 4DVAR fail to demonstrate the desired performance despite its outstanding suitability for remote sensing observations (for an example see Marécal and Mahfouf, 2002). Park and Droegemeier (1999, 2000) found the phase error to contribute more to the total error in the forecasts of developing convective systems than the amplitude error for domain integrated accumulated rainfall.

Another method for data assimilation is physical initialization (PI) which was introduced by Krishnamurti et al. (1991). The authors showed that PI entails the assimilation of observed rain rates in a numerical prediction model. Haase et al. (2000) and later Milan et al. (2008) extended the original PI to a scheme called PIB (Physical Initialization Bonn). To give an example for a typical application of such an assimilation scheme the PIB is explained in more detail and applied to one case to better perceive the impact of such a scheme.

The basic idea of PI and PIB is that the improvement of precipitation forecasts depends strongly on the coupling of humidity and wind fields in the atmosphere and it assumes that updrafts connected with horizontal humidity flux convergence in the lower part of the cloudy column lead to rain formation (compare also Wilson and Schreiber, 1986; Cotton and Anthes, 1989). The PIB connects the observation space directly with the model space in such a way that both the model and the observed precipitation at every grid point for every time step is taken into account. In case the difference between model precipitation and analyzed precipitation exceeds 20%, the profiles of vertical wind, specific water vapour, cloud water content and the cloud ice content are modified. The threshold is a rough guess of the uncertainty of the precipitation estimate. The modification consists of the following steps:

- For every grid point with analysed precipitation above 0.1 mm/h a single column cloud/precipitation model is utilised to modify the simulated cloud base and top heights, the vertical wind profile, and the humidity profile.
- At grid points with analysed precipitation below 0.1 mm/h PIB reduces the water vapour content, the cloud water content and the cloud ice content based on the information from the satellite data.
- In areas where precipitation data are not available, the model fields are not modified.

An advantage of PIB compared to other assimilation methods is the short assimilation window: Only a few time steps are necessary to achieve acceptable results.

The scheme is applied for one forecast by means of the COSMO-model (described above) during the COPS experiment, where the radar-based RADOLAN (radar online adjustment, Bartels et al. (2004)) data are used to estimate surface precipitation. The results are compared to another assimilation technique, the Latent Heat Nudging (LHN). This scheme is based on the work of Manobianco et al. (1994) and on the successive application from Jones and Macpherson (1997). The principal idea is to correct the model's latent heating at each time step by an amount calculated from the difference between observed and modelled precipitation. Practically this scheme nudges (Anthes, 1974; Hoke and Anthes, 1976) the model temperature profile to the estimated temperature profile (using a saturation adjustment technique).

For the chosen example the weather is characterised by an upper low pressure system near the western part of Brittany in the early hours of the day. Its cold front moves slowly eastwards while the convective activity weakens until noon. In the radar data (Fig. 2.5) an area with convective activity moves across southern and central Germany. During this event the DWD rain gauge network in Germany shows strong precipitation, e.g. the 6 h rain gauge accumulation was 48 mm at Cologne-Bonn Airport (in the middle of the model region). This case is interesting to see if and how PIB improves a poor model forecast. The operational COSMO prediction in the time range between 0 and 6 UTC indicates no or only little precipitation. At the end of the assimilation window, the PIB run reproduces the cold front, although with some overestimation of precipitation in the middle of western Germany.

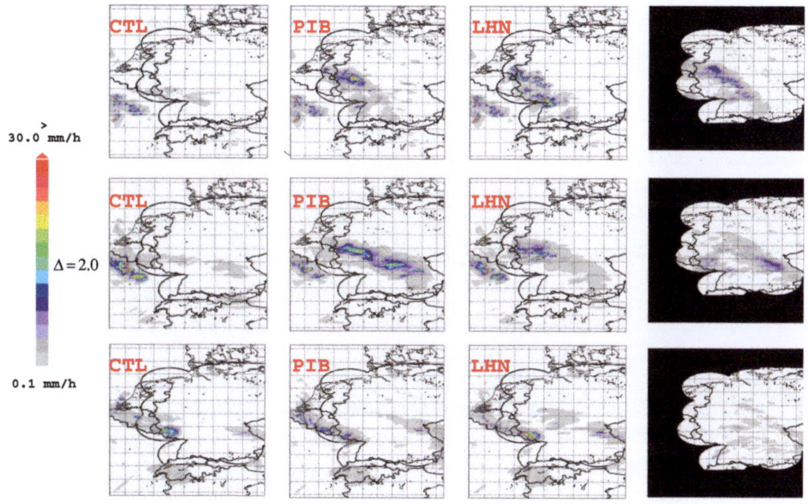

Fig. 2.5 Hourly accumulated precipitation. Results at the end of the assimilation window, *upper* row, after 3 h, *middle* row, and after 6 h of free forecast, *lowest* row

The LHN run creates convective cells instead of a contiguous field and also has the tendency to overestimate precipitation. After 3 h of forecast both LHN and PIB overestimate precipitation but only the PIB run keeps the frontal structure of the rain field with strong precipitation in eastern Germany. After another 3 h of forecast the cold front has passed the model area: The precipitation in the eastern part of Germany cannot be attributed to the assimilation.

The forecast is quantitatively evaluated by a fuzzy logic system based on the work of Ebert (2008) using the Equitable Threat Score (ETS) (see more detailed description below). The ETS is both spatially scaled and calculated for a range of thresholds. The basic idea is that for high resolution forecasts the verification should also take into account the area mean and not only a grid point value. In a consistent forecast the ETS score should decrease with a threshold increase and it should increase when enlarging the spatial scale. Figure 2.6 clearly shows that the PIB assimilation scheme is a suitable tool. It can also seen, however, that PIB still has a location problem, because the ETS values rise strongly when analyzing the different scales.

In contrast to variational assimilation methods explained above non-variational data assimilation systems, for example ensemble Kalman filter methods

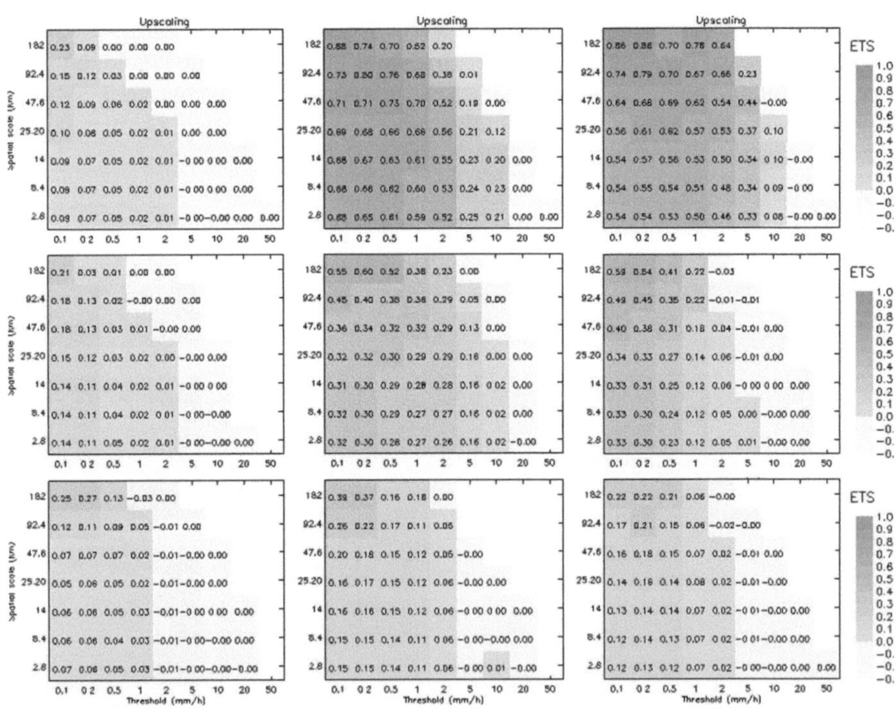

Fig. 2.6 Equitable Threat Score (ETS) for hourly accumulated precipitation. Results at the end of the assimilation window, *upper* row, after 3 h, *middle* row, and after 6 h of free forecast, *lowest* row

(Houtekamer and Mitchell, 1998; Anderson, 2001) implicitly calculate ensemble-based estimates of the flow-dependent forecast error as part of the solution and give a systematic way to calculate the time evolution of the forecast-error statistics according to the dynamics of the forecast model. One of the advantages of the ensemble Kalman filter approach is the fact that it is embarrassingly parallel (Evensen, 1994). This affords the opportunity to construct larger ensembles on an operational basis (compare Section 2.5). One important point to mention is that every assimilation scheme has to be implemented in a model-consistent way to not introduce larger disturbances (e.g. strong gravity waves) which would result in degradation of the weather forecast. In addition a model-consistent assimilation of precipitation and the corresponding fields of humidity can reduce the lack in the hydrological variables results in the so called spin up (spin-down) problem (Krishnamurti et al., 1993), i.e. a hydrological imbalance between precipitation and evaporation.

Although there is no doubt that data assimilation techniques have improved over the last decades there are certain limitations of the developed and applied techniques that lead to low performance gains of utilized NWP models of the atmosphere. Optimal Interpolation (OI) and all versions of the Kalman filter, variational algorithms rest on the assumption of Gaussian error distributions. But especially convective cloud processes with their highly nonlinear features and short life-times do not comply with this assumption. A theoretically sound treatment of non-Gaussian error characteristics provides Van Leeuwen (2001) in the context of an ensemble Kalman smoother by introducing high complexity and poorly known PDFs. On an ad hoc basis this problem has so far been addressed in practise at ECMWF by Andersson et al. (1998) for radiosonde data assimilation, in Simmons et al. (1999) for stratospheric humidity analysis, and in Andersson et al. (2000) for water vapour sensitive radiance observations. A means to ease the problem for water vapour has been devised by Dee and da Silva (2003), who introduced a pseudo-relative humidity, where the mixing ratio is scaled by the background saturation mixing ratio. Nevertheless, discontinuities cannot be removed. Discontinuities due to modelled water phase transitions not only violate the assumption of Gaussian error characteristics, but also render tangent linear approximations (TLA) and hence adjoint calculations questionable (Douady and Talagrand, 1990). Park and Droegemeier (1997) found the TLA highly sensitive to the amount of input modification; they found e.g. total TLA failure close to regime changes, with the largest errors near cloud tops. In a model study with artificial parameter perturbations, Park (1999) found predictability limits beyond 140 min. Fillion and Mahfouf (2003) re-examined diagnostic and prognostic cloud schemes, aiming at smoothing by suitable simplifications for a better applicability in the variational context. The Tiedtke (1989) operational convection scheme at ECMWF has been subject to a modification towards reduced complexity, which is less prone to non-linearities, and making assimilation of space based observation more feasible. The lack of differentiability (lack of smoothness) also leads to TLA errors due to the restricted vertical and horizontal extent of cloud formation (location problem) and finite lifetime (timing problem). It is partly true that these errors can be significantly reduced by well analysed synoptic scale fields with advanced data assimilation methods, especially on the meso-Gamma-scale, particularly in regard to the vertical extent of convective systems. However, even sophisticated data assimilation algorithms cannot exclude severe misplacement errors in the timing and horizontal location of convective systems.

Difficulties in using meso-scale forecasted fields often stem from conditions, where convection is properly developed in the model, but improperly positioned. This may be due to erroneous propagation velocities, phase errors, or wrong positioning of the embedding meso-scale structure. Attempts to address the problem using techniques originally developed for image processing have been made by several researchers. Hoffman et al. (1995) defined and estimated distortion errors of forecasted fields, which were extended to SSM/I precipitable water fields by Hoffman and Grassoti (1996). Mehrkorn et al. (2003) further developed this method to decompose analysis and forecast errors into phase error, amplitude error, and residual error, calling their procedure Feature Calibration and Alignment (FCA). This method was applied only to large-scale two-dimensional meteorological fields. In addition to quantification and separation of the forecast errors in general, such methods also hold promise for correcting phase errors, and may be applicable to high-resolution fields. Alexander et al. (1998) employed a "digital warping" method, which was adopted from animated motion picture production.

In addition with ever-increasing resolution of NWP models, the nonlinearities, non-Gaussian PDFs, flow-dependent and poorly-known balance within the systems are likely to become so strong that the mentioned approaches will not provide enhanced weather forecast. Therefore there is clear need to further develop ensemble-based assimilation strategies.

2.4 Reasons for Uncertainties

In general, the above described models and related data assimilation systems – as usually employed in meso-scale weather prediction – will be influenced by three main sources of uncertainty:

- Initial conditions: uncertainty due to structures not seen by the observing system or estimated by the data assimilation scheme, due e.g. to limited resolution or other instrument limitations
- Physical parameterisations: uncertainty resulting from the model formulation of convection, cloud microphysical, planetary boundary layer, or other processes
- Boundary conditions: uncertainty in the synoptic and meso-scale environment that influences the limited area model through the boundary conditions obtained from a global model

The uncertainty in initial conditions is of particular interest due to the nonlinearity of the system and the fact that small errors in the initial conditions can grow very fast and make all forecasts to diverge from the truth after a certain time. The initial state of the atmosphere is by observations alone *largely underdetermined* because~10^7 regularly distributed model variables are constrained usually by only~10^5 irregularly distributed observations.

The physical parameterizations for the different model formulations of convection, cloud microphysical, planetary boundary layer, land surface or other processes, form the second large source of uncertainty, since the related processes are characterized by extremely complex inner patterns, structures and processes that act at different time and space scales. As a consequence the state variables are strongly heterogeneous and variable in space. In addition the mentioned processes interact on the different scales in time and space as described in Fig. 2.7. This makes the correct description of the system and the related forecast extremely challenging. In this context it is crucial to note that with new developed parameterisations for the described processes any deviation from the former (operational) setup of a NWP model needs to be validated first in order to optimally adapt the other "screws" in the model.

The typical model setup for NWP by *nesting* of models of different resolution, as described above, forms the third source of uncertainty (compare also Fig. 2.7). The needed information is available on a coarser model domain and the spatial resolution of the different fields from these domains is often insufficient to capture all scales relevant for the smaller domain. To address this problem, downscaling algorithms were and still have to be developed, that produce 3-dimensional fields with a higher

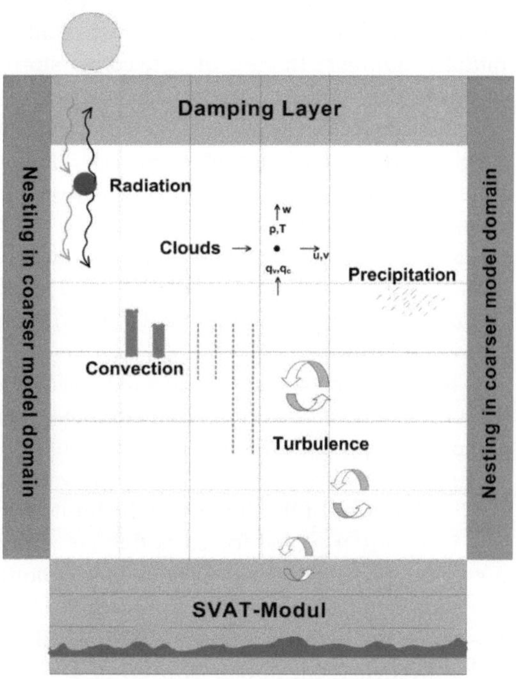

Fig. 2.7 Sketch of the related processes of convection, cloud microphysical, planetary boundary layer, land surface that are characterized by extremely complex inner patterns, structures and processes that act at different time and space scales (adapted from F. Ament, private communication)

resolution from such available coarse domains. The produced 3-dimensional fields should add subscale variability to the coarse field of the right amount and with the right spatial correlations.

2.5 Handling of Uncertainties

As stated in the introduction, only ensemble forecasts seem to be able to solve the general predictability problem. The main question is how to produce a probability density function (PDF) from the ensemble forecast with known deficiencies of the ensemble, by mapping the uncertainties in initial and boundary conditions as well as uncertainty in physical parameterizations. Decisive for the quality of an ensemble forecast is thus the extent to which the discussed uncertainties are embedded in the ensemble generation. In effect the ensemble should be created in such a way that the ensemble results can be transformed into a PDF of the future atmospheric state. A number of methods have been proposed to generate ensembles that reflect the described sources of uncertainty. In this context it is important to take into account the different time scales for NWP.

In medium range ensemble prediction, the most important source of uncertainty is probably the *initial condition,* and a number of methods have been developed to generate perturbations that reflect this uncertainty. These include lagged and scaled lagged average forecasting (Hoffman and Kalnay, 1983; Ebisuzaki and Kalnay, 1991, respectively), the breeding method (Toth and Kalnay, 1997), and a singular vector (SV) approach (Molteni et al., 1996, Buizza, 1997). The latter methods in particular attempt to identify the most rapidly growing linear perturbations of the initial conditions. In case of convective storms, methods based on linear growth of errors may be less appropriate, however. The timescale for growth – about an hour for convective systems – is short compared to the forecast length, which may range from several hours up to several days, and the errors may have long-since saturated. This is more analogous to seasonal prediction than the classical problem of medium-range weather forecasting. In this case, a more random distribution of initial conditions, or a stochastic forcing term, may be required to explore the state variable space, with predictability coming via the surface and lateral boundary conditions, rather than preferred rapidly growing structures during the initial part of the forecast.

The most obvious way to account for *boundary condition* uncertainty is to use a set of boundary conditions generated by a global ensemble forecasting system. A brute-force approach that uses every member of the global ensemble is likely to be inefficient, however, since much of the variability in the global ensemble may be confined outside of the domain of the limited-area model. One solution to this problem is to cluster global forecasts that are similar in the target region, and use only a single representative set of boundary conditions for each cluster. This approach has been implemented in the Consortium for Small-scale MOdelling Limited-area Ensemble Prediction System (COSMO-LEPS) developed in Italy using the former Lokal Modell (LM, now COSMO) of DWD (Molteni et al., 2001, Tibaldi et al.,

2003). It was found in this system that most of the variability in the 51 member ECMWF EPS for a region centred on the Alps can be retained by as few as five members.

Various studies concerning the effects of *model errors* have also been published. Further improvements for the ensemble approach can be expected by additionally applying error regression techniques on the ensemble members, which was demonstrated with great success by Krishnamurti et al. (1999). Shin and Krishnamurti (2003) tested super-ensemble approaches based on three different composition methods: multi-cumulus scheme, multi-model scheme, and a multi-analysis scheme, which achieved a QPF improvement of about 20%, with the multi-model configuration being the most effective. Verlinde and Cotton (1993) investigated the possibility and computational costs of parameter optimisation with intermittent observations, which are formulated as functions of prognostic variables, as given by reflectivity and liquid water path. Hou et al. (2001) investigated the ensemble approach in the framework of the Storm and Meso-scale Ensemble Experiment (SAMEX) with resolutions down to 3 km. Apart from the beneficial impact of multi-model ensembles, the authors found improved results with perturbations to model physics parameterisations. A detailed investigation of the relative contributions of initial value versus model physics perturbations for forecast skill is given in Stensrud et al. (2000), demonstrating that both act primarily on different time scales: model physics variance acts much faster during the first 12 h than initial conditions. A promising approach for parameterized convection is presented by Grell and Dévényi (2002). Ensembles are generated by a suite of different parameterisation schemes, for which domain-averaged precipitation rates show significant improvements with respect to single ensemble members. An effect is exploited here which makes use of better statistics, first demonstrated by Leith (1974). Grell and Dévényi (2002) showed the beneficial impact of Bayesian PDF construction with assimilated observations, to reduce diagnostically the precipitation bias.

Relatively new is the approach of using stochastic parameterisations; the uncertainty of defining parameterisations, which are usually on very restricted measurements data on complex or sub-scale physical processes, is exploited by a random selection of parameterisation constants. Different selections, e.g. for spatial or temporal domains, will lead to different results when running the model in configurations with different selection criteria. Preliminary tests on the meso-scale by Theis (2005)[2] exploiting the current rather simple ECMWF approach, have, however, not lead to sufficient ensemble spread. Bright and Mullen (2002), however, found that introducing a simple stochastic cumulus parameterisation in a mesoscale model significantly increased the spread of the ensemble, but did not affect the bias. Craig et al. (2005) describe a stochastic convection scheme based on a physical model of small-scale convective behaviour, i.e. a canonical ensemble of convective plumes,

[2] PdD thesis by Susanne Theis, Meteorological Institute of University Bonn. Stochastic parameterisations were introduced to the LM in a similar way, as it is tested currently at ECMWF. The results on the prediction of temperature and precipitation were evaluated against measurements.

that in single-column tests is able to reproduce the convective variability found in a cloud-resolving simulation.

Ongoing research is dealing with ensemble-based data assimilation systems for the convective scale (1–3 km model mesh-size; forecast time up to 24 h). For these applications the above mentioned uncertainties as well as increasing nonlinearities due to even higher resolutions have to be taken into account. Until now two different approaches with different strength and weaknesses appear to be suitable to enhance forecast capabilities for the convective scale:

- Different methods that make use of the Kalman filter
- Different methods that belong to the group of particle filters, where the model PDF is represented by a discrete set of model states

It is far beyond the scope of this document to give a detailed overview of the related techniques. Instead the reader is referred to the different publications for the given examples. For the first group of methods that rely on the Kalman filter the Local Ensemble Kalman Filter (LEKF, see Ott et al., 2004) and the Local Ensemble Transform Kalman Filter (LETKF, see Hunt et al., 2007) look promising for applications on the convective scale. The most important features of both schemes are that (i) they assimilate all observations that may affect the analysis at a given grid point simultaneously and (ii) they obtain the analysis independently for each model grid point (Szunyogh et al., 2008). The LETKF also introduces changes that improve the computational efficiency of the algorithm and adds flexibilities that are beneficial when non-local observations, such as satellite radiances, are assimilated. When only local observations are assimilated the LETKF and the LEKF schemes produce identical results (Szunyogh et al., 2008).

In the LETKF (Hunt et al., 2007), an ensemble of forecasts is used to represent a situation-dependent background error covariance. This information is deployed in three steps taking into account the observations and their errors:

1. The analysed state is computed from the ensemble mean forecast.
2. The analysis covariance is calculated.
3. The analysis perturbations are determined as a suitable linear combination of the background perturbations (i.e. the deviations of the individual ensemble forecasts from the ensemble mean forecast).

Localisation by using only observations in the vicinity of a certain grid point, requires the ensemble to represent uncertainty in only a rather lower-dimensional local unstable sub-space. The method provides an algorithm how to make smooth transitions between different linear combinations of the ensemble members in different regions. This should allow keeping both the ensemble size and the imbalances reasonably small (source: http://cosmo-model.cscs.ch/content/tasks/priorityProjects/kenda/default.htm: last visited 25.01.2010).

The group of particle filters, where the model PDF is represented by a discrete set of model states are seen as an alternative and very elegant way to be applied on the convective scale. A typical particle filter consists of the following steps (see e.g. van Leeuwen, 2009 and references therein for a detailed description):

1. Take an ensemble together with a prior probability density function (PDF)
2. Compute the "distance" of each member to the available observations. Direct and indirect observations (e.g. radar, satellite) with complex observation operators can be used, as no direct inversion of these operators is needed.
3. Based on the distances, Bayesian statistics is used to calculate the posterior PDF, reflecting the weights the individual members should get in the next steps.
4. Use the posterior PDF to re-sample the members. Members with small weights are discarded and the importance of members with large weight will be increased. In the simplest form, several copies of them will be generated.
5. Cycle to the next analysis time, typically 3 h later, and repeat steps 1–3.

There are two potential areas of concern. One is the ensemble size needed to realistically sample the PDF. They might become too high to become operationally feasible. The other concern is filter divergence. In the pure form, observations never influence the model states – they only weigh the member importance; thus all members might drift away from reality.

The method offers a number of advantages over traditional assimilation schemes. These include the possibility to handle highly nonlinear processes and to deal with non-Gaussian probability density functions. Also, there is no need to know and formulate balance conditions between the analysis increments for the mass-, wind-, and humidity fields, which is still an unsolved problem for the convective scale. The related technique has, however, some problems with ensemble collapse. Thus special care must be taken in spreading by varying the duplicated ensemble members. Van Leeuwen (2009) lists a number of resampling methods to eventually tackle the problem of filter divergence and also gives first ideas to relax the boundary between particle filtering and other solution methods (e.g. the ensemble Kalman filter) for data assimilation problems

For all the mentioned methods it is crucial to note that short range Numerical Weather Prediction (NWP) are typically limited by the time the model needs to establish its hydrological cycle, the so called *spin-up time,* as already mentioned above. This limitation can be tackled by Precipitation Radar (PR) data for improving precipitation nowcasting (usually up to a few hours) based on PR data alone, by analysing the motion and intensity trends in precipitation fields in the past hours and project these changes into the future (e.g. KONRAD in Germany (Lang et al., 2003), GANDOLF in the United Kingdom (Pierce et al., 2000), and TITAN (Dixon and Wiener, 1993) and TREC (Rinehart and Garvey, 1978) in Colorado. Nowadays the needed data are available from many national Radar networks almost immediately after the measurement process with horizontal resolutions of kilometres and temporal resolution of minutes, which means that these techniques can be applied for larger regions.

Currently enhanced nowcasting algorithms by means of an ensemble of Radar-based forecasts are also under development. Therefore an ensemble of predictions is generated taking the free parameters concerning initial track direction and intensity tendency into account. In such a probabilistic framework the potential user of a weather forecast obtains information on the likelihood of a range of weather states and developments. Such an ensemble can be very useful for predicting floods in flash flood prone catchments of low mountain ranges, i.e. in relatively fast responding catchments without utilizing any NWP model. Furthermore the velocity of the precipitation field can be analyzed in a statistical manner to gain deeper knowledge about typical velocities and intensity trends. The Radar-based nowcasting techniques can also form the basis for enhanced data assimilation procedures by utilizing theses extrapolation techniques to produce an extended ensemble forecast as was experienced by Krishnamurti et al. (1997) by utilizing the PI method for hurricane ensemble forecasts. In such a way a very robust system is set up that can be reliably applied to any convective or other case. A major advantage of such a system would be the short length of the assimilation window: Only a few time steps are necessary to achieve acceptable results and therewith it does not rely on large computing resources. In addition such a combined approach could be very useful to overcome the above mentioned problem of ensemble collapse due to particle divergence by utilizing information from the radar-based forecast together with PIB for precipitating or at least cloudy situations in which the particles are confronted with the observations before the actual measurement time to guide them in the right direction (guided SIR; van Leeuwen (2009)).

2.6 Verification and Applications

One of the central tasks related to the above described weather prediction model systems is, to quantify the quality of the produced forecasts, and to demonstrate that such forecasts do indeed enable better decisions compared to those made using alternate information like e.g. deterministic forecasts combined with model-output-statistics (MOS systems) using a collection of long-term observations of similar weather situations. The verification procedure should also guide forecasters and potential users on which products to trust most and use.

The most intuitive way to judge forecast performance is to use human judgment by looking at time series or maps of the latest forecast. This is a difficult or even impossible task in case of utilizing larger ensembles. To judge in detail how good or bad a deterministic forecasts or a probabilistic forecast is, different verification tools already exist or are in development. In general, verification in the meteorological sense is the comparison of the forecast to the true atmospheric state or in a more practical sense, the estimation and analysis of the common probability density function between forecast and observation (Murphy and Winkler, 1987; Murphy, 1991). A general overview of verification methods can for example be found in Stanski et al. (1989) or more recently by Toth et al. (2003) in Jolliffe and Stephenson (2003).

Even for a deterministic forecast it is not easy to prove the real value using traditional verification statistics. For any model that produces a high-resolution forecast it is very difficult to match precisely observed small-scale features. The case of a feature being forecast but offset in space and/or time from the observations, resulting in a "double penalty" of observed-but-not-forecast and forecast-but not- observed, is more prevalent at high resolution than at low resolution Ebert (2008). Even when the forecast captures the large-scale weather well, the small-scale errors easily dominate the total error using simple methods (e.g. Mass et al., 2002).

Ebert (2008) describe "Fuzzy" verification techniques which require that the forecasts are in *approximate* agreement with the observations, meaning that forecasts are close in space, time, intensity, and/or other aspects. This is especially important for the forecast of precipitation due to the fact that at some scales the features may be unpredictable, for example, intermittent convective rainfall (Ebert, 2008). The techniques described by Ebert (2008) provide guidance on the spatial or temporal scales of all crucial model output for deterministic NWP.

For ensemble forecasts current research develops enhanced verification techniques focussing on the probabilities or uncertainties of the ensemble forecast. A fundamental way to analyze an ensemble forecast and to answer the question how well the ensemble spread of the forecast represents the uncertainty of the observations is to calculate the ensemble spread and the related forecast errors (e.g. the Rank histogram, see Talagrand et al., 1997). Other methods like the Relative operating characteristics (ROC) to discriminate between forecasted events and non-forecasted events became already standard verification tools. These techniques do also pinpoint sources of error in the generation of ensemble forecast systems.

2.7 Outlook

For flood risk management, forecasts from the above described weather prediction model systems should provide constructive information for optimal management and planning in the case of flooding. To tackle the meteorological and hydrological model uncertainties, rigorous meteorological/hydrological/hydraulic modeling together with computationally highly efficient algorithms has to be performed. One crucial aspect to note is the sample size for these rare events. There may be no statistical significance reached by means of traditional verification and therewith even the application of ensemble weather forecasts might be only the necessary prerequisite.

There is also a growing interest by the NWP community in predicting discharge with their models. Grasselt et al. (2008) extended the Land Surface Parameterization TERRA_ML of the operational NWP model system COSMO of DWD by a routing scheme and tested the impact on soil moisture and discharge evolution. Such an approach could lead to operational flood forecast. In this context the limited sample size of most NWP ensemble systems, where the ensemble size comprises usually less than 50 members, is an open issue for flood forecasting.

Acknowledgements We acknowledge large contributions to this text from the research proposal of the DAQUA PI-team (G. Craig, H. Elbern, D. Leuenberger, C. Simmer, W. Wergen) in the framework of the Priority Programme 1167 of the German Science Foundation (DFG).

References

Alexander GD, Weinman JA, Schols JL (1998) The use of digital warping of microwave integrated water vapor imagery to improve forecasts of marine extratropical cyclones. Mon Weather Rev 126:1469–1496

Anderson JL (2001) An ensemble adjustment filter for data assimilation. Mon Weather Rev 129:2884–2903

Andersson E, Fisher M, Munro R, McNally A (2000) Diagnosis of background errors for radiances and other observable quantities in a variational data assimilation scheme, and the explanation of a case of poor convergence. Q J R Meteorol Soc 126:1455–1472

Andersson E, Haseler J, Undén P, Courtier P, Kelly G, Vasiljevic D, Brankovic C, Cardinali C, Gaffard C, Hollingsworth A, Jakob C, Janssen P, Klinker E, Lanzinger A, Miller M, Rabier F, Simmons A, Strauss B, Thépaut J-N, Viterbo P (1998) The ECMWF implementation of three dimensional variational assimilation (3D-Var). Part III: experimental results. Q J R Meteorol Soc 124:1831–1860

Anthes RA (1974) Data assimilation and initialization of hurricane prediction models. J Atmospheric Sci 31:702–719

Bartels H, Weigl E, Reich T, Lang P, Wagner A, Kohler O, Gerlach N (2004) Routineverfahren zur Online-Aneichung der Radarniederschlagsdaten mit Hilfe von automatischen Bodenniederschlagsstationen. – Abschlussbericht, Selbstverlag, Deutscher Wetterdienst Offenbach

Bougeault P (2005) The ECMWF data assimilation and forecasting system. Presentation at the celebration "100 Jahre Meteorologisches Observatorium Lindenberg", 12–16 Oct, Lindenberg, Germany

Bouttier F, Courtier P (1999) Data assimilation concepts and methods. Meteorological Training Course Lecture Series. ECMWF, Reading

Bright DR, Mullen SL (2002) Short-range ensemble forecasts of precipitation during the southwest monsoon. Weather Forecasting 17:1080–1100

Bryan GH, Wyngaard JC, Fritsch JM (2003) Resolution requirements for the simulation of deep moist convection. Mon Weather Rev 131:2394–2416

Buizza R (1997) Potential forecast skill of ensemble prediction and spread and skill distributions of the ECMWF ensemble prediction system. Mon Weather Rev 125:99–119

Cotton WR, Anthes RA (1989) Storm and cloud dynamics. Academic, San Diego, CA, 863 pp

Courtier (1997) Dual formulation of four-dimensional variational assimilation. Q J R Meteorol Soc 123:2449–2462

Craig GC, Cohen BG, Plant RS (2005) Statistical mechanics and stochastic convective parameterisation. Workshop on Representation of sub-grid processes using stochastic dynamic models, ECMWF, Reading, England

Dee DP, da Silva A (2003) The choice of variable for atmospheric moisture analysis. Mon Weather Rev 131:155–171

Dirmeyer PA, Schlosser CA, Brubaker KL (2009) Precipitation, recycling, and land memory: an integrated analysis. J Hydrometeorol 10:278–288

Dixon M, Wiener G (1993) TITAN: thunderstorm identification, tracking, analysis, and nowcasting – a radar-based methodology. J Atmospheric Oceanic Technol 10(6):785–797

Douady D, Talagrand O (1990) The impact of threshold processes on variational assimilation. Preprints, international symposium on assimilation of observations in meteorology and oceanography, WMO, Clermont-Ferrand, pp 486–487

Ebert EE (2008) Fuzzy verification of high resolution gridded forecasts: a review and proposed framework. Meteorol Appl 15:51–64

Ebisuzaki W, Kalnay E (1991) Ensemble experiments with a new lagged average forecasting scheme. WMO Report #15

Ehrendorfer M (1994a) The Liouville Equation and its potential usefulness for the prediction of forecast skill. Part I: Theory. Mon Weather Rev 122:703–713

Ehrendorfer M (1994b) The Liouville equation and its potential usefulness for the prediction of forecast skill. Part II: Applications. Mon Weather Rev 122:714–728
Evensen G (1994) Sequential data assimilation with a nonlinear quasi-geostrophic model using Monte-Carlo methods to forecast error statistics. J Geophys Res 99(C5):10143–10162
Fillion L, Mahfouf JF (2003) Jacobians of an operational prognostic cloud scheme. Mon Weather Rev 131:2838–2856
Grasselt R, Schüttemeyer D, Warrach-Sagi K, Ament F, Simmer (2008) Validation of TERRA-ML with discharge measurements. Meteorol Z 17(6):763–773
Grell GA, Dévényi D (2002) A generalized approach to parameterizing convection combining ensemble and data assimilation techniques. Geophys Res Lett 29:1693
Grell GA, Dudhia J, Stauffer DR (1994) A description of the fifth-generation Penn State/NCAR Mesoscale Model (MM5). NCAR/TN-398+STR, 122 pp
Haase G, Crewell S, Simmer C, Wergen W (2000) Assimilation of radar data in mesoscale models: physical initialization and latent heat nudging. Phys Chem Earth 25:1237–1242
Hoffman RN, Grassotti C, (1996) A technique for assimilating SSM/I observations of marine atmospheric storms: tests with ECMWF analyses. J Appl Meteorol 35:1177–1188
Hoffman RN, Kalnay E (1983) Lagged average forecasting, an alternative to monte-carlo forecasting. Tellus 35A:100–118
Hoffman RN, Liu Z, Louis JF, Grassotti C (1995) Distortion representation of forecast errors. Mon Weather Rev 123:2758–2770
Hoke JE, Anthes RA (1976) The initialization of numerical models by a dynamic-initialization technique. Mon Weather Rev 104(12):1551–1556
Hou DC, Kalnay E, Droegemeier KK (2001) Objective verification of the SAMEX ′98 ensemble forecasts. Mon Weather Rev 129:73–91
Houtekamer PL, Mitchell HL (1998) Data assimilation using an ensemble Kalman filter technique. Mon Weather Rev 126:796–811
Hunt BR, Kostelich EJ, Szunyogh I (2007) Efficient data assimilation for spatiotemporal chaos: a local ensemble transform Kalman filter. Physica D 230:112–126
Jolliffe IT, Stephenson DB (2003) Forecast verification. A practitioner′s GUIDE in atmospheric science. Wiley, New York, NY, 240 pp.
Jones C, Macpherson B (1997) A latent heat nudging scheme for the assimilation of precipitation data into an operational mesoscale model. Meteorol Appl 4:269–277
Kalnay E (2003) Atmospheric modeling, data assimilation and predictability. Cambridge University Press, Cambridge, 341 pp.
Krishnamurti TN, Bedi HS, Ingles K (1993) Physical initialization using SSM/I rain rates. Tellus 45A:247–269
Krishnamurti TN, Correa-Torres R, Rohaly G, Oosterhof D, Surgi N (1997) Physical initialization and hurricane ensemble forecasts. Weather Forecasting, 12:503–514
Krishnamurti TN, Kishtawal CM, LaRow TE, Bachiochi DR, Zhang Z, Williford CE, Gadgil S, Surendran S (1999) Improved weather and seasonal climate forecasts from multimodel superensemble. Science 285:1548–1550
Krishnamurti TN, Xue J, Bedi HS, Ingles K, Oosterhof D (1991) Physical initialization for numerical weather prediction over the tropics. Tellus 43A:53–81
Lang P, Plörer O, Munier H, Riedl J (2003) KONRAD – Ein operationelles Verfahren zur Analyse von Gewitterzellen und deren Zugbahnen, basierend auf Wetterradarprodukten. Berichte des Deutschen Wetterdienstes 222. Technical Report. DeutscherWetterdienst: Offenbach
Leith CE (1974) Theoretical skill of Monte-Carlo forecasts. Mon Weather Rev 102:409–418.
Lorenz EN (1963) Deterministic nonperiodic flow. J Atmospheric Sci 20:130–141
Macpherson B (2001) Operational experience with assimilation of rainfall data in the Met Office Mesoscale model. Meteorol Atmospheric Phys 76:3–8
Manobianco J, Koch S, Karyampudi V, Negri A (1994) The impact of assimilating satellite-derived precipitation rates on numerical simulations of the ERICA IOP 4 cyclone. Mon Weather Rev 122:341–365

Marécal V, Mahfouf J-F (2002) Four-dimensional variational assimilation of total column water vapour in rainy areas. Mon Weather Rev 130:43–58

Mass CF, Ovens D, Westrick K, Colle BA (2002) Does increasing horizontal resolution produce more skillful forecasts? Bull Am Meteorol Soc 83:407–430

Mehrkorn Th, Hoffman RA, Grasotti Ch, Louis J-F (2003) Feature calibration and alignment to represent model forecast errors: empirical regularization. Q J R Meteorol Soc 129: 195–218

Milan M, Venema V, Schüttemeyer D, Simmer C (2008) Assimilation of radar and satellite data in mesoscale models: a physical initialisation scheme. Meteorol Z 17(6):887–902

Molteni F, Buizza R, Marsigli C, Montani A, Nerozzi F, Paccagnella T (2001) A strategy for high-resolution ensemble prediction. I: definition of representative members and global-model experiments. Q J R Meteorol Soc 127:2069–2094

Molteni F, Buizza R, Palmer TN, Petroliagis T (1996) The ECMWF ensemble prediction system: methodology and validation. Q J R Meteorol Soc 122:73–119

Murphy AH (1991) Probabilities, odds, and forecasts of rare events. Weather Forecasting 6: 302–307

Murphy AH, Winkler RL (1987) A general framework for forecast verification. Mon Weather Rev 115:1330–1338

Ott E, Hunt BR, Szunyogh I, Zimin AV, Kostelich EJ, Corazza M, Kalnay E, Patil DJ, Yorke JA (2004) A local ensemble Kalman Filter for atmospheric data assimilation. Tellus 56A:415–428

Park SK (1999) Nonlinearity and predictability of convective rainfall associated with water vapor perturbations in a numerically simulated storm. J Geophys Res Atmos 104(D24):31575–31587

Park SK, Droegemeier KK (1997) Validity of the tangent linear approximation in a moist convective cloud model. Mon Weather Rev 125:3320–3340

Park SK, Droegemeier KK (1999) Sensitivity analysis of a moist 1-D Eulerian cloud model using automatic differentiation. Mon Weather Rev 127:2128–2142

Park SK, Droegemeier KK (2000) Sensitivity analysis of a 3D convective storm: implications for variational data assimilation and forecast error. Mon Weather Rev 128:140–159

Pierce CE, Hardaker PJ, Collier CG, Haggett CM (2000) GANDOLF: a system for generating automated nowcasts of convective precipitation. Meteorol Appl 7:341–360

Rinehart RE, Garvey ET (1978) Three-dimensional storm motion detection by conventional weather radar. Nature 273:287–289

Shin DW, Krishnamurti TN (2003) Short- to medium-range superensemble precipitation forecasts using satellite products: 1. Deterministic forecasting. J Geophys Res 108(D6):8383

Simmons AJ, Untch A, Jakob C, Kallberg P, Unden P (1999) Stratospheric water vapour and tropical tropopause temperatures in ECMWF analyses and multi-year simulations. Q J R Meteorol Soc 125:353–386

Stanski H, Wilson LJ, Burrows WR (1989) Survey of common verification methods in meteorology. WMO World Weather Watch Technical Report 8, 114 pp.

Stensrud DJ, Bao JW, Warner TT (2000) Using initial condition and model physics perturbations in short-range ensemble simulations of mesoscale convective systems. Mon Weather Rev 128:2077–2107

Sun J, Crook NA (1997) Dynamical and microphysical retrieval from doppler radar observations using a cloud model and its adjoint. part 1: model development and simulated data experiments. J Atmospheric Sci 54:1642–1661

Szunyogh I, Kostelich EJ, Gyarmati G, Kalnay E, Hunt BR, Ott E, Satterfield E, Yorke JA (2008) A local ensemble transform Kalman filter data assimilation system for the NCEP global model. Tellus 60A:113–130

Talagrand O, Vautard R, Strauss B (1997) Evaluation of probabilistic prediction systems, 1999. Proceedings of the ECMWF workshop on predictability, 20–22 October 1997, Reading, pp 372

Theis S (2005) Deriving probabilistic short-range forecasts from a deterministic high-resolution model. PhD-Thesis, Bonn University, Bonn

Tibaldi S, Paccagnella T, Marsigli C, Montani A, Nerozzi F (2003) Short-to-medium range limited area ensemble prediction: the LEPS system, Quaderno Tecnico ARPA-SMR, 13/2003

Tiedtke M (1989) A comprehensive mass flux scheme for cumulus parameterization in large-scale models. Mon Weather Rev 117(8):1779–1800

Toth Z, Kalnay E (1997) Ensemble forecasting at NCEP and the breeding method. Mon Weather Rev 125:3297–3319

Toth Z, Talagrand O, Candille G, Zhu Y (2003) Probability and ensemble forecasts. In: Jolliffe IT, Stephenson DB (eds) Forecast verification: a practitioner's guide in atmospheric science. Wiley, West Sussex, pp 137–163

Van Leeuwen PJ (2001) An ensemble smoother with error estimates. Mon Weather Rev 129: 709–728

van Leeuwen PJ (2009) Particle filtering in geophysical systems. Mon Weather Rev 137: 4089–4114

Verlinde J, Cotton WR (1993) Fitting microphysical observations of nonsteady convective clouds to a numerical model: an application of the adjoint technique of data assimilation to a kinematic model. Mon Wea Rev 121:2776–2793

Wilhelmson RB, Wicker LJ (2001) Severe storm modeling. Severe convective storms. Meteorol Monogr 28(50), Am Meteorol Soc:123–166

Wilks DS (1995) Statistical methods in the atmospheric sciences. Academic, San Diego, CA, 467 pp

Wilson JW, Schreiber WE (1986) Initiation of convective storms at radar-observed boundary-layer convergence lines. Mon Weather Rev 114:2516–2536

Wulfmeyer V, Behrendt A, Bauer H-S, Kottmeier C, Corsmeier U, Blyth A, Craig G, Schumann U, Hagen M, Crewell S, Di Girolamo P, Flamant C, Miller M, Montani A, Mobbs S, Richard E, Rotach MW, Arpagaus M, Russchenberg H, Schlüssel P, König M, Gärtner V, Steinacker R, Dorninger M, Turner DD, Weckwerth T, Hense A, Simmer C (2008) The convective and orographically-induced precipitation study: a research and development project of the world weather research program for improving quantitative precipitation forecasting in low-mountain regions. Bull Am Meteorol Soc 89(10) 1477–1486

Zupanski M, Zupanski D, Parrish D, Rogers E, DiMego G (2002) Four-dimensional variational data assimilation for the Blizzard of 2000. Mon Weather Rev 130:1967–1988

Zus F, Grzeschik M, Bauer H-S, Wulfmeyer V, Dick G, Bender M (2008) Development and optimization of the IPM MM5 GPS slant path 4DVAR system. Meteorol Z 17:867–885

Chapter 3
Interpolation of Precipitation for Flood Modelling

Uwe Haberlandt

Abstract This chapter discusses possibilities for the spatial estimation of rainfall for flood modelling. It is assumed, that some high time resolution point measurements from a network of recording rainfall gauges are available as basic information. Conventional and geostatistical methods are presented for the spatial interpolation of the point measurements to raster cells and areas. The focus here is on the application of stationary and non-stationary geostatistical methods. The latter are able to utilise additional information e.g. from daily non-recording stations, weather radar and elevation for the estimation of mean areal rainfall. Alternatively, methods for conditional spatial simulation of precipitation are discussed. Those simulation approaches preserve the high spatial variability of rainfall and can be used for uncertainty assessment. Two examples regarding flood modelling are presented at the end of the chapter. The first one deals with interpolation of hourly rainfall using radar as additional information. The second one compares the application of precipitation data obtained from interpolation and simulation for rainfall-runoff modelling.

Contents

U. Haberlandt (✉)
Institute for Water Resources Management, Hydrology and Agricultural Hydraulic Engineering, Leibniz Universität Hannover, Hannover, Germany
e-mail: haberlandt@iww.uni-hannover.de

A.H. Schumann (ed.), *Flood Risk Assessment and Management*,
DOI 10.1007/978-90-481-9917-4_3,

3.1 Introduction

For distributed hydrological modelling of floods precipitation data with high-resolution in space and time are needed. The spatial resolution of recording precipitation networks for hourly or shorter time step data is seldom sufficient. For example the network of recording rain gauges in Germany operated by the German Weather Service (DWD) over a period of at least 10 years has a density of about one station per 2,000 km^2. Much denser networks exist usually for non-recording gauges; e.g. the network density for daily measurements in Germany is about one station per 90 km^2. Alternatively to point observations, weather radar data, which have a very high spatial resolution are increasingly used as input for hydrological modelling (from Ehret et al., 2008; Kim et al., 2008; Segond et al., 2007). However, there is often a large space-time variable bias in radar rainfall estimates (Krajewski and Smith, 2002). To provide optimal input for flood modelling, the best strategy is certainly a combination of all available information on rainfall and applying sophisticated interpolation methods.

Quite a number of modern interpolation methods have been proposed for rainfall (Dubois et al., 1998). Besides geostatistical approaches, such as ordinary kriging, external drift kriging and co-kriging (Carrera-Hernandez and Gaskin, 2007; Goovaerts, 2000; Lloyd, 2005), other techniques based on splines (Hutchinson, 1998a, b) or genetic algorithms (Demyanov et al., 1998; Huang et al., 1998) have been applied. This article focuses on geostatistical interpolation approaches. Non-stationary geostatistical methods are able to consider additional information in the interpolation procedure. Time invariant elevation data are most frequently used as supplementary information (Goovaerts, 2000; Hevesi et al., 1992a, b). However, the extra value of the elevation information depends on the time step and generally decreases with increasing time resolution. Using time variant additional information allows the combination of point measurements from rain gauges with other data sources like radar (Goudenhoofdt and Delobbe, 2008; Haberlandt, 2007; Seo et al., 1990a, b), satellite data (Grimes et al., 1999) or results from numerical weather prediction models (Haberlandt and Kite, 1998).

Interpolation provides a smooth mapping of the variable under consideration with minimum error variance. Disadvantage is the loss of variability, i.e. small values will be over- and large values underestimated. This may lead to errors, especially if the data are used for modelling of floods with high non-linearity between input and output. One possibility to preserve this variability is conditional spatial simulation of precipitation providing several equal probable realisations. Subsequent runs of a hydrological model using these different realisations of precipitation would then allow an assessment of uncertainty from the precipitation input and provide a less biased model result (Bliefernicht and Bárdossy, 2008; Haberlandt and Gattke, 2004). There are different methods available for spatial stochastic simulation of random variables like Gaussian sequential simulation, turning bands method, simulated annealing etc. (see e.g. Deutsch and Journel, 1992; Goovaerts, 1997). The peculiarity of precipitation simulation is the space-time character of precipitation which needs to be preserved in the simulation process.

The chapter is organised as follows. In Section 3.2 the idea of the interpolation procedure and the two most common conventional interpolation methods are presented. In Section 3.3 geostatistical methods are introduced, including the statistical model, variograms as well as the most common simple and advanced interpolation methods. Section 3.4 discusses briefly the validation of interpolation methods. In Section 3.5 conditional simulation is presented. At the end of the Chapter 2 examples regarding rainfall estimation for flood modelling are briefly discussed, one for rainfall interpolation in Section 3.6 and one for rainfall simulation in Section 3.7.

3.2 Interpolation Principle and Conventional Methods

The objective of spatial interpolation is the estimation of a variable, in our case precipitation, for unobserved locations. Usually the interpolation is done on a regular raster, often with subsequent spatial averaging of raster values over larger areas like subbasins to calculate mean areal precipitation as input for hydrological modelling. The general estimator for all interpolation methods can be written as

$$Z^*(u) = \sum_{i=1}^{n} \lambda_i \cdot Z(u_i), \tag{3.1}$$

where $Z^*(u)$ is the estimated value at the unknown location u, $Z(u_i)$ are the known values at observed points u_i in the neighbourhood and λ_i the weights with $i = 1, \ldots, n$. The estimation of the weights λ_i is different for different interpolation methods.

The simplest and frequently applied *nearest neighbour* (NN), or *Thiessen* method uses only the precipitation measurement Z at the nearest point u_k for estimating precipitation Z^* at an unknown point u, such that

$$\lambda = 1 \quad \text{for } d(u, u_k) = \min d(u, u_k), \tag{3.2}$$

with $d(u, u_i)$ denoting the Euclidean distance between the points u and u_i.

The *inverse distance weighting* (IDW) method is often applied for interpolation using the four nearest neighbours, one in each quadrant $Q_j, j = 1, \ldots, 4$ around the unknown location u. The weights are calculated as

$$\lambda_j = \frac{1 \Big/ d_j^2(u, u_{kj})}{\sum_{j=1}^{4} 1 \Big/ d_j^2(u, u_{kj})} \quad \text{for } d_j(u, u_{kj}) = \min_{i \in Q_j} d_j(u, u_i). \tag{3.3}$$

Most often the squared distance is used for IDW. However other exponents are also possible for d.

The conventional methods NN and IDW cannot account for the spatial persistence structure of a special target variable like precipitation. They rely solely on the geometric configuration of the network, i.e. the estimated weights are the same for any variable under consideration if the same network is considered. Advantage of the NN and IDW methods are their simplicity. The latter one can often be applied with sufficient performance when the network density is high enough.

For flood modelling it is required to apply these methods on hourly or shorter time-step data. Often the recording station network is very poor. To augment the sparse networks it recommended including rainfall data from the denser daily station networks. Note, that this is only possible if no operational forecast is required. The simplest way for the disaggregation of daily data $Z_d(u)$ at location u say to hourly data $Z_h(u,t)$ with $t = 1, \ldots, 24$ at the same location is to apply direct scaling using the intensities from the nearest neighbour u_k with hourly data:

$$Z_h^*(u,t) = Z_h(u_k,t) \cdot \frac{Z_d(u)}{\sum_{t=1}^{24} Z_h(u_k,t)}. \tag{3.4}$$

After disaggregating the data from the daily network, they can be applied together with the hourly data using e.g. NN or IDW for interpolation.

3.3 Geostatistical Interpolation

3.3.1 Statistical Model

Geostatistics uses the concept of *random functions* (RF) for the description of spatial persistence, interpolation and simulation of the variables under consideration. A random function $Z(u)$ is the set of all random variables (RV) $Z(u_i)$ at the points u_i in the domain D. A *regionalised variable* $z(u)$ is one realisation of the RF, e.g. the set of all observations for one state at all points u_i in D. For the complete description of the RF a n-dimensional distribution function $F[Z(u_1), \ldots, Z(u_n)]$ would be necessary. The problem is that the sample size is usually not sufficient to estimate F. Available is only one value at each point u_i for the estimation of the RV and only one realisation for the estimation of the RF. A solution to this problem would be replacing unavailable replications at a point by observations at other points from the whole domain. This requires certain statistical assumptions which are discussed in the following.

One assumption is the hypothesis of *second order stationarity* which consists of two conditions:

1. The expected value of the RF $Z(u)$ is constant all over the domain D

$$E[Z(u)] = m \quad \text{for all } u \in D. \tag{3.5}$$

2. The covariance of two random variables depends only on the distance vector h between the two point and not on the specific locations u and $u + h$

$$C(h) = E[(Z(u+h) - m) \cdot (Z(u) - m)] \quad u \in D. \tag{3.6}$$

The function $C(h)$ is called the *covariance function.* This hypothesis requires also that all points in D have the same finite variance. This is not always met.

A weaker assumption is the *intrinsic hypothesis*, which requires:

1. The expected value of the RF $Z(u)$ is constant all over the domain D, Eq. (3.5).
2. The variance of the increment $Z(u+h) - Z(u)$ of two RV's depends only on the distance vector h between the two points and not on the specific locations

$$\gamma(h) = \frac{1}{2} Var[Z(u+h) - Z(u)] \quad u \in D. \tag{3.7}$$

The function $\gamma(h)$ is called the *semivariogram*, or simply *variogram* as in the following text.

3.3.2 Variograms

The variogram describes the spatial persistence of the RV under study. The *experimental variogram* can be estimated from the observed sample with

$$\gamma^*(h) = \frac{1}{2n(h)} \sum_{i=1}^{n(h)} [z(u_i) - z(u_{i+h})]^2, \tag{3.8}$$

where $n(h)$ is the number of data pairs separated by the distance vector h. If the data are observed on a regular raster a sufficient subsample for different distances h is usually available. In case of irregular observations classes for distances and angels have to be built as follows in order to have enough values for averaging:

$$\left|u_i - u_j\right| \leq |h| \pm \Delta h \quad \text{and} \quad angle\ \varphi \leq \varphi \ \pm \Delta\varphi. \tag{3.9}$$

For robust variogram estimation the number of available data point pairs per class should be $n(h) \geq 30$.

For interpolation (Kriging) a *theoretical variogram model* is required, to obtain variogram values for any distance vector. The standard variogram models are the nugget effect model, the spherical model, the exponential model, the Gaussian model and the power model (Goovaerts, 1997). The power model is the only one of them without sill. Also linear combinations of those permissible models can be used. A very common combination is that of a nugget effect and a spherical model

$$\gamma(h) = c_0 + \begin{cases} c\left(\frac{3}{2}\frac{h}{a} - \frac{1}{2}\frac{h^3}{a^3}\right) & \text{if } h \leq a \\ c & \text{otherwise} \end{cases} \tag{3.10}$$

where a is the range, c is the partial sill and c_0 is the nugget variance. If the variogram depends only on the length of the vector h and not on the direction the random function is said to be *isotropic*. Otherwise *anisotropy* needs to be considered e.g. by applying different variograms for different directions or by data transformations (see Isaaks and Srivastava, 1989). In practice small sample sizes often prevent the consideration of anisotropy.

The parameters of the theoretical variogram models are inferred based on the experimental variogram. Manual (visual) and automatic variogram fitting is possible. Manual variogram fitting is labor intensive but expert knowledge can be considered. Automatic fitting is much faster but a "blind" method.

Considering that usually time series of precipitation need to be interpolated, the fitting of the theoretical model $\gamma(h)$ to the experimental one $\gamma^*(h)$ has to be done for each time step anew. If done manually this can be a very time consuming procedure. Two different approaches are suggested in this case. The first one is based on an average experimental variogram

$$\bar{\gamma}^*(h) = \frac{1}{m \cdot 2n(h)} \sum_{t=1}^{m} \frac{1}{s_t^2(Z)} \sum_{i=1}^{n(h)} \left[Z(u_i, t) - Z(u_{i+h}, t)\right]^2, \tag{3.11}$$

where m is the number of time steps t and s_t^2 is the variance for each time step. In this case only one variogram is finally obtained, for which the fitting of the theoretical model can be done manually. The second approach is an automatic fitting procedure, which provides a specific variogram $\gamma_t(h)$ for each time step t such that the weighted sum of squares between the experimental and the theoretical variogram values for $k = 1, \ldots, K$ lags becomes a minimum (Cressie, 1985):

$$\sum_{k=1}^{K} \frac{n_t(h_k)}{\gamma_t^2(h_k)} \cdot \left[\gamma_t^*(h_k) - \gamma_t(h_k)\right]^2 \rightarrow Min. \tag{3.12}$$

This gives more importance to the smaller lags and the ones computed from more data pairs. The minimisation of Eq. (3.12) can be done e.g. using the Nelder and

Mead optimisation method (see e.g. Press et al., 1992) assuming that $s_t^2 = c_0 + c$. Thus, only two parameters, the range a and the ratio c_0/c need to be estimated for the theoretical model providing a sufficient robust optimisation procedure.

The sensitivity of the variogram estimation on the interpolation performance is often not very high. However, if the error variance is needed e.g. for uncertainty assessment or simulation a good variogram estimation is necessary.

3.3.3 Ordinary Kriging (OK)

Ordinary Kriging (OK) (Matheron, 1971) is the best known geostatistical interpolation method. For using OK the requirements of the intrinsic hypothesis have to be met (Eqs. 3.5 and 3.7). For estimation of values *Z*(*u*) at unknown points Eq. (3.1) is applied. The weights λ are calculated using the OK kriging system

$$\sum_{j=1}^{n} \lambda_j \gamma(u_i - u_j) + \mu = \gamma(u_i - u) \quad i = 1,\ldots,n \sum_{j=1}^{n} \lambda_j = 1, \tag{3.13}$$

where γ are the variogram values and μ is a Lagrange parameter. Equation (3.13) is obtained by minimising the estimation variance

$$\sigma^2(u) = -\sum_{j=1}^{n}\sum_{i=1}^{n} \lambda_j \lambda_i \gamma(u_i - u_j) + 2\sum_{i=1}^{n} \lambda_i \gamma(u_i - u) \tag{3.14}$$

considering the variogram, the network configuration and ensuring that all weights add to one (unbiasedness condition). In praxis, the stationarity conditions from Eqs. (3.5) and (3.7) can often be satisfied by selecting a small enough search radius to restrict the domain *D*.

3.3.4 Simple Kriging (SK)

For *Simple Kriging* the expected value *m*(*u*) needs to be known (compared to OK) in the whole domain *D*. The hypothesis of second order stationarity applies, which requires a finite variance. Only variograms with sill are permitted. For estimation of *Z*(*u*) Eq. (3.1) can be used if the expected value *m*(*u*) is subtracted from all data before. The weights λ are calculated using the SK kriging system

$$\sum_{j=1}^{n} \lambda_j C(u_i - u_j) = C(u_i - u) \quad i = 1,\ldots,n \tag{3.15}$$

where C are the covariances. Equation (3.15) is obtained by minimising the estimation variance

$$\sigma^2(u) = -\sum_{j=1}^{n}\sum_{i=1}^{n} \lambda_j \lambda_i C(u_i - u_j) + C(0) - 2\sum_{i=1}^{n} \lambda_i C(u_i - u). \tag{3.16}$$

For SK the expected value $m(u)$ needs not to be constant in the domain D. So, SK can also be applied for instationary random functions, when a model for estimating $m(u)$ is available.

3.3.5 Residual Kriging (RK)

Residual Kriging (RK) is a non-stationary method which allows the consideration of additional information (e.g. topography) to explicitly estimate the trend $m(u)$ of the target variable $Z(u)$. RK is applied usually in the following steps:

1. Estimate the trend $m(u)$ using regression:

$$Z(u) = \alpha + \sum_{j=1}^{k} \beta_j Y_j(u) + R(u) = m(u) + R(u). \tag{3.17}$$

2. Subtract the trend from the original data:

$$R(u) = Z(u) - m(u) \quad \text{with } \mathrm{E}[R(u)] = 0. \tag{3.18}$$

3. Apply the residuals for interpolation e.g. using OK:

$$R^*(u) = \sum_{i=1}^{n} \lambda_i R(u_i). \tag{3.19}$$

4. Add the trend back to the interpolated residuals:

$$Z^*(u) = R^*(u) + m(u). \tag{3.20}$$

This method is statistically not fully consistent, since regression requires independent residuals but kriging relies on dependent residuals (to estimate the variogram). As a work around an iterative procedure could be used (Neuman and Jacobson, 1984). However, in praxis the above method is often successfully applied without iteration.

3.3.6 External Drift Kriging (EDK)

External Drift Kriging (EDK) (Ahmed and De Marsily, 1987) allows the implicit treatment of non-stationary random functions $Z(u)$, so Eq. (3.5) does not have to be met. The approach uses an additional variable $Y(u)$, which is linearly related to $Z(u)$. The conditional expected value of Z is then

$$E\left[Z(u)|Y(u)\right] = a + bY(u) \tag{3.21}$$

where a and b are unknown constants. The linear estimator is equivalent to Eq. (3.1). The Kriging system for calculating the weights γ is now derived by minimising the estimation variance Eq. (3.14), considering both the unbiasedness condition and the additional variable given in Eq. (3.21) as

$$\begin{gathered} \sum_{j=1}^{n} \lambda_j \gamma(u_i - u_j) + \mu_1 + \mu_2 Y(u_i) = \gamma(u_i - u) \quad i = 1, \ldots, n \\ \sum_{j=1}^{n} \lambda_i = 1 \\ \sum_{j=1}^{n} \lambda_i Y(u_j) = Y(u) \end{gathered} , \qquad (3.22)$$

where μ_1 and μ_2 are two Lagrange parameters. From Eq. (3.22) it can be seen that the constants a and b have been eliminated, but the additional variable Y has to be given for all target points u and for all other points where Z is known. In praxis it is sufficient to estimate the variogram using the regional data $Z(u)$. EDK can be used as an alternative to the better known but more demanding co-Kriging approach, where variograms of Y and cross variograms between Z and Y are required additionally. EDK can be extended to the multivariate case using several additional variables.

When EDK is applied for the spatial interpolation of a whole time series of precipitation, the weights are estimated independently for each time step of the series using Eq. (3.22). The coefficients a and b denoted in Eq. (3.21) will implicitly change not only in space (as compared to RK), but also for each time step. Thus, with EDK, it is possible to consider a space-time variable connection between precipitation as the primary variable and any additional information. This feature is quite important when radar data are used as an additional variable since it enables the imitation of a space-time variable Z-R relationship between rainfall and reflectivity (see Section 3.6).

3.4 Validation of Interpolation Methods

Objective of the validation exercise is to test in general the functioning of the geostatistical model and in special the interpolation performance of the selected kriging method. The geostatistical model includes the variogram(s), the type of kriging and the search strategy. There are two basic approaches for validation

1. *Jack-knife*. This is resampling without replacement, i.e., when one data set is re-estimated from another non-overlapping data set. This requires a large enough sample, which can be divided into two subsets.
2. *Cross-validation*. The observed data are dropped one at a time and re-estimated from some of the remaining neighbouring data. After re-estimation the datum is replaced again with its original value.

Cross-validation is usually preferred because the sample size is seldom large enough for jack-knife. The observed values $Z(u)$ are then compared with the interpolated ones $Z^*(u)$ using different performance measures. Some useful criteria with complementary information are:

the *bias*

$$Bias = \frac{1}{n}\sum_{i=1}^{n}\left[Z^*(u_i) - Z(u_i)\right], \tag{3.23}$$

the *relative standard error* normalised with the observed average

$$RSE = \frac{1}{\bar{Z}} \cdot \sqrt{\frac{1}{n}\sum_{i=1}^{n}[Z^*(u_i) - Z(u_i)]^2} \tag{3.24}$$

and the *coefficient of correlation*

$$Cor = \frac{Cov\left[Z(u), Z^*(u)\right]}{\sqrt{Var\left[Z(u)\right] \cdot Var\left[Z^*(u)\right]}}. \tag{3.25}$$

3.5 Simulation Methods

With simulation the spatial variability of the target variable can be preserved. Spatial simulation of precipitation provides equal probable realisations which can be applied to assess the uncertainty of precipitation and resulting flood flows through rainfall-runoff modelling. There are different methods available for spatial stochastic simulation of precipitation which are mostly known from the geostatistical literature like sequential simulation, turning bands method, simulated annealing etc. (see e.g. Deutsch and Journel, 1992; Goovaerts, 1997). The main difference of precipitation simulation to parameter simulation is the space-time character of precipitation which needs to be preserved in the simulation process.

Two general cases of simulation can be distinguished: *conditional simulation*, which is directly honoring the observations at the known locations and *unconditional simulation*, which is not reproducing the observations. Although geostatistical simulation methods can be used in both cases the main application here is conditional simulation. Methods for unconditional simulation are often considered as stochastic models which are discussed in Chapter 7 of the book. The two simulation methods sequential simulation and simulated annealing are briefly described in the following.

3.5.1 Sequential Simulation

In sequential simulation values of the variable $Z(u)$ are drawn from its conditional distribution, which is constructed using data from the neighbourhood and previously simulated values. Usually the following steps apply:

1. The set of points to be simulated $S = \{y_1, ..., y_N\}$ and the set of available observed points $B = \{x_1, ..., x_n\}$ are defined.
2. A point y_k is selected at random from the set S.
3. The conditional distribution function $F(Z(y_k)|x_1, \ldots, x_n)$ for the variable at point y_k given the observed points B is estimated.
4. A value $z^*(y_k)$ from $F(.)$ is drawn at random.
5. The point y_k is removed from the simulation set S and added to the observation set B.
6. Steps 1–6 are repeated, until all points from set S have been generated.

The central point is the estimation of the conditional distribution function $F(Z(y_k)|x_1, \ldots, x_n)$ for the unknown location y_k in step 3. The two most common ways for the assessment of $F(.)$ are the Gaussian and the Indicator approach outlined below (for details see e.g. Deutsch and Journel, 1992; Goovaerts, 1997).

For *sequential Gaussian simulation* first the Normal score transformation is applied on the original data. Then simple kriging (SK) can be used with observed and previously simulated points from the neighbourhood providing $E[Z(y_k)]$ as interpolation results and $Var[Z(y_k)]$ as estimation variance. This way for each location y_k an estimation of the conditional Gaussian distribution $F(Z(y_k)|x_1, \ldots, x_n)$ is possible.

For *sequential Indicator simulation* first binary indicator transformations are applied on the original data for different thresholds indicating non-exceedance and exceedance probabilities. The successive interpolation of all those indicators provides then an approximation of the conditional distribution function $F(Z(y_k)|x_1, \ldots, x_n)$ at location y_k.

This still leaves the problem of space-time sequential simulation as required for precipitation. On way to do this is to consider time as third dimension in the interpolation procedure. There are only few publications dealing with this question. Some ideas can be found in De Cesare (2002) or Nunes and Soares (2005).

3.5.2 Simulated Annealing

Simulated annealing can be considered as a discrete optimisation technique in which an initial image is gradually perturbed so long as to match a set of constraints like the reproduction of the probability distribution and the spatial covariance estimated from the observed sample data. Information about the theoretical background of simulated annealing is given by Aarts and Korst (1989) and about practical geostatistical aspects by Goovaerts (1997). The procedure can be applied with the following steps:

1. A start raster is defined. The observations are assigned to the nearest grid points each and random values are generated for all other grid points from the sample histogram.

2. An objective function O is defined and its initial value calculated, e.g. using observed $\gamma(h)$ and simulated $\gamma^*(h)$ variogram values:

$$O = \sum_{h_i} \left[\frac{\gamma^*(h_i) - \gamma(h_i)}{\gamma(h_i)} \right]^2 \tag{3.26}$$

3. A start annealing temperature $T_a = T_0$ and a reduction factor R $(0 < R < 1)$ are selected.
4. Two pixels are chosen at random and their values swapped. The objective function (Eq. 3.26) is recalculated providing O_{new}.
5. The change is accepted with the probability:

$$P_{accept} = \begin{cases} 1, & \text{if } O_{new} \leq O_{old} \\ \exp\left(\frac{O_{old}-O_{new}}{T_a}\right), & \text{if } O_{new} > O_{old} \end{cases} \tag{3.27}$$

6. The steps 4 and 5 are repeated M times.
7. The annealing temperature is reduced $T_{a+1} = T_a \times R$.
8. The steps 6 and 7 are repeated until O becomes smaller than a prescribed value or if O does not decrease anymore for a certain number of iteration steps.

Step 5 is essential for the optimisation not to stop at any local minimum but to find a good solution near the global minimum. The annealing temperature regulates the probability of negative changes. The lower T_a the less likely is the acceptance of a negative change. The convergence is ensured in reducing the annealing temperature T_a each time after M iteration steps.

Simulated annealing is a very flexible method since it allows the incorporation of any desired rainfall characteristic in the objective function. However, the method is numerically very demanding. An example for considering space-time simulation of precipitation is given in Section 3.7.

3.6 Example for Rainfall Interpolation

This example discusses the optimal interpolation of hourly precipitation exemplarily for the heavy rainfall event of 10–13 August 2002 in parts of the Elbe River basin (Haberlandt, 2007). For interpolation, multivariate External Drift Kriging (EDK) using additional information from weather radar (Radar), daily precipitation in form of the total event sum (Pev) and elevation (El) has been used. The main focus was on the optimal combination of point surface precipitation and weather radar data. The EDK method considers implicitly a space-time variable relationship between hourly point rainfall and all additional information, honoring the variability of the Z–R relationship. For the investigation, hourly precipitation from 21 recording stations, daily precipitation from 281 non-recording stations, and radar data from 3 locations were used.

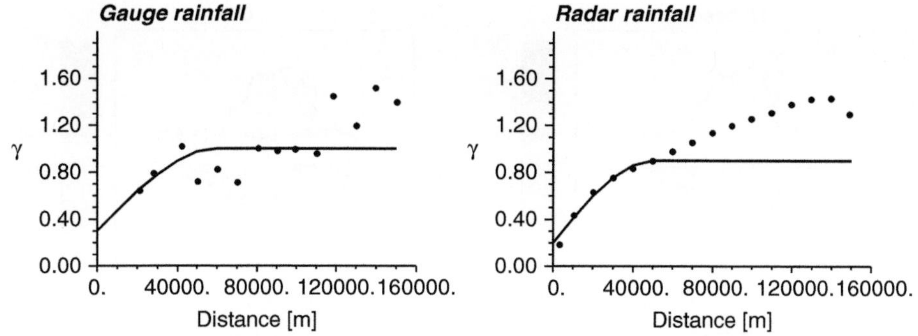

Fig. 3.1 Average experimental and fitted theoretical standardised variograms characterising hourly rainfall for the August 2002 extreme rainfall event over the Elbe river basin

Variograms are inferred based on automatic estimation (Eq. 3.12), manual fitting of a mean variogram (Eq. 3.11) using gauge and radar rainfall with and without considering anisotropy. Comparing interpolation results has revealed only little sensitivity to the selected variogram estimation method. Figure 3.1 shows exemplarily two variograms, one fitted using hourly gauge rainfall and the other fitted using hourly radar data. The fitted spherical models (Eq. 3.10) are similar. The scatter in the experimental gauge based variogram results from the small sample size. In both variograms the values increase almost permanent with increasing distance. This indicates a drift in the data, which has been neglected here but is considered in the interpolation through EDK.

Figure 3.2 shows the results of a step-wise cross validation for the interpolation of hourly precipitation using different methods with correlation as the performance criterion. The performance of the EDK approach is clearly better than for any univariate reference method. If no radar data are available the best method is EDK using Pev as an additional variable. In this case, including elevation has no significant

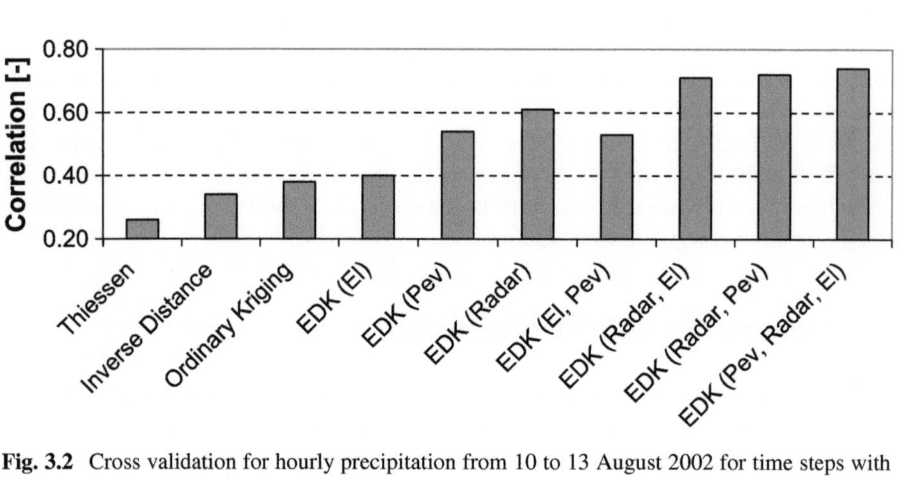

Fig. 3.2 Cross validation for hourly precipitation from 10 to 13 August 2002 for time steps with P > 1 mm/h (from Haberlandt et al., 2007)

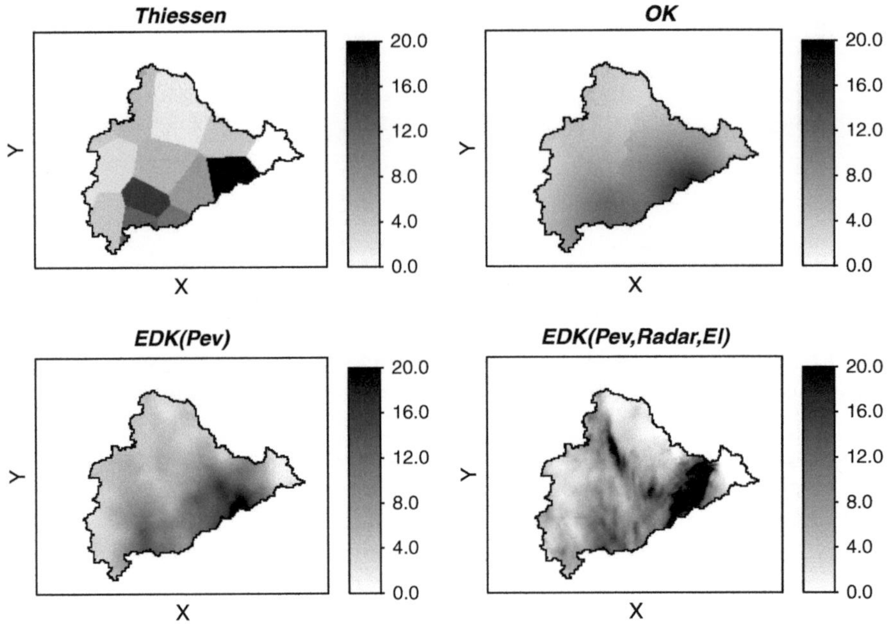

Fig. 3.3 Spatial distribution of precipitation in parts of the Elbe river basin in mm/h for the 12th of August 2002 from 6 to 7 h estimated using different interpolation methods

impact. However, when radar data are available elevation becomes significant and a model using Radar and El is recommended. This model is almost as good as the model using all three variables and has the advantage of being applicable for operational forecasts.

Figure 3.3 illustrates the spatial distribution of precipitation for one selected hour of the event for the four interpolation methods Thiessen, OK, EDK with the event sum Pev and EDK with Pev, Radar and El as additional variables. The result from the Thiessen interpolation clearly shows the polygons separating the geometric regions of influence for each available station. The application of OK based on the small sample size of 21 stations leads to a very smooth map. When using EDK with Pev more spatial structure becomes visible. The application of EDK with Radar, Pev and El as additional information improves the spatial representation of precipitation significantly. This map shows more precipitation variability, higher extremes and the typical anisotropic rainfall patterns.

3.7 Example for Rainfall Simulation

In this example conditional spatial simulation of precipitation is presented using simulated annealing (Haberlandt and Gattke, 2004). The simulated annealing algorithm is designed here such that it honors the measured values at their locations, reproduces the histogram and the variogram of the observed sample data. To consider in addition the space-time persistence of precipitation the simulated maps are forced to correlate with the interpolated maps, which preserve the correlation in time. The required correlation is estimated from cross validation using OK. This way, the similarity between interpolated and simulated maps depends on the uncertainty of the interpolation.

The technique has been tested exemplarily for the 107 km^2 Lippe river basin using daily precipitation data from two advective storms with durations of about 10 days. A simple conceptual hydrological model (FGM-IWK, Plate et al., 1988) has been forced by interpolated precipitation using ordinary kriging and 100 realisations of simulated precipitation using simulated annealing in comparison.

Figure 3.4 shows a comparison between interpolated and simulated precipitation together with prescribed and simulated variograms exemplary for 1 day of one event.

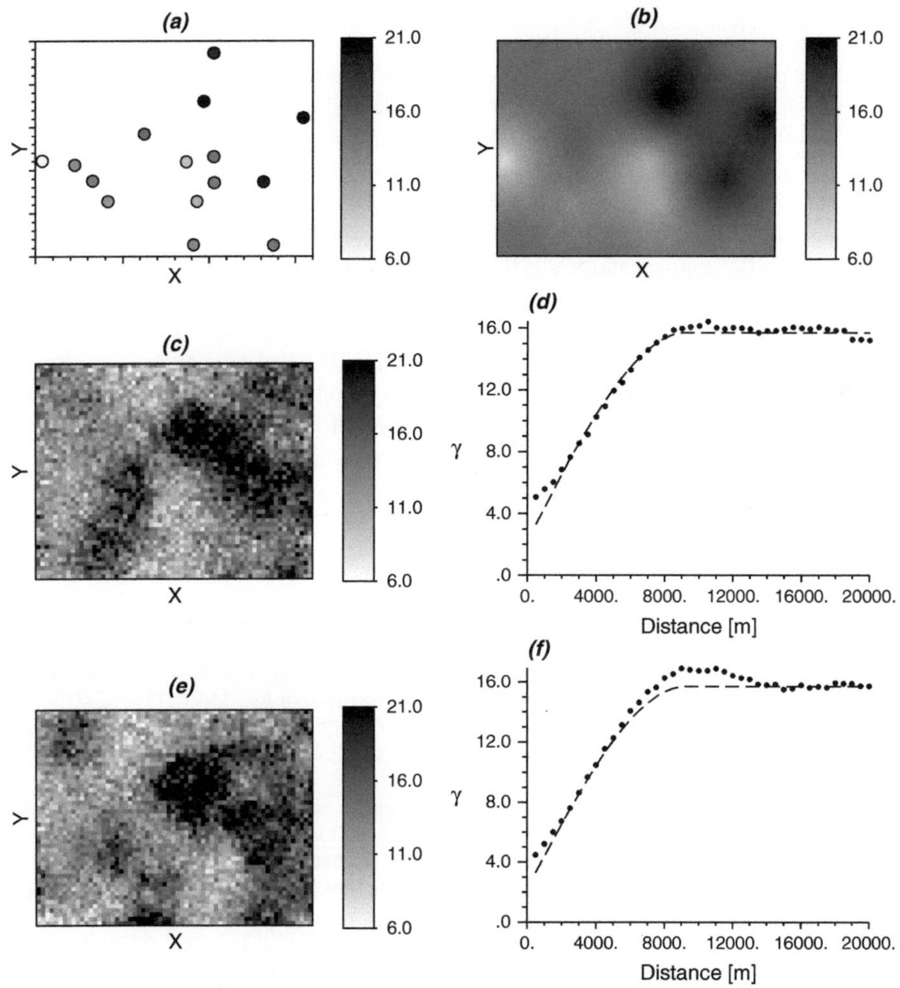

Fig. 3.4 Observed data (a), interpolated map (b) and two simulated realisations of precipitation (c, e) in mm/d with prescribed and simulated variograms (d, f) for March 5, 1998

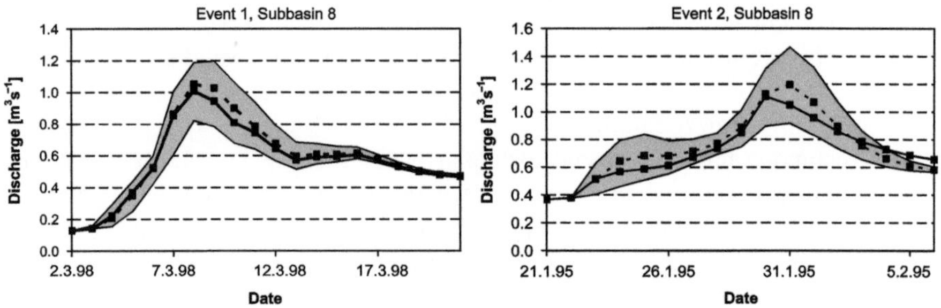

Fig. 3.5 Modeled discharge using interpolated precipitation (*solid line*) and simulated precipitation (average: *dashed line*, range from 100 realisations in *grey*) (from Haberlandt et al., 2007)

The smooth interpolation map is clearly different from the two simulated maps, which are expressing significantly more variability but are still conserving the major patterns. Also, the simulated realisations reproduce the prescribed variograms quite well. For this time step the simulation is conditioned with a correlation of 0.38 on the OK interpolation map.

Figure 3.5 compares simulated runoff for one selected sub-basin using interpolated and simulated precipitation for both storm events. The relatively wide range of simulated flows gives an idea of the uncertainty resulting from the uncertain spatial rainfall distribution. However, averaged hydrographs based on simulated rainfall do not deviate very much from the hydrographs based on interpolated rainfall. The reason is, on the one hand, the relatively low dynamics of advective precipitation events, and on the other hand, the application of a quite simple almost linear hydrological model. Further work is required to extend the study to convective storms based on smaller time steps and using a more physically based model.

References

Aarts E, Korst J (1989) Simulated annealing and boltzmann machines: a stochastic approach to combinatorial optimization and neural computing. Wiley, New York, NY

Ahmed S, De Marsily G (1987) Comparison of geostatistical methods for estimating transmissivity using data on transmissivity and specific capacity. Water Resour Res 23(9):1717–1737

Bliefernicht J, Bárdossy A (2008) Stochastic simulation of hourly precipitation fields for extreme events on the rivers Freiberger Mulde, Oberer Main, and Fränkische Saale (in German). HyWa 52(4):168–172

Carrera-Hernandez JJ, Gaskin SJ (2007) Spatio temporal analysis of daily precipitation and temperature in the Basin of Mexico. J Hydrolo 336(3–4):231–249

Cressie N (1985) Fitting variogram models by weighted least squares. Math Geol 17:563–586

De Cesare L, Myers DE, Posa D (2002) FORTRAN programs for space-time modelling. Comput Geosci 28(2):205–212

Demyanov V, Kanevski M, Chernov S, Savelieva E, Timonin V (1998) Neural network residual kriging application for climatic data. J Geogr Inf Decis Anal 2(2):215–232

Deutsch CV, Journel AG (1992) GSLIB: geostatistical software library and user's guide. Oxford University Press, New York, NY, 340 pp

Dubois G, Malczewski J, Cort MD (eds) (1998) Spatial interpolation comparison 97. J Geogr Inf Decis Anal 2(1–2):1–10, Special Issue

Ehret U, Götzinger J, Bárdossy A, Pegram GGS (2008) Radar-based flood forecasting in small catchments, exemplified by the Goldersbach catchment, Germany. Int J River Basin Manage 6(4):323–329

Goovaerts P (1997) Geostatistics for natural resources evaluation. Oxford University Press, New York, Oxford, 483 pp

Goovaerts P (2000) Geostatistical approaches for incorporating elevation into the spatial interpolation of rainfall. J Hydrol 228(1–2):113–129

Goudenhoofdt E, Delobbe L (2008) Evaluation of radar-gauge merging methods for quantitative precipitation estimates. Hydrol Earth Syst Sci 13:195–203

Grimes DIF, Pardo-Iguzquiza E, Bonifacio R (1999) Optimal areal rainfall estimation using raingauges and satellite data. J Hydrol 222:93–108

Haberlandt U (2007) Geostatistical interpolation of hourly precipitation from rain gauges and radar for a large-scale extreme rainfall event. J Hydrol 332(1–2):144–157

Haberlandt U, Gattke C (2004) Spatial interpolation vs. simulation of precipitation for rainfall-runoff modelling – a case study in the Lippe river basin In: Webb B, Acreman M, Maksimovic C, Smithers H, Kirby C (eds) Hydrology science and practice for the 21st century, vol 1. British Hydrological Society, London, pp 120–127

Haberlandt U, Kite GW (1998) Estimation of daily space-time precipitation series for macro-scale watershed modelling. Hydrol Processes 12(9):1419–1432

Haberlandt U, Ebner von Eschenbach A-D, Belli A, Gattke C (2007) Space–time representativity of precipitation for rainfall–runoff modelling: experience from some case studies. Proceeding of Symposium HS2004 at IUGG2007, Perugia. IAHS Publ. No. 313, pp 61–69

Hevesi JA, Flint AL, Istok JD (1992a) Precipitation estimation in mountainous terrain using multivariate geostatistics. Part I: structural analysis. J Appl Meteorol 31:661–676

Hevesi JA, Flint AL, Istok JD (1992b) Precipitation estimation in mountainous terrain using multivariate geostatistics. Part II: Isohyetal maps. J Appl Meteorol 31:677–688

Huang Y, Wong P, Gedeon T (1998) Spatial interpolation using fuzzy reasoning and genetic algorithms. J Geogr Inf Decis Anal 2(2):204–214

Hutchinson MF (1998a) Interpolation of rainfall data with thin plate smoothing splines: I Two dimensional smoothing of data with short range correlation. J Geogr Inf Decis Anal 2(2): 139–151

Hutchinson MF (1998b) Interpolation of rainfall data with thin plate smoothing splines: II Analysis of topographic dependence. J Geogr Inf Decis Anal 2(2):152–167

Isaaks EH, Srivastava RM (1989) Applied geostatistics. Oxford University Press, New York, NY

Kim BS, Kim BK, Kim HS (2008) Flood simulation using the gauge-adjusted radar rainfall and physics-based distributed hydrologic model. Hydrol Processes 22(22): 4400–4414

Krajewski WF, Smith JA (2002) Radar hydrology: rainfall estimation. Adv Water Resour 25 (8–12):1387–1394

Lloyd CD (2005) Assessing the effect of integrating elevation data into the estimation of monthly precipitation in Great Britain. J Hydrol 308(1–4):128

Matheron G (1971) The theory of regionalized variables and its applications. Les Cahiers du Centre de Morphologie Mathématique, Fasc 5

Neuman SP, Jacobson EA (1984) Analysis of nonintrinsic spatial variability by residual kriging with application to regional groundwater levels. Math Geol 16(5):499–521

Nunes C, Soares A (2005) Geostatistical space-time simulation model for air quality prediction. Environmetrics 16(4):393–404

Plate EJ, Ihringer J, Lutz W (1988) Operational models for flood calculations. J Hydrol 100: 489–506

Press WH, Teukolsky SA, Vetterling WT, Flannery BP (1992) Numerical recipes in Fortran 77. The art of scientific computing. Cambridge University Press, New York, NY

Segond M-L, Wheater HS, Onof C (2007) The significance of spatial rainfall representation for flood runoff estimation: a numerical evaluation based on the Lee catchment, UK. J Hydrol 347(1–2):116–131

Seo D-J, Krajewski WF, Bowles DS (1990a) Stochastic interpolation of rainfall data from rain-gages and radar using co-kriging: 1. Design of experiments. Water Resour Res 26(3):469–477 (89WR02984)

Seo D-J, Krajewski WF, Bowles DS (1990b) Stochastic interpolation of rainfall data from rain-gages and radar using co-kriging: 2. Results Water Resour Res 26(5):915–924 (89WR02992)

Chapter 4
Framing Uncertainties in Flood Forecasting with Ensembles

Andreas H. Schumann, Yan Wang, and Jörg Dietrich

Abstract Under the assumption of rational decision making flood forecasts produce economic benefits only if their application reduces the uncertainties of future developments. In general a forecasting system cannot provide the exact future value of the predictand. There are many different sources of uncertainties. In this chapter tools are presented which can be used to characterize them. It is focused on ensemble methods. Ensembles do not only provide descriptions of uncertainties. They can be combined with data assimilation to produce "best guess" forecasts based on the Bayes theorem. The forecasting system can be adapted indirectly to the actual state of information by selecting an ensemble member which is in good agreement with observed data. A widely used approach, based on this methodology, is the Ensemble Kalman Filter (EnKF). The EnKF can be used to update state variables of hydrological models. In addition meteorological ensembles forecasts and parameter ensembles are discussed. A case study, demonstrating the applicability of ensemble forecasts, closes this chapter.

Contents

A.H. Schumann (✉)
Ruhr-University Bochum, Chair of Hydrology and Water Management, Bochum, Germany
e-mail: andreas.schumann@rub.de

A.H. Schumann (ed.), *Flood Risk Assessment and Management*,
DOI 10.1007/978-90-481-9917-4_4,

4.1 Introduction

Flood forecasting is a prerequisite for operational flood risk management. Flood emergency management which is based on forecasts can prevent damages and losses or at least reduce them. Often the economic effect of emergency management results from the decisions to start countermeasures (alerts, evacuation etc.). If such measures are not required then they produce only costs, otherwise benefits. False alerts are critical not only by their economic aspects but also from the psychological point of view, as they undermine the confidence in the forecast system. Here it is essential that a forecast is "good". Murphy (1993) specifies a good meteorological forecast by three aspects: consistency, quality and value. With "consistency" the degree to which a forecast corresponds to the forecaster's best judgement about the situation is characterised. The quality of a forecast describes the agreement of the forecast conditions with the observed conditions during the valid time of the forecast. The value of a forecast is the degree to which a forecast helps users to realise some incremental economic and other benefits.

The economic benefits of forecasts depend on the following aspects. Actions for operational flood damage prevention have to be started in relationship to an expected but future unknown value of a predictand *y* (water stage, discharge etc.). If the economic effect of a certain operational decision can be described with a utility function *U*(*y*), which depends on the hydrological variable *y*, its expected value *E* can be estimated as the integral value of the probability density function (pdf) of the predictand *f(y)*:

$$E(U_t) = \int_{o}^{\infty} U(y) \cdot f(y_t)\, dy \qquad (4.1)$$

Rational decision making requires that the total uncertainty about a hydrologic predictand can be quantified in terms of a probability distribution, conditional on all available information and knowledge (Krzysztofowicz, 1999). A discussion of this "predictive uncertainty" in flood forecasting is given by Todini (2010). It is used here to specify the forecasting problem. In Eq. (4.1) the index *t* characterises the fact that the predictand has to be known in the future with $t > t_O$ (t_O is the present time). The difference between *t* and t_O is the forecasting time T_F. If the actual value of the predictand y_{tO} is known and a forecast is available for y_t, then the economic effect of this forecast will be defined by the differences between the pdf's $f(y_t)$ and $f(y_t|y'_t)$, where y'_t is the forecasted value of y_t. The pdf $f(y_t|y'_t)$ describes the predictive uncertainty. It characterises the uncertainty that a value y_t occurs if the value y'_t is forecasted. The integration of additional information in the form of a forecast aims at sharpening this posterior pdf. If a flood forecast $y'(t)$ is available, the economic effects of a decision should be estimable with a higher degree of certainty. The main task of a forecasting model prediction is to reduce uncertainty in the decision making process. It has to be accepted that a model does not provide the future value of the predictand as it involves many uncertainties. These uncertainties are induced by the model choice, the limited representation of hydrological processes, the errors of input and output data and other impacts. The hydrological model represents here the knowledge about the hydrological system, its actual status and future developments. Unfortunately the predictive uncertainty $f(y_t|y'_t)$ can be characterised empirically only by analyses of differences between observed and forecasted values. With regard to rare floods such assessments are not possible. Instead attempts to characterise predictive uncertainties are based on re-forecasts of observed floods. However, the uncertainty of simulations $f(y'_t|y_t)$ of an observed value y_t, which was characterised by Todini (2010) as "validation" uncertainty, is different from the predictive uncertainty $f(y_t|y'_t)$. In Fig. 4.1 the differences between $f(y'_t|y_t)$ and $f(y_t|y'_t)$ are shown schematically.

In reality the predictive uncertainty increases with forecast time T_F (Fig. 4.2). Re-forecasts of observed flood events of the past can be used to analyse the different sources of uncertainties regarding their impact on the forecasted values (Fig. 4.3).

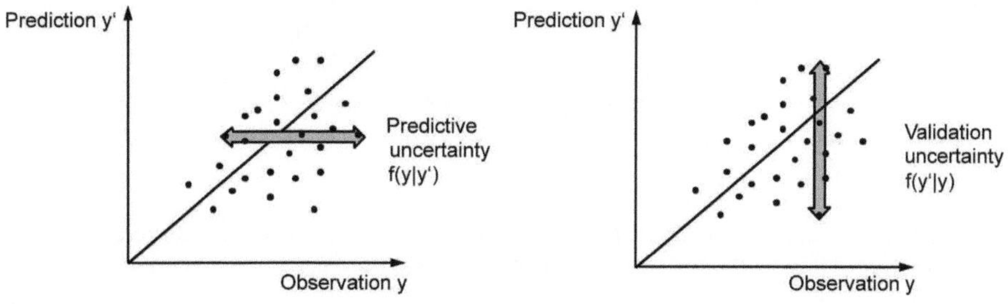

Fig. 4.1 Predictive uncertainty (*left*), describing the variance of observed values if a prediction is available, and validation uncertainty (*right*) characterising the uncertainties of predictions of an observed value

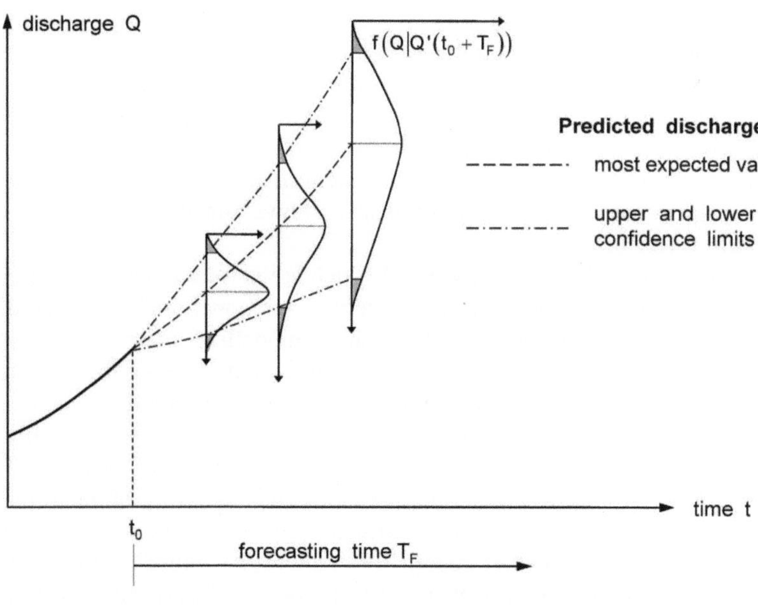

Fig. 4.2 Conditional PDFs of the predictand Q (e.g. discharge) derived from uncertain forecasts Q'

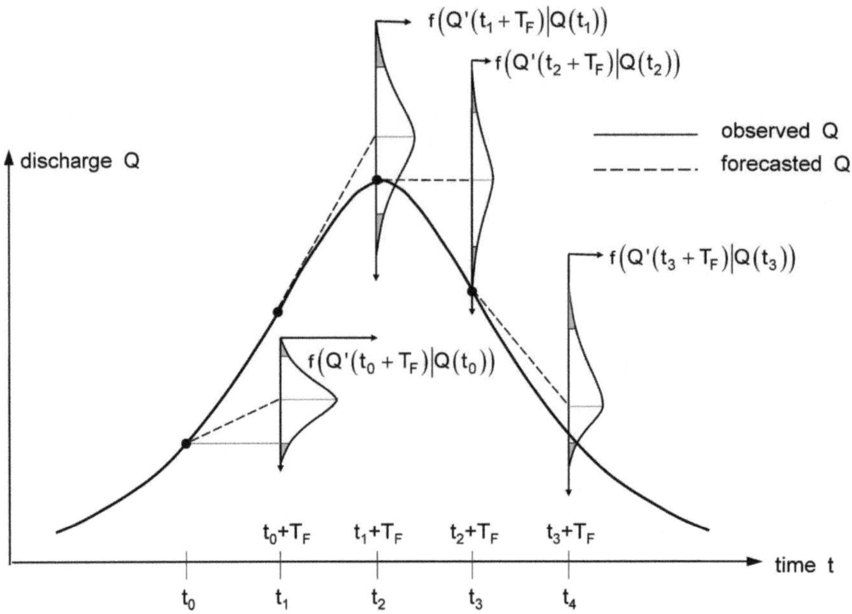

Fig. 4.3 Uncertainty of simulations characterised by conditional PDFs of the predicted value $Q'(t+T_F)$ depending on the hydrological situation characterised here by the observed value Q at the forecasting point t

Ensembles are tools to specify uncertainties of input data, of parameters or even of models. Needless to say, these analyses will not result in narrowing the pdf of forecasts $f(y'_t|y_t)$. The benefit of applying ensemble forecasts consists in the characterisation of validation uncertainty, as it is shown schematically in the Figs. 4.1 and 4.3. The variance of forecasts is temporal variable, as the different components of uncertainties interact in different ways at different time steps. Ensembles demonstrate that forecasts are based on uncertain assumptions of future developments or on presumed behaviour of a hydrological system. If the spread of ensemble forecasts does not include the real development it shows that the expectations about uncertainties are insufficient.

In the following tools will be discussed which are helpful to characterise and to compensate modelling (validation) uncertainties. As shown in Fig. 4.1, predictive uncertainty is related with, but not equivalent to modelling uncertainty. Forecasts of many different events in the past with the same model, data collecting system and within the same range of input and output data would be needed to characterise predictive uncertainties empirically. Obviously this is not possible. Normally only differences between observed and re-forecasted values for some observed events can be used to review a flood forecasting system. The credibility of a hydrological simulation model is usually assessed by estimating the goodness of fit of the model output in comparison to observed records in a calibration and/or validation period. To evaluate a flood forecast model it has to be operated in a quasi "forecasting mode", where records of input variables (predictors) are used to compute a record of the predictand in the same way as it would be done operationally. Especially precipitation forecasts should be provided in a similar way as this would be done during a flood event in real time to ensure a realistic evaluation. However, this is often not possible as meteorological forecasts can not be provided for flood events of the past by re-forecasting with the same modelling systems which are used today.

In the operational mode analyses of differences between observed and forecasted values from previous time steps can be applied to sharpen the pdfs of predicted values. Ensemble forecasts can be coupled with data assimilation for this purpose. This will be shown exemplary in the case study later on in this chapter for a German flood forecast model. Here ensembles are used which specify the impacts of uncertainties of state variables, input data (characterised with ensembles of predicted rainfall) and model parameters.

4.2 Sources of Uncertainties

In Systems Theory a model is composed of seven different components: system boundary, inputs u, initial states x_O, parameters θ, structure, states x and outputs y (Liu and Gupta, 2007). During simulations the input, state and output variables change with time. The structure of the resulting State- Space- model is shown in Fig. 4.4. The model itself can be separated into two parts (Brammer and Siffling, 1994). One part, characterized by an operator *f()*, describes the evolution of the state variable x_t with time, depending on the state at the end of the previous time interval x_{t-1}, the input u_t during this time interval and the set of model parameters θ. The output y_t is estimated from the state variable x_t with the second component of the model, the transformation function $h(x_t)$. Differentiating a model in these

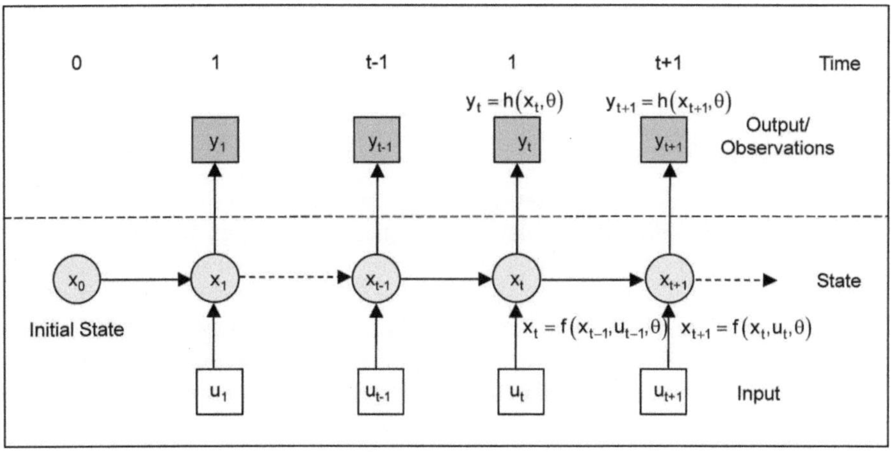

Fig. 4.4 Structure of simulations with the State-Space-model

parts, the input – output relationships are described in an indirect way. In hydrological applications the input variable u_t is e. g. precipitation and output y_t is runoff and the state variable x_t could be e. g. the soil moisture content.

Uncertainties of modelling can be categorized into three primary types (Liu and Gupta, 2007):

1. *Structural errors of the model.* Models are imperfect approximations to the complex reality, based on assumptions and simplifications.
2. *Model parameter uncertainties.* Model parameters are conceptual representations of the properties of a real system. As the model is only an approximation of reality, parameters are uncertain per se. Often parameters have to be estimated in an indirect way by calibration, comparing the results of simulation with observations. The limited amount of information which can be used for their estimation is a main source of parameter uncertainties. The widely applied deterministic calibration techniques often ignore uncertainties of measured data and structural errors of models.
3. *Data errors.* Data errors generate uncertainties in hydrologic predictions through model inputs. In many cases calibration problems may result from observational errors. Liu and Gupta (2007) differentiate data errors between errors of measurements and representativeness errors due to scale incompatibility between the measured variable (e.g. precipitation measured at a point) and the corresponding model input variable (e.g. areal precipitation).

Structural, parameter and data uncertainties collectively lead to uncertainties in hydrologic predictions of model outputs and states. Ensemble based methods are attempts to specify the different sources of forecast uncertainties. They can be applied to compare the relative importance of the different components of uncertainties with the aim to handle them in operational flood risk management.

4.3 Treatment of Uncertainties in Operational Flood Forecasts Using Ensemble Methods

4.3.1 Overview

As described before, there are several reasons why forecasts will be uncertain. In reality the uncertainties of simulations will become obvious with time in comparisons with observations. If such analyses are implemented in operational mode during a flood event it gives flood managers options to control the simulation process, e. g. in cases where the forecasts show systematic biases or if initial state variables of simulations seem to be inappropriate. These activities can be supported by ensemble forecasts, which demonstrate the sensitivity of simulation results in relationship to the uncertainties of input and state variables or model parameters. The value of a prediction is limited if the corresponding uncertainties are not understood, quantified and reduced. Often sampling (or ensemble) methods are used to quantify the uncertainty in model output. Samples from the assumed pdfs of uncertain input data, initial state variables and model parameters are taken to run the model forward several times. As result of such an analysis the uncertainties in model outputs from a large sample of predictions can be described with statistical means. Monte Carlo simulations can be applied to account for uncertainties associated with the parameters of a probability model that Bayesian methods handle natively. The limitations of this approach are: the pdfs of uncertain characteristics and variables are not known; they vary in time and depend on the actual hydrological situation. In contradiction ensembles are based on a lower number of data points and involve many assumptions about the model, model parameters and other characteristics which are not modified. Among the different sources of uncertainties only some can be considered within the forecasting ensemble to limit the complexity of such analyses in real-time. An ensemble forecasting system does not completely represent the uncertainties of models. One basic assumption which is often used is that each result within the forecast ensemble has the same probability. Neither the total variation (which would result from a large amount of possible combinations of uncertain aspects of modelling) nor a differentiation in more or less probable forecasts can be represented in this way. However, a probabilistic evaluation of the outcome would be useful in many applications. Thus it is challenging to develop ensemble generation mechanisms which do not only result in a large number of model outcomes, but also represent the probabilistic assessment of the variables under consideration.

Ensembles can be applied in the operational mode to merge data and information in real time that is provided by imperfect models and uncertain data. Their application aims to select such combinations of parameters, initial states and forecasted input values which adapt the model output optimal to observations. Here the Bayes' theorem can be applied to assimilate data and to select the most adequate members from the ensemble. A prototype of this methodological approach is the ensemble Kalman filter (EnKF), which is widely used for updating state variables in hydrological models. It will be presented below as an option to adapt the state variables in accordance with observed output variables. Assimilation of observed data and Bayesian inference can be applied to produce a "best guess" forecast. Uncertainties of model parameters can be characterised also by ensembles. As mentioned before, the third group of uncertainties are uncertainties of input data. Here ensemble forecasts of precipitation, which are described in Chapter 2, can be used to characterise these uncertainties. In the following the different options for the applications of ensembles are discussed in greater detail.

4.3.2 Updating of State Parameters by Data Assimilation Using the Ensemble Kalman Filter (EnKF)

The EnKF is based on the Kalman filter (KF), which was developed for optimal control of linear systems. Here the generic discrete-time – stochastic-dynamic system which can be derived from the State-Space model (Fig. 4.4) is used to describe the methodology of the KF (Moradkhani et al., 2005). The state vector x_t, an n-dimensional vector representing the state variables of the system at time t, is expressed in the form of

$$x_{t+1} = f(x_t, u_t, \theta) + \omega_t, \quad \omega_t \sim N(0, \Sigma_t^m). \tag{4.2}$$

The nonlinear propagator *f*( ) contains the model input vector u_t, the state vector at the previous time step x_t and the time-invariant model parameter vector θ. The model error is considered by a dynamical noise term ω_t with covariance Σ_t^m. It represents all uncertainties related to the model structure and the forcing data. This stochastic forcing term flattens the probability distribution of the states. Simplified the model error can be represented as a stochastic perturbation. The output variables of the model y'_t are functions of the state variables and the parameters characterizing the model:

$$y'_{t+1} = h(x^*_{t+1}, \theta) + v_{t+1}, \quad v_{t+1} \sim N(0, \Sigma^y_{t+1}) \tag{4.3}$$

Where the propagator *h*( ) relates the state variables to the measured variables. It yields the expected value of the prediction given the true model states x^* and parameters. All sources of errors in the observation are reflected by v_{t+1} which will be assumed to be Gaussian distributed. It denotes the error deviation of the observations. At each measurement time an output observation y_t becomes available and the output forecast error z_t can be computed

$$z_t = y_t - y'_t = y_t - h(x^f_t, \theta) \tag{4.4}$$

where x^f_t is the forecasted state vector and not the true one. The Kalman Filter KF is now applied to update the state vector

$$x^u_t = x^f_t + K_t(y_t - h(x^f_t, \theta)) \tag{4.5}$$

where x^u_t is the updated state and K_t denotes the so-called Kalman gain. K_t depends on the size of the measurement and model errors. The recursive transition to the next state is

$$x^f_{t+1} = f(x^u_t, u_t, \theta). \tag{4.6}$$

The application of the KF is limited to the linear case. In case of nonlinear dynamics the state vector can be linearized to use the Extended Kalman filter (EKF) (Jazwinski, 2007). An alternative is the Ensemble Kalman filter (EnKF) (Evensen, 1994). It is based upon ensemble generations of model trajectories from which the time evolution of the probability density of the model states and related error covariances are estimated. The approximation of the forecast state error covariance matrix (a priori) is made by propagating an ensemble of model states using updated states (ensemble members) from the previous time step. Thus the EnKF propagates an ensemble of n state vectors in parallel. Each state vector represents one realisation of generated model replicates. A description of its functionality is given by Moradkhani et al. (2005). The model forecast is computed for each ensemble member as follows:

$$x_{t+1}^{i-} = f(x_t^{i+}, u_t^i, \theta, t) + \omega_t^i \quad i = 1, \ldots, n \tag{4.7}$$

where x_{t+1}^{i-} is the *i*th ensemble member forecast at time $t + 1$ and x_t^{i+} is the *i*th updated ensemble member at time *t*.

In addition to representing the additive process noise, EnKF represents the multiplicative model errors through forcing data perturbations, adding the noise ζ_t^i with covariance Σ_t^u to the forcing data u_t^i each time step *t*

$$u_t^i = u_t + \zeta_t^i \quad \zeta_t^i \sim N(0, \Sigma_t^u) \tag{4.8}$$

The error covariance matrix associated with the forecasted (a priori) estimate can be calculated as the ensemble covariance matrix:

$$\begin{gathered} \boldsymbol{P}_{t+1}^{-} = E(\boldsymbol{X}_{t+1}\boldsymbol{X}_{t+1}^{T}) = \tfrac{1}{n-1}\boldsymbol{X}_{t+1}\boldsymbol{X}_{t+1}^{T} \\ \textit{where } \boldsymbol{X}_{t+1} = [x_{t+1}^{i-} - \bar{x}_{t+1}^{-}, \ldots, x_{t+1}^{n-} - \bar{x}_{t+1}^{-}] \quad \textit{and} \\ \bar{x}_{t+1}^{-} = E(x_{t+1}^{i-}) = \tfrac{1}{n}\Sigma_{i=1}^{n} x_{t+1}^{i-} \end{gathered} \tag{4.9}$$

After updating all of the ensemble members the updated (a posteriori) error covariance can be estimated similarly.

With the assumption of the forecasted states x_{t+1}^{i-} the observation y_{t+1} is used to obtain the posterior estimatex_{t+1}^{i+}. According to the standard KF a linear correction equation is used to update forecasted state ensemble members:

$$x_{t+1}^{i+} = x_{t+1}^{i-} + K_{t+1}(y_{t+1}^{i} - y'^{i}_{t+1}) \tag{4.10}$$

where y_{t+1}^i is the *i*th trajectory of the observation replicates generated by adding the noise of η_{t+1}^i with covariance Σ_{t+1}^y to the actual observation:

$$y_{t+1}^i = y_{t+1} + \eta_{t+1}^i, \quad \eta_{t+1}^i \sim N(0, \Sigma_{t+1}^y). \tag{4.11}$$

Observations are treated in the EnKF as random variables by generating an observation ensemble with a mean equal to the actual observation at each time and a predefined covariance. The *i*th predictive variable y'^{i}_{t+1} is estimated with the model *h( )* using the *i*th a priori state variable

$$y'^{i}_{t+1} = h(x_{t+1}^{i-}, \theta) \tag{4.12}$$

The Kalman gain matrix is estimated in the ensemble case as

$$K_{t+1} = \Sigma_{t+1}^{xy-}\left[\Sigma_{t+1}^{xy} + \Sigma_{t+1}^{y}\right]^{-1} \tag{4.13}$$

where Σ_{t+1}^{yy} is the forecast error covariance matrix of the prediction y'^{i}_{t+1} and Σ_{t+1}^{xy-} is the forecast cross covariance of the state variables $x^{i}_{t+1}-$ and y'^{i}_{t+1}.

The main advantage of the EnKF compared with KF consists in the possibility to estimate the Kalman gain *K* without an observation transition operator after linearization of the observation model *f( )*. The estimation of the a priori model error covariance $P^{?}$ is not needed to estimate *K*. However, the quality of the ensemble generation method, the forecast model and the analysis scheme influence the performance of the EnKF. The variance of the noises introduced to the input and output variables can be selected proportional to the magnitude of the variables (Moradkhani et al., 2005).

4.3.2.1 Meteorological Ensemble Forecasts

Numerical weather predictions are essential for flood forecasts with lead times longer than the runoff concentration time of the catchment of interest. As described in Chapter 2 the accuracy of the precipitation forecast is often not sufficient. Precipitation forecasts form the end of a meteorological process chain which bridges different spatial and temporal scales. The non-linear and complex atmospheric system cannot be predicted exactly. Over the last decades ensemble prediction systems were developed to produce multiple weather predictions for the same location and time (Cloke and Papenberger, 2009). Ensembles specify the differences between several forecasts which are derived from variations of model parameters, initial and/or boundary conditions or conceptualisations of process models. Different types of ensembles can be classified according to the generating mechanisms:

- single system ensembles are specified by perturbation of uncertain initial and boundary conditions or different model components (e.g. convection schemes in atmospheric models or perturbation of model parameters),
- multi model ensembles combine simulations for the same location and time derived from different models (e.g. Georgakakos et al., 2004 or Ajami et al., 2007),
- lagged average ensembles are specified combining actual forecasts with earlier forecasts if forecast times are overlapping.

Several meteorological ensemble prediction systems (EPS) are operational at the global or regional scale (e.g. ECMWF, MSC, NCEP) (Buizza et al., 2005). One example for an EPS is COSMO-LEPS, a Limited Area Ensemble Prediction System (Molteni et al., 2001) for the medium range (3–5 days lead time). It was developed within the COSMO (Consortium for Small-scale Modelling) to improve the predictability of extreme weather events in Central and Southern Europe. The added value of the system consists in joining the skill of the ECMWF EPS to depict the possible evolution scenarios with the capability of the COSMO limited area model to improve the descriptions of local meteorological processes. It runs on a daily basis using 10 km grid spacing and 40 vertical layers, starting at 12UTC and with a forecast range of 132 h. Driven by a cluster of ensemble forecasts which is derived from 102 members of the EMCWF EPS a forecast ensemble with 16 members is provided as a physical ensemble.

The short-range SRNWP-PEPS (Denhard and Trepte, 2006) is a “Poor Mans Ensemble Prediction System” which combines up to 23 deterministic forecasts from 21 national meteorological services with a lead time of 2 days. It can be seen from case studies and probabilistic verification for Germany (Trepte et al., 2006) that this ensemble is a valuable tool for severe weather forecasting. A major benefit of this multi-model EPS is the possibility to compare the behaviour of all operational European limited area models.

Hydrological applications of meteorological ensemble forecasts started in the late 1990s and are still subject of ongoing research (e.g. Verbunt et al., 2006; Komma et al., 2007; Reed et al., 2007; Diomede et al., 2008). The HEPEX project (Hydrological Ensemble Prediction Experiment, Schaake et al., 2007) promotes the development of ensemble stream flow forecast systems. The European Floods Alert System (EFAS) which is under development at the Joint Research Centre of the European Commission (Thielen et al., 2009) uses the ECMWF ensemble with 51 members and a forecast time up to 10 days, but also COSMO- LEPS with 16 members for 5 days forecasts. It provides flood information for the medium to long-range at large scale river basins being relevant for decisions at national or EU level.

The application of meteorological ensemble forecasts in flood forecasts is connected with several problems. One problem consists in the limited options to validate such systems. The meteorological reasons of floods are manifold but extreme flood events are rare. Thus the performance of meteorological ensemble systems can be evaluated only with relatively small samples of limited representativeness with regard to the large variety of flood inducing meteorological situations. Other problems result from the fact that EPS are still under development and can be tested only occasionally under operational conditions. Re-forecasts of flood events of the past are costly and restricted by limited data availability. In particular if nested regional models have to be used for re-forecasting, a specification of initial and boundary conditions of the atmosphere is needed which could be provided by modelling at the global scale only.

4.3.2.2 Utilisation of Parameter Ensembles

Simulated discharges result normally from an overlay of several process models, which can not be calibrated separately as observations for the different runoff components are missing. In the mathematical sense such models are over-parameterised. The calibration results in a large number of equifinal sets of parameters. Equifinality means that different combinations of parameters could result in similar model outputs with small differences in model efficiencies (Beven and Freer, 2001). Other problems which are causing non-uniqueness of parameters are the utilization of different quantitative measures to specify the adaptation of simulations to observations or the limited amount of data which is available for calibration. This data problem is aggravated as hydrological processes are non-linear and extreme floods are rare. Nevertheless hydrological models have to be extrapolated to extreme situations.

The operational online recalibration of model parameters during a flood event is critical as the information about the input variables is often limited. However, it is possible to use an ensemble with different sets of parameters which were derived during the model calibration process in advance. The variability of flood generation processes, which can be estimated from observed events in the past, could be considered in this way. The application of a parameter ensemble demands parallel simulations with the same model but different sets of parameters. This results in an ensemble of forecasts. It can be evaluated against observed values. According to the principle of Bayesian inference this can be done operationally with the likelihood which has to be specified for each parameter set (Box and Tiao, 1992). This likelihood can be estimated from the temporal development of differences between observed and simulated data. Here the methodological approach of Bayesian Model Averaging (BMA) (Vrugt and Robinson, 2007; Ajami et al., 2007) can be applied.

Bayesian model averaging (BMA) (Hoeting et al., 1999) is a probabilistic technique which estimates a consensus prediction from competing predictions using a weighting with likelihood measures. As the name implies it is frequently applied in multi-model simulations. The likelihood measure (weight) for each member model is based on the success frequency of the predictions that an individual model has made within the observations. For this reason, BMA weights are a measure of individual model performance. BMA has been applied successfully in meteorology (Raftery et al., 2005, Chapter 2). A description can be found in Ajami et al. (2007) which is used here to specify this methodological approach.

A quantity y is the observed output variable which has to be forecasted by a set of K models $M = [M_1, M_2, \ldots M_K]$. The $pk(y'_k|M_k, u, y)$ is the posterior distribution of y'_k which represents the quantity to be forecasted under model M_k, given a discrete data set u (input data) and y (observed outputs). The posterior distribution of the BMA prediction is

$$p(y'_{BMA}|M_1, \ldots, M_k, u, y) = \sum_{k=1}^{K} p(M_k|u, y) \cdot p_k(y'_k|M_k, u, y) \tag{4.14}$$

where $p(M_k|u, y)$ is the posterior probability of model M_k, characterising the likelihood of model M_k being the correct model. It represents a weight to combine the outputs of the several models $w_k = p(M_k|u, y)$ with $\Sigma_{k=1}^{K} w_k = 1$. The second multiplicator is represented by the normal distribution with the mean equal to the prediction y'_k made by model M_k and standard deviation σ_k. The BMA prediction for y is denoted here with y'_{BMA}. It is characterized by

$$E(y'_{BMA}|y'_1, \ldots y'_k, u, y) = \sum_{k=1}^{K} w_k \otimes y'_k \tag{4.15}$$

and

$$Var(y'_{BMA}|y'_1 \ldots y'_K, u, y) = \sum_{k=1}^{K} w_k \left(y'_k - \sum_{i=1}^{K} w_i y'_i \right)^2 + \sigma^2 \tag{4.16}$$

where σ^2 is the variance of the time series shaped based on one of the model predictions (ensembles) being the best at each time step. As shown above, a BMA prediction is the weighted average of predictions in which the weights are likelihoods that an individual model is correct. The uncertainty of the BMA forecast can be characterized by the variance between models (first term) and the weighted deviations simulated from observed values from the beginning of the flood event until time T:

$$\sigma^2 = \frac{1}{T} \sum_{t=1}^{T} \sum_{k=1}^{K} w_k (y_t - y'_{k,t})^2 \tag{4.17}$$

The first term of (4.16) informs about the spread-skill relationship as it represents the variance between the ensemble forecasts. The second term characterises the internal variance of the optimal (BMA) forecast and the observed values. This can be done with a maximum-likelihood approach where the likelihood function is formulated as follows:

$$L(w_1, \ldots w_k, \sigma^2) = \sum_{t=1}^{T} \log \left(\sum_{k=1}^{K} w_k \otimes p(y_t|y'_{kt}) \right) \tag{4.18}$$

The expectation-maximization algorithm (Dempster et al., 1977) can be applied to estimate the weights (Ajami et al., 2007).

In a multi-parameter ensemble a single model M is applied with different parameter sets θ_k. Thus the simulations depend on the parameters set θ_k only:

$$y'_k = M(u, \theta_k). \tag{4.19}$$

If the residuals e can be assumed to be additive, then the observed quantity will be

$$y = M(u, \theta) + e(\theta). \tag{4.20}$$

The parameter set θ is here a probabilistic variable with the posterior probability distribution $P(\theta\,|u, y)$, which is conditional on the observed data u and y. According to Bayes statistics $P(\theta\,|u, y)$ is proportional to the product of likelihood function and prior distribution function:

$$P(\theta\,|y) \propto P(\theta) \otimes L(y\,|\theta) \tag{4.21}$$

Assuming that the residuals $e(\theta)$ are additive, uncorrelated and normal distributed with mean equal zero and a constant but unknown variance σ_y the likelihood of a parameter set describing the observed data y over the number of time steps (T) can be estimated as follows (Box and Tiao, 1992):

$$L\left(\theta, \sigma_y \,|u, y\right) = \frac{1}{\sigma_y^T} \exp\left(-\frac{1}{2\sigma_y^2}\left(\sum_{t=1}^{T} \left(e\left(\theta\right)_t\right)^2\right)\right) \tag{4.22}$$

As normal distributed residuals can not be assumed, Thiemann et al. (2001) suggested the one-to-one transformation of the outputs $z = g(y)$ such that the errors in the transformed output space $\nu = g(y) - g(y')$ are independent, each having an exponential power density. With regard to the heteroscedastic variance of output errors these authors applied the Box-Cox transformation $z = (y^{\lambda} - 1)/\lambda$ in a case study. Using the exponential power distribution (Box and Tiao, 1992) the conditioned probability of the residuals ν can be estimated with a fixed shape parameter $\beta$and the standard deviation of residuals σ which is unknown but constant with respect to time:

$$p\left(\nu \,|\sigma, \beta\right) = \frac{\omega(\beta)}{\sigma} \cdot \exp\left[-c(\beta)\left|\frac{\nu_t}{\sigma}\right|^{2/(1+\beta)}\right] \tag{4.23}$$

$$\text{with } c(\beta) = \left\{\frac{\Gamma[\frac{3}{2}(1+\beta)]}{\Gamma[\frac{1}{2}(1+\beta)]}\right\}^{1/(1+\beta)}$$

$$\text{and } \omega(\beta) = \frac{\{\Gamma[\frac{3}{2}(1+\beta)]\}^{1/2}}{(1+\beta)\{\Gamma[\frac{1}{2}(1+\beta)]\}^{3/2}} \tag{4.24}$$

where β is the shape parameter ($-1 < \beta \leq 1$), a measure of kurtosis, σ the standard deviation of residuals and ν the residual at time t.

With assumption of this error model the conditional density of the transformed outputs observed at times $t = 1, \ldots, T$ will be (Thiemann et al., 2001):

$$p\left(z \,|u; \theta, \sigma, \beta\right) = \left[\frac{\omega(\beta)}{\sigma}\right]^T \exp\left(-c(\beta)\sum_{t=1}^{T}\left|\frac{\nu_t(\theta)}{\sigma}\right|^{2/(1+\beta)}\right) \tag{4.25}$$

Assuming that θ and σ are independent Thiemann et al. computed a maximum likelihood estimate of σ_t at the current time step with respect to σ

$$\sigma_T(\theta)^{2/(1+\beta)} = \frac{T-1}{T}\sigma_{T-1}(\theta)^{2/(1+\beta)} + \frac{1}{T}\frac{2c(\beta)}{(1+\beta)}\left|\nu_T(\theta)\right|^{2/(1+\beta)} \tag{4.26}$$

The posterior density of θ can be estimated with the following recursive formulation

$$p\left(\theta \,|u, z_{T+1}, z; \beta\right) \propto N_T(\theta)\, p\left(\theta \,|u, z; \beta\right) \tag{4.27}$$

where

$$N_T(\theta) = \frac{1}{\sigma_T(\theta)} \exp\left[-c(\beta)\left|\frac{\nu_T(\theta)}{\sigma_T(\theta)}\right|^{2/(1+\beta)}\right]. \tag{4.28}$$

During a flood event more and more data will become available from observations. With an updating of posterior densities the differentiation between parameter sets within the ensemble can be adapted stepwise.

4.4 Case Study: Ensembles as a Part of a Flood Forecast System for the Mulde River Basin

The different ensemble based methods presented above were applied in a case study for the Mulde River basin (Schumann, 2009; Dietrich et al., 2009, 2008). The Mulde River basin is located at the northern part of the Ore Mountains which forms the natural borderline between Germany and the Czech Republic. The watershed has an area of around 7,400 km^2. The basin is formed by several parallel sub-basins, draining from South to North. Especially for west-cyclonic rainfall events, which have caused extreme floods in the past, the uncertainty of precipitation forecasts in location, time and volume is crucial. To develop an operational flood management system, the hydrological model ArcEGMO (Becker et al., 2002) was adapted to the needs of operational flood forecasting. Three ensemble based methods were applied to consider the uncertainties of initial state variables, forecasted meteorological inputs and model parameters:

- the EnKF,
- two meso-scale EPS and one lagged average EPS to consider the uncertainties of precipitation forecasts and
- a parameter ensemble, combined with Bayesian inference to select the best parameterisation considering the difference between forecasts and observations.

The model ArcEGMO is a GIS-based modular modelling system containing several sub-models. It is a conceptual model, however several parameters have a physical meaning and some of them can be derived directly from a geodatabase. The watersheds are partitioned into homogenous units (hydrotopes), for which the vertical and horizontal processes are simulated. Distribution functions for essential parameters like saturated conductivity, field capacity, etc. are applied to consider the spatial variability of land use and soil related properties. The model concept is shown schematically in Fig. 4.5. With regard to the geomorphological structure of the river basin, the model was differentiated horizontally into uplands and hillslopes. Subsurface flow is modelled with a system of linear reservoirs. A second, fast responding reservoir is activated if the water level in the soil reservoir exceeds a threshold (Becker et al., 2002). A sensitivity analysis (Wang et al., 2007) revealed

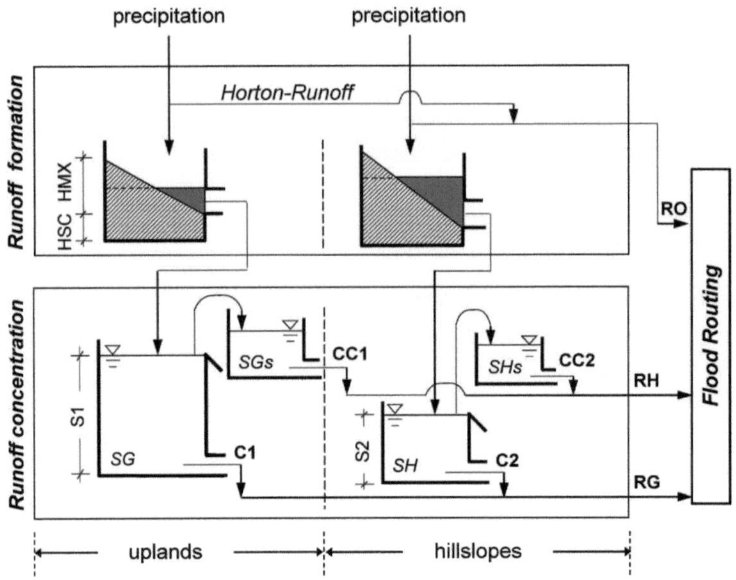

Fig. 4.5 Conceptual diagram of the hydrological model ArcEGMO (Becker et al., 2002) as it was applied to the Mulde River Basin using a subdivision into two different hydrological systems (uplands and hillslopes)

that the model parameters for the fast reservoirs are most relevant for the calibration of runoff in the headwaters during the summer season.

The model can be used in a water balance simulation mode with a daily time step continuously or event-based in a flood mode with hourly time steps. The system state of the water balance mode is used as initial state of the flood event mode. The flood model can be operated parallel with different parameter sets and different meteorological inputs (e.g. from rainfall ensemble forecasts). So the computational demand is reduced to one set of model initialization runs including a state update procedure and a large number of semi-automated process simulations are executed for a limited number of time-steps. This strategy allows an efficient integration of ArcEGMO into the computation and optimization of a probabilistic flood forecast chain.

Uncertainty in initial conditions (e.g. soil moisture, content of groundwater storage) is reduced by an update of the state variable with the Ensemble Kalman Filter. The efficiency of this approach is shown with the example of 1 year simulations with daily time step in the water balance mode (Fig. 4.6). The application of EnKF improves particularly simulations of recession periods. This is advantageous for flood forecasts where the initial state has to be assumed from these simulations.

The efficiency of EnKF is shown for the example of a single flood event in Fig. 4.7. At the left the differences between storage contents of simulations with and without updating by application of EnKF are shown. At the right, the impact of an improved approximation of the initial storage content is presented.

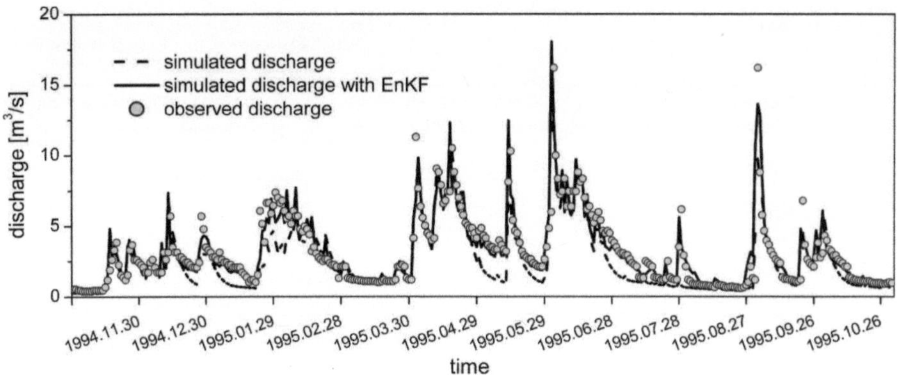

Fig. 4.6 Runoff simulation for 1 year (daily time steps) with and without an update of state variables by EnKF (year 1995, gauge Zoeblitz)

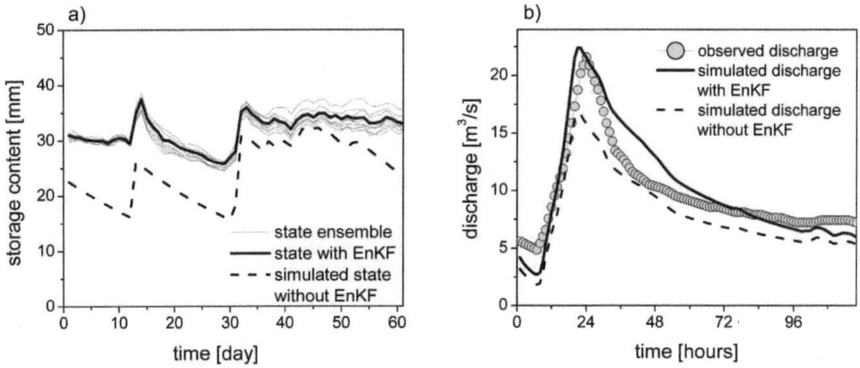

Fig. 4.7 Update of the initial state of the daily model (*left*) and the effect of an update of the initial state for flood simulations (*right*)

The EnKF was applied in this case study only for estimation of the initial state at the beginning of a flood event. It was used for updating the storage content of the continuous model which defines this initial state. Other ensemble methods which were applied in this case study were meteorological ensemble forecasts and parameter ensembles.

A meteorological ensemble system was developed which combines medium-range forecasts provided by COSMO-LEPS (3–5 days lead time), short-range forecasts from SRNWP-PEPS (1–2 days lead time) and very short-range forecasts from the COSMO-DE model (Doms and Förstner 2004) with 21 hours lead time and horizontal resolution of 2.8 km. The COSMO-DE model runs from different initialization times were combined to generate a lagged average forecast (LAF) ensemble. The ensemble simulation strategy is shown in Fig. 4.8. It differentiates between forecasts in medium and short ranges.

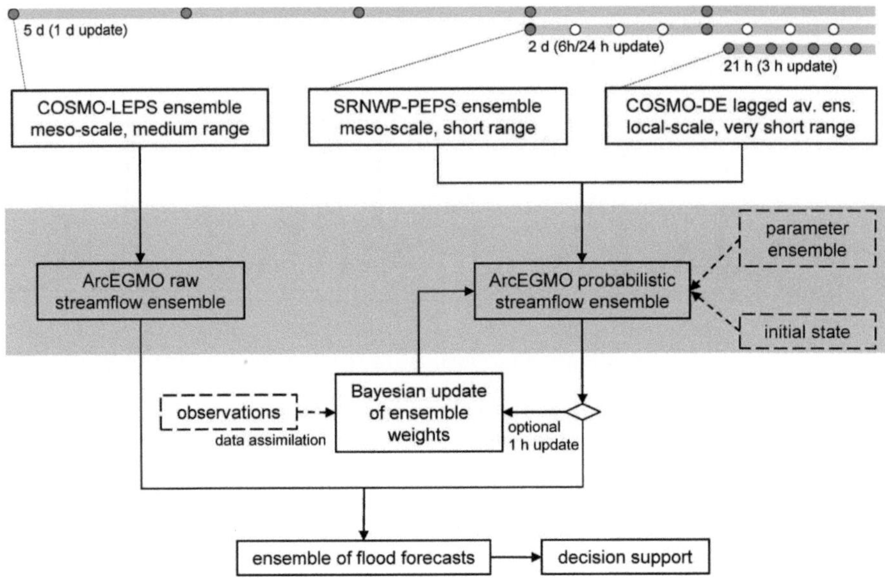

Fig. 4.8 Flow chart of the ensemble system applied in the Mulde River case study

Medium range forecasts have lead times of 3–5 days. These forecasts are provided for reservoir management and early warnings previous to a potential extreme flood event. The hydrological model ArcEGMO was applied with a default parameter set, which proved to be efficient for simulations of observed flood events in the past. It can be assumed that the hydrological uncertainty is explicitly lower than meteorological uncertainties of medium-range precipitation forecasts. These meteorological uncertainties are mainly represented by the spread of forecasts provided by COSMO-LEPS. An example for re-forecasts with COSMO-LEPS is given in Fig. 4.9.

Short to very short-range forecasts are useful for issuing flood alerts and initiating flood defence measures. Here decision makers need more detailed information about relevant criteria like peak time, peak discharge and possible inundation areas. Updates of meteorological forecasts are provided more often and observed data can be assimilated. For forecasts in short range the hydrological uncertainty becomes more important. Here the parameter ensemble was applied in combination with the COSMO-DE lagged average ensemble. Thus the hydrological model provides updated forecasts in 1–3 hourly intervals. Uncertainties in initial conditions were reduced by application of the EnKF as an iterative update procedure, which adjusts

Fig. 4.9 Sequence of re-forecasts ensembles for the flood in 2002, gauge Wechselburg at the Zwickauer Mulde River, a tributary of the Mulde River. COSMO-LEPS hindcasts were initialized at 09/08, 10/08 and 11/08/2002 at 12:00 UTC and processed by the model Arc-EGMO. The *dotted line* shows the observed flood wave

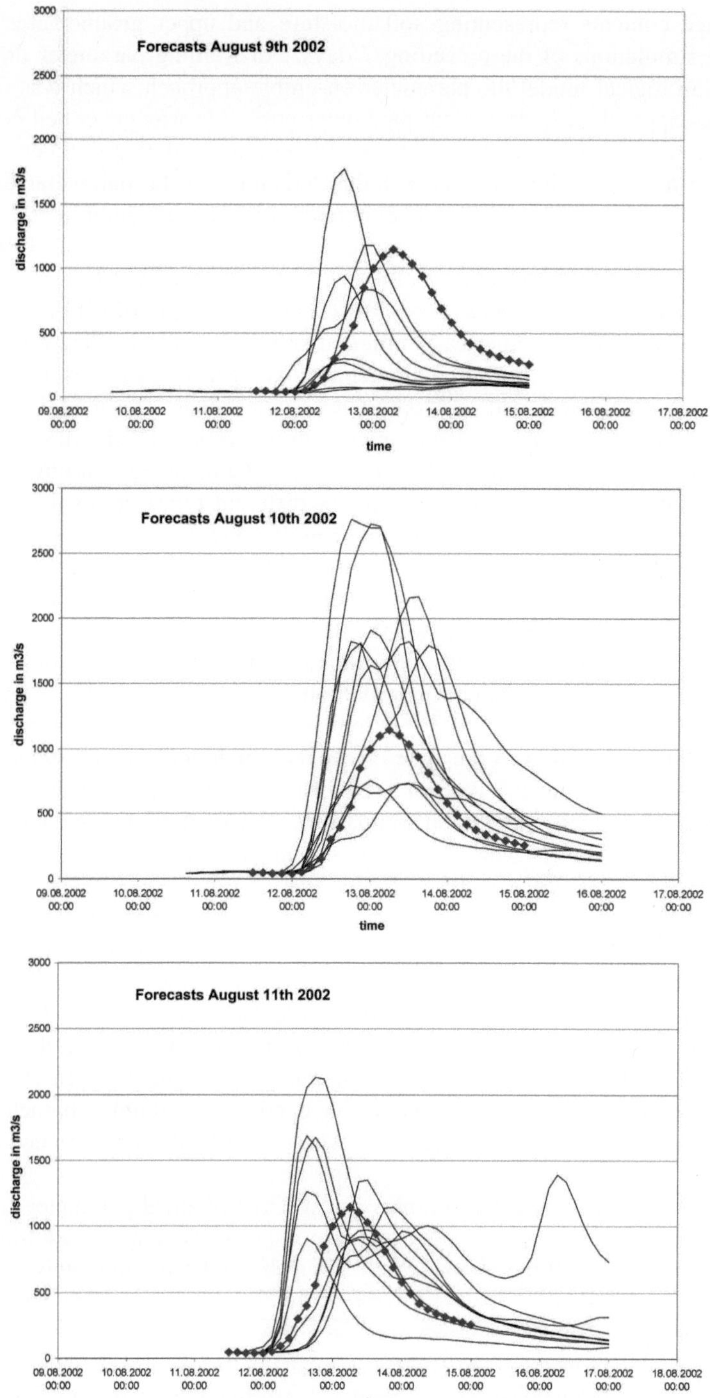
Forecasts August 9th 2002
discharge in m3/s
3000
2500
2000
1500
1000
500
0
09.08.2002 00:00
10.08.2002 00:00
11.08.2002 00:00
12.08.2002 00:00
13.08.2002 00:00
14.08.2002 00:00
15.08.2002 00:00
16.08.2002 00:00
17.08.2002 00:00
time
Forecasts August 10th 2002
discharge in m3/s
3000
2500
2000
1500
1000
500
0
09.08.2002 00:00
10.08.2002 00:00
11.08.2002 00:00
12.08.2002 00:00
13.08.2002 00:00
14.08.2002 00:00
15.08.2002 00:00
16.08.2002 00:00
17.08.2002 00:00
time
Forecasts August 11th 2002
discharge in m3/s
3000
2500
2000
1500
1000
500
0
09.08.2002 00:00
10.08.2002 00:00
11.08.2002 00:00
12.08.2002 00:00
13.08.2002 00:00
14.08.2002 00:00
15.08.2002 00:00
16.08.2002 00:00
17.08.2002 00:00
18.08.2002 00:00
time

Fig. 4.9 (Contd.)

the storage contents representing soil moisture and upper groundwater storages based on simulations of the preceding 7 days. For framing parameter uncertainty of the hydrological model the parameter ensemble approach which was presented above was applied. A hydrological parameter ensemble was generated combining model parameter sets, which proved to be efficient in simulating flood events in the calibration and validation periods. For the derivation of the parameter ensemble, optimization, stochastic methods and expert knowledge were combined. In the first step multi-criteria optimizations with different algorithms (Yapo et al., 1998; Vrugt et al., 2003) were used to estimate efficient parameter sets which were capable to simulate observed flood events. By selection and weighting of different objective functions a large set of numerically efficient solutions could be derived. Based on these results a large number of parameter sets was generated to cover the feasible range of the parameters with a Monte Carlo experiment. The model efficiencies based on these parameter sets differ between flood events. Nevertheless it was possible to specify a common default parameter set. The parameter ensemble contains sets of parameters, which were efficient for different types of floods. To specify these dependencies of parameters on the runoff conditions the floods were categorised into two groups. One group contains events which induced large floods with return periods of more than 50 years, the other ones "normal" floods with smaller return periods. Even though only two floods of the category "large" were available it became obvious that the quality of simulations for these events would be lower than for normal floods if the common default parameter set would be applied. At the beginning of an operational flood forecast the parameter ensemble members were weighted according to the type of events which could be expected a priori. If no information about the type of event was available, all parameter sets were equally weighted. During the flood simulations, observed discharge data were used to estimate the residuals which specified the likelihoods as was described above for the Bayesian inference method. In this way it became possible to update the weights of the ensemble member if new information became available. An example for this approach, based on the re-analysis of the disastrous 1954 flood event, is shown in Fig. 4.10. It starts with equal likelihoods at the beginning of the flood event (Fig. 4.10a). After 24 h observed discharge data were considered. A subset of the parameter set shows higher likelihoods and will be preferred in the following forecasts (Fig. 4.10b). After 52 h preference is given to another subset of parameters based on new observed discharges. A band of uncertainty characterises the differences between those parameter sets where the likelihoods exceeded a certain threshold.

When the meteorological ensembles and the hydrological parameter ensembles are combined within an operational flood forecast model chain, the resulting stream flow ensemble has 520 members, which has to be updated completely every 12 h (when all systems deliver new data) and partly every 3 h (when the COSMO-DE lagged average ensemble receives a new member).

Obviously the large numbers of forecasts can not be provided to decision makers directly without aggregation of the results. Unfortunately at the time of the case study a probabilistic differentiation between the members of the meteorological

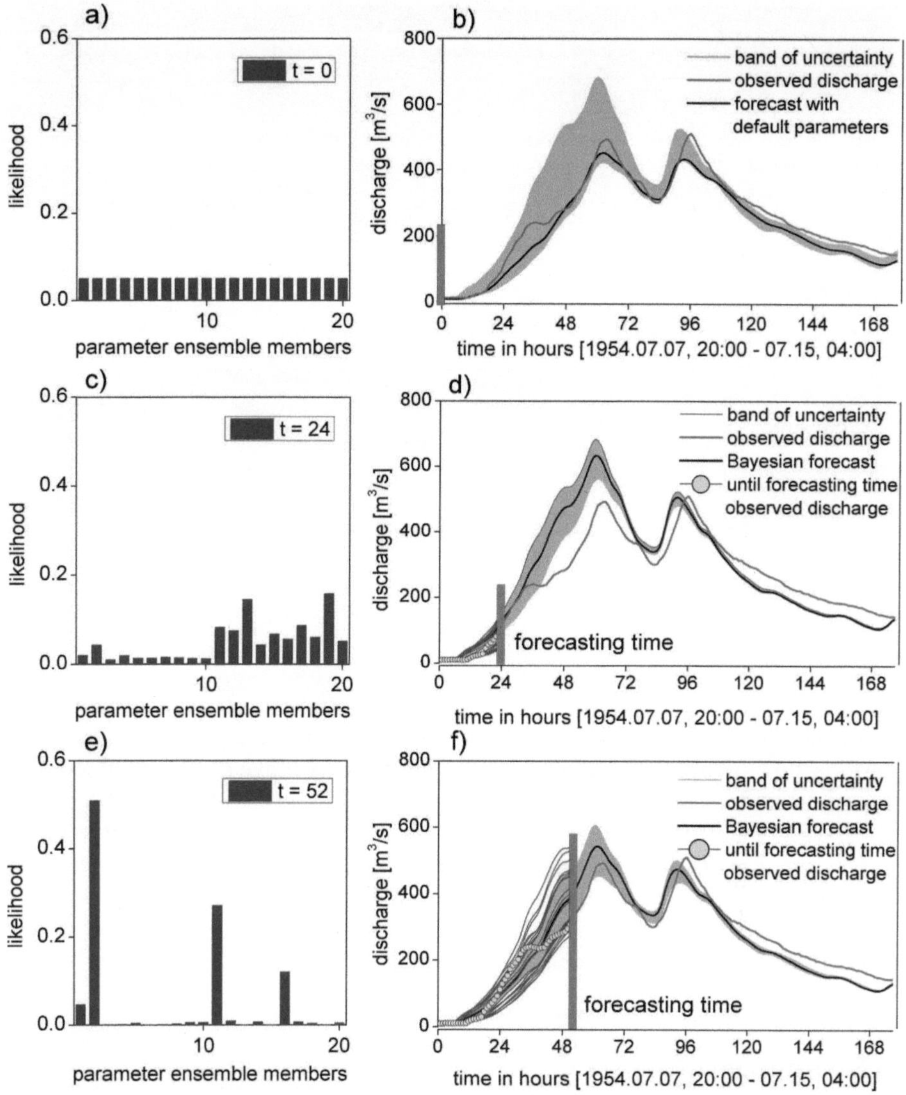

Fig. 4.10 Example for applications of the Bayesian inference to specify and reduce uncertainties of parameters by likelihoods, derived from assimilation of observed discharge values

ensemble was not available. However, a frequency analysis of ensemble forecast exceeding critical thresholds is possible but not appropriate from a probabilistic point of view. In medium range forecasts a large variety within the ensemble exists even if the number of forecast is limited.

For operational flood risk management in Germany flood alert levels are specified for runoff gauges. If the water level is exceeding these levels, damage preventing actions are planned. To support this system a post-processing tool which converted

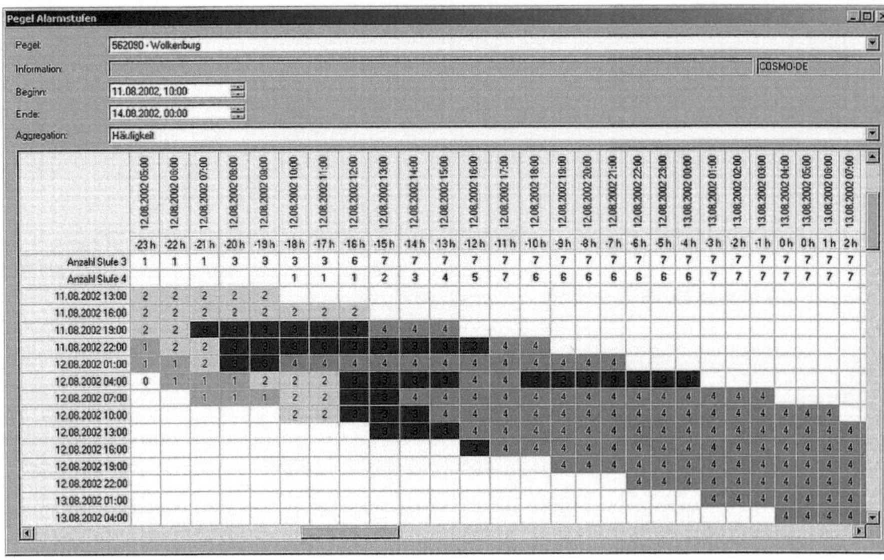

Fig. 4.11 Persistency diagram, describing the frequency of flood alert levels, forecasted with ensembles which were provided every 3 h (rows). The different gray shadows visualise the flood alert levels

simulated discharges into water levels was developed to integrate ensemble forecasts in this system. The predicted water levels were linked with the flood alert system to assess the actual risk of an exceedance of flood alert levels. After discussions with practitioners about their needs, alert persistency charts were developed, which visualize the frequency and duration (columns, background) of predicted alert levels (numbers 0–4) from ensembles, depending on the time of forecasts (rows, Fig. 4.11). These charts show the frequencies, tendencies and persistence of critical water levels. Details are given in Dietrich et al. (2009).

4.5 Summary

Flood forecasting is the attempt to specify future developments on the base of incomplete information. The uncertainties of these forecasts limit their practical value. Many attempts to specify these uncertainties are based on ensembles. Ensembles provide multiple forecasts based on variations of input variables, initial states, model physics or parameter values. They visualise the impacts of incomplete knowledge of these characteristics. In contrast to Monte Carlo-simulations they do not cover the whole range of uncertainties, but specify the main sources of uncertainties by scenarios. The added value of this approach consists in options to specify the current knowledge about future developments, to assimilate data which underline the plausibility of assuming a certain initial state, parameter set or meteorological forecast which is competing with other assumptions. Thus the range of uncertainty can be confined and a forecasted ensemble of future developments can be reduced based on an improved knowledge base.

References

Ajami NK, Duan Q, Sorooshian S (2007) An integrated hydrologic Bayesian multimodel combination framework: confronting input, parameter, and model structural uncertainty in hydrologic prediction. Water Resour Res 43:W01403. doi:10.1029/2005WR004745

Becker A, Klöcking B, Lahmer W, Pfützner B (2002) The hydrological modelling system ARC/EGMO. In: Singh VP, Frevert DK (eds) Mathematical models of large watershed hydrology. Water Resources Publications, Littleton, CO, 891 pp

Beven KJ, Freer J (2001) Equifinality, data assimilation, and uncertainty estimation in mechanistic modelling of complex environmental systems. J Hydrol 249:11–29

Box GEP, Tiao GC (1992) Bayesian inference in statistical analysis, Wiley Classics Library Edition. Wiley-Interscience, Wiley, New York, NY

Brammer K, Siffling G (1994) Kalman-Bucy-Filter: Deterministische Beobachtung und stochastische Filterung. Oldenbourg Verlag

Buizza R, Houtekamer PL, Toth Z, Pellerin G, Wei M, Zhu Y (2005) A comparison of the ECMWF, MSC, and NCEP Global Ensemble Prediction Systems. Mon Weather Rev 133: 1076–1097

Cloke HL, Pappenberger F (2009) Ensemble flood forecasting: a review. J Hydrol 375(3–4): 613–626

Dempster AP, Laird NM, Rubin DB (1977) Maximum likelihood from incomplete data via the *EM* algorithm. J Roy Stat Soc B 34:1–38

Denhard M, Trepte S (2006) Calibration of the European multi-model ensemble SRNWPPEPS, 2nd THORPEX international science symposium, WMO/TD No. 1355, WWRP/THORPEX No. 7

Dietrich J, Schumann AH, Redetzky M, Walter J, Denhard M, Wang Y, Pfützner B, Büttner U (2009) Possibilities and drawbacks of ensemble techniques in framing the uncertainty of extreme flood forecasts. Nat Hazards Earth Syst Sci 9:1529–1540

Dietrich J, Trepte S, Wang Y, Schumann AH, Voß F, Hesser FB, Denhard M (2008) Combination of different types of ensembles for the adaptive simulation of probabilistic flood forecasts: hindcasts for the Mulde 2002 extreme event. Nonlinear Processes Geophys 15:275–286

Dietrich J, Denhard M, Schumann AH (2009) Can ensemble forecasts improve the reliability of flood alerts? J Flood Risk Manage 2:232–242

Diomede T, Davolio S, Marsigli C, Miglietta MM, Moscatello A, Papetti P, Paccagnella T, Buzzi A, Malguzzi P (2008) Discharge prediction based on multi-model precipitation forecasts. Meteorol Atmospheric Phys 101:245–265

Doms G, Förstner J (2004) Development of a kilometre-scale NWP-system: LMK. In: Doms G, Schättler U, Montani A (eds). COSMO Newsletter, consortium for small scale modelling, accessible via http://www.cosmo-model.org/ vol 4.:pp 168–176

Evensen G (1994) Sequential data assimilation with a nonlinear quasi-geostrophic model using Monte {Carlo} methods to forecast error statistics. J Geophys Res 99(C5):10.143–10.162

Georgakakos KP, Seo DJ, Gupta H, Schake J, Butts MB (2004) Characterizing streamflow simulation uncertainty through multimodel ensembles. J Hydrol 298:222–241

Hoeting JA, Madigan D, Raftery AE, Volinsky CT (1999) Bayesian model averaging: a tutorial. Stat Sci (4):382–417

Jazwinski AH (2007) Stochastic processes and filtering theory. Dover, Mineola, NY

Komma J, Reszler C, Blöschl G, Haiden T (2007) Ensemble prediction of floods – catchment non-linearity and forecast probabilities. Nat Hazards Earth Syst Sci 7:431–444

Krzysztofowicz R (1999) Bayesian theory of probabilistic forecasting via deterministic hydrologic model. Water Resour Res 35(9):2739–2750
Liu Y, Gupta HV (2007) Uncertainty in hydrologic modeling: toward an integrated data assimilation framework. Water Resour Res 43:W07401. doi:10.1029/2006WR005756
Molteni F, Buizza R, Marsigli C, Montani A, Nerozzi F, Paccagnella T (2001) A strategy for high-resolution ensemble prediction, part I: definition of representative members and global model experiments. Q J R Meteorol Soc 127:2069–2094
Moradkhani H, Sorooshian S, Gupta HV, Houser PR (2005) Dual state-parameter estimation of hydrological models using ensemble Kalman filter. Adv Water Resour 28:135–147, Elsevier Ltd.
Murphy AH (1993) What is a good forecast? An essay on the nature of goodness in weather forecasting. Weather Forecasting, 8:281–293
Raftery AE, Gneiting T, Balabdaoui F, Polakowski M (2005) Using Bayesian model averaging to calibrate forecast ensembles. Mon Weather Rev 133:1155–1174
Reed S, Schaake J, Zhang Z (2007) A distributed hydrologic model and threshold frequency-besed method for flash flood forecasting at ungauged locations. J Hydrol 337:402–420
Schaake JC, Hamill TM, Buizza R, Clark M (2007) HEPEX: the hydrological ensemble prediction experiment. Bull Am Meteorol Soc 88(10):1541–1547
Schumann A (Hrsg.) (2009) Entwicklung integrativer Lösungen für das operationelle Hochwassermanagement am Beispiel der Mulde, Schriftenreihe Hydrologie/ Wasserwirtschaft Ruhr- Universität Bochum, Heft 24
Thielen J, Bartholmes J, Ramos M-H, de Roo A. (2009) The European flood alert system – Part 1: concept and development. Hydrol Earth Syst Sci 13(2):125–140
Thiemann M, Trosset M, Gupta H, Sorooshian S (2001) Bayesian recursive parameter estimation for hydrologic models. Water Resour Res 37(10):2521–2535
Todini E (2010) Predictive uncertainty in flood forecasting and emergency management. Proceeddings of the 17th Congress of the Asia and Pacific Division of the IAHR, Auckland, New Zealand, February 24–27 accepted paper, personal communication
Trepte S, Denhard M, Göber M, Anger B (2006) SRNWP-PEPS: some results of verification, Second THORPEX international science symposium, WMO/TD No. 1355, WWRP/THORPEX No. 7
Verbunt M, Zappa M, Gurtz J, Kaufmann P (2006) Verification of a coupled hydrometeorological modelling approach for alpine tributaries in the Rhine basin. J Hydrol 324:224–238
Vrugt JA, Robinson BA (2007) Treatment of uncertainty using ensemble methods: comparison of sequential data assimilation and Bayessian model averaging. Water Resour Res 43:WO1411. doi: 10.1029/2005WR004838
Vrugt JA, Gupta HV, Bouten W, Sorooshian S (2003) A shuffled complex evolution metropolis algorithm for optimization and uncertainty assessment of hydrologic model parameters. Water Resour Res 39(8):1201
Wang Y, Gattke C, Schumann A (2007) Reducing the uncertainty of flood forecasts using multi-objective optimization algorithms for parameter estimation, in Quantification and reduction of predictive uncertainty for sustainable water resources management, IAHS Publ. 313, Perugia
Yapo PO, Gupta HV, Sorooshian S (1998) Multi-objective global optimization for hydrologic models. J Hydrol 204(1–4):83–97

Chapter 5
Design of Artificial Neural Networks for Flood Forecasting

Johannes Cullmann and Gerd H. Schmitz

Abstract This chapter provides an overview of currently available neural network design with the purpose of a timely warning for operational flood risk management, considering the need to evaluate the uncertainty of the forecast. Neural network models are very effective with regard to their computational requirements and provide new options for operational scenario analysis and ensemble forecasts. Here the widely used "multi layer feed forward network" is compared to an alternative, the "polynomial neural network". A new training strategy permits to discriminate between input vectors. This method opens a way to reflect physical facts by means of input vectors in neural models, i.e. the neural model is portraying the rainfall runoff process on the basis of process understanding and physical boundary conditions of the considered catchment.

Contents

5.1 The Challenge of Flood Forecasting

Rain-induced floods represent one of the most common and dangerous natural hazards for small to medium sized catchments. Floods in this chapter are defined as

J. Cullmann (✉)
Federal Institute of Hydrology, Koblenz, Germany
e-mail: cullmann@bafg.de

A.H. Schumann (ed.), *Flood Risk Assessment and Management*,
DOI 10.1007/978-90-481-9917-4_5,

high flow events which cause or threaten damage (NWS/NOAA, 2007). Often these floods are a consequence of severe rainstorms of short duration but high rainfall intensities. Regarding the total runoff volume, they are often much smaller than long-lasting floods in large river systems which affect vast inundation areas. Nonetheless, due to the critical discharge velocities and steep gradients in the rising limb of hydrographs, such floods pose a serious threat to human life.

In the effort of anticipating flood related damages to human lives and economic goods, structural flood retention is the most frequently applied measure. It is a very effective way to reduce flood impacts on human society. However, timely flood forecasts are essential to control technical flood retention structures in an optimal way. Structural flood retention requires large investments if rare flood events are used in hydrological design. Nonetheless, there is no absolute protection from extremely rare events by means of technical measures.

Especially for fast reacting watersheds more efficient flood forecasting systems are of primary importance. Prolonging forecasting and warning times enables the affected people to safeguard their belongings as well as their lives in case of a devastating flood event.

Assuring a lead time with a span that allows for an effective reaction to the forecast requires taking into consideration a quantitative precipitation forecast and a detailed and comprehensive description of rainfall runoff processes. Therefore rainfall-runoff models are the means of choice for flood forecasting. If forecasting is based on measured rainfall, the gain in lead-time is a function of various processes, e.g. interception, surface water storage, soil storage, surface water travel times and flood wave propagation in river channels. When the lead-time is insufficient for effective flood mitigation and management measures- which is generally the case in small and steep catchments – radar-based now-casting and/or quantitative precipitation forecast can be applied to prolongate the forecasting time significantly. Here, the hurdle is the inclusion of the precipitation forecast uncertainty. Online evaluation of this uncertainty requires an extremely fast forecasting model which could be used e.g. for ensemble forecasts very efficiently.

With regard to these problems, the two main focal points of recent research in flood forecasting are:

- Classical hydrological modelling approaches which rely on routing, rainfall runoff modelling and ensemble forecasts in the context of flood forecasting.
- Data-driven models which comprise artificial neural networks (ANN) and probabilistic methods. They are being investigated as a fast and reliable means of flood forecasting.

In this chapter we focus on ANN. The main contribution of this work consists in a description of artificial neural networks which are capable to represent all important rainfall-runoff processes in fast reacting watersheds in the context of rain-induced floods.

The approach presented here takes full advantage of artificial intelligence in the form of neural networks. In the following sections neural networks based strategies are explained in detail for readers which are not familiar with this topic.

5.2 Representation of Rainfall-Runoff Processes with Artificial Neural Networks

Trying to compensate for some of the inconveniences of highly sophisticated numerical approaches (computational effort, data requirements), a considerable amount of research has been invested during the last 15 years for adapting the theory of ANN to flood modelling and forecasting. This was mainly motivated by the principal advantage of neural nets: Once they are trained they are extremely easy to use and outperform classical models by far in terms of computational time requirements and simulation speed. General aspects concerning artificial neural networks and their role in hydrology are concisely reviewed in (ASCE, 2000a, b). Apart from this fundamental work, a vast number of detailed publications describe the advances in the field of applying neural nets to hydrological modelling:

Hjelmfelt and Wang (1993) developed a neural network based on the unit hydrograph theory. Using linear superposition, a composite runoff hydrograph for a watershed was developed by the appropriate summation of unit hydrograph ordinates. Smith and Eli (1995) applied a back propagation neural network model to predict peak discharge and peak time for a hypothetical watershed. In their study, non-linear reservoir models generate data sets for training and validation. Hsu et al. (1995) propose straight forward predictions of stream flow employing a three-layer network. Hsu et al. (1997) further developed the potential of recurrent neural approaches in the context of hydrological modelling. Dawson and Wilby (1998) used a three-layer back propagation network to determine runoff from the catchments of the rivers Ambers and Mole in England. The two catchments are of nearly equal size (about 140 km^2). Observed flow data and mean historical rainfall data serve as inputs in their study. Their results show that nets perform similar to an existing model with less input information. Zealand et al. (1999) described the potential of neural nets for short term forecasting of stream flow. Their work explored the capabilities of artificial neural networks and compared their performance to conventional approaches used to forecast stream flow. Dawson and Wilby (2001) provide a general overview about neural networks and their application with respect to rainfall runoff modelling.

After an introduction of the most widely used class of ANN – the multi layer feed forward net – polynomial neural networks are explained. The focus here lies on the structure of both net types as well as their specific training procedures.

5.2.1 Multi Layer Feed Forward Nets

Multi layer feed forward nets (MLFN) are widely applied for rainfall-runoff modelling. They have been introduced into the flood forecasting context in the early 1990s. Lately, more attention has been paid to MLFN. Hu et al. (2001) use this technique for river stage forecasting. Imrie et al. (2000) test different activation functions for the output layers of MLFN. Cullmann et al. (2006) promoted this net type in their flood forecasting approach. This study evaluates MLFN performance in the context of rainfall-induced flood forecasting.

5.2.1.1 Principles of Multi Layer Nets

The principal function (Fig. 5.1) of multi layer nets is based on the neurons in hidden and output layers. They transform their respective inputs to outputs through two separate stages. First, for each neuron, all single inputs are multiplied by the corresponding weight (w) and secondly, the total sum of these products plus a constant known as bias (b) yields the node output in the hidden and output layers.

An elementary neuron with n inputs is shown in the lower part of Fig. 5.1. Each input is weighted with an appropriate weight w. The sum of the weighted inputs and the bias forms the input to the transfer function f (.). Any differentiable transfer function f (.) is suited to generate output. The output S from the jth node in the hidden layer – after the summation operation – is defined as follows:

$$S_j = \sum_{j=1}^{n} W_{ji} X_i + b_j \tag{5.1}$$

with: n = number of elements in the vector
X_i = the input signal from ith node
W_{ji} = MLFN weights
b_j = MLFN bias

Then the net output Y_j from the jth output node is:

$$Y_j = f(S_j) \tag{5.2}$$

5.2.1.2 Structure of Multi Layer Neural Networks

The network geometry determines the number of connection weights and the way how these are arranged. Generally, a fixed number of hidden layers is defined and the number of nodes in each layer is chosen afterwards. It can been shown that three layer networks (1 input – 1 hidden – 1 output layer) with sigmoid transfer functions in the hidden layer and linear transfer functions in the output layer can approximate virtually any function of interest to any degree of accuracy, provided that a sufficient number of neurons are available in the hidden layer (Hornik et al., 1989).

The number of nodes in the input layer is determined by the number of input vectors, whereas the number of nodes in the output layer has to be equal to the number of model outputs. The critical aspect is the choice of the number of nodes in the hidden layers and hence the number of connection weights. The importance of finding a balance between a sufficient number of free parameters (weights) to guarantee the desired representation of the function which has to be approximated, and too many free parameters, which will result in an over-fitting, is well known. This issue has

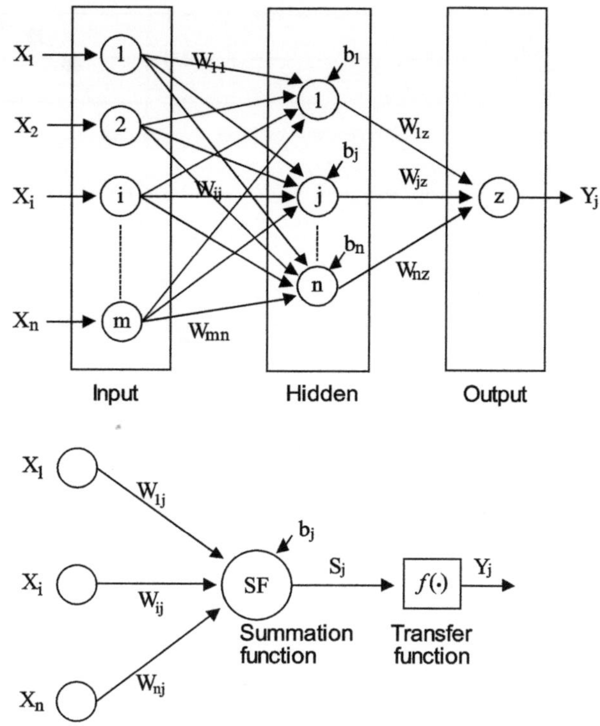

Fig. 5.1 Typical three-layer network structure (*top*) of an ANN and operating scheme for a single node *j* (*bottom*)

been discussed widely in the literature (e.g. Maren et al., 1990; Rojas, 1996). The number of hidden layer nodes influences the performance of a network significantly. With too few nodes, the network approximates the system output poorly. The ANN will over-fit the training data if too many nodes are applied. Consequently, an optimal hidden layer geometry reduces both the computational effort which is needed for a training of the ANN and ensures the best possible generalization performance, avoiding the problem of over-fitting.

Until now no general theory exists for determining optimal network geometry. In the case study which is presented here three layers feed forward networks are used. Following Hornik et al. (1989), this type ensures the required net performance in the flood forecasting context. The number of input neurons equals the number of input vectors, while one fixed node represents the model output.

The transfer function converts the effective incoming signal of node j, S_j (Eq. 5.1) into the output signal (Y_j). Multi layer nets typically use sigmoid transfer functions in the hidden layers. These functions are often called "squashing" functions, since they compress an infinite input range into a finite $(-1;\ 1)$ output range. Here non-linear bipolar tan-sigmoid transfer functions are used for hidden layer nodes (Eq. 5.2; Fig. 5.2), whereas linear transfer functions are characterising the output layer node.

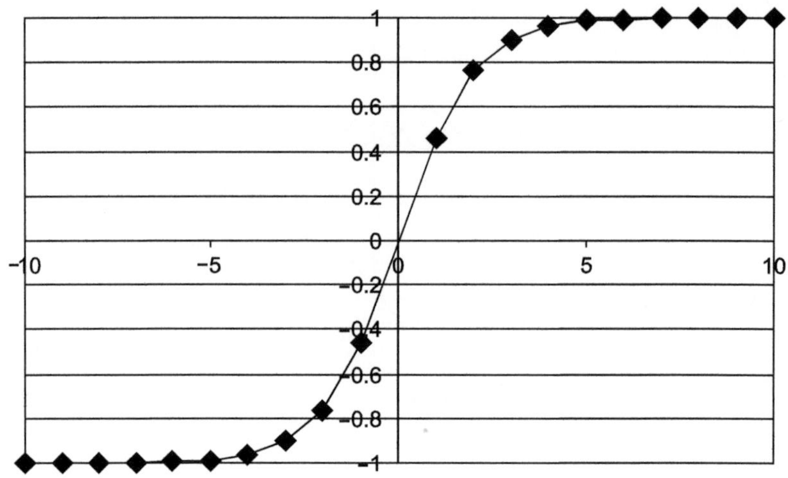

Fig. 5.2 tan-sigmoid function (Eq. 5.3)

$$f(S_j) = \frac{2}{1 + \exp^{-S_j}} - 1 \tag{5.3}$$

with: S_j = input
$f(S_i)$ = output of the transfer function.

The characteristics of the tan-sigmoid function are:

- Upper and lower bounds exist;
- it is monotonically increasing;
- it is continuous and differentiable everywhere.

The simple linear transfer function used for the output layer of the MLFN is expressed as:

$$f(S_j) = S_j \tag{5.4}$$

with: S_j = input

5.2.1.3 Training of Multi Layer Nets

The optimisation of the multi layer network weights is known as "training" or "learning". This process can be compared to the parameter estimation phase of conventional hydrological modeling. The aim is to find a global solution for a highly non-linear optimisation problem (White, 1989). Therefore, the general theory of non-linear optimisation seems to be the method of choice (Battiti, 1992). The suitability of a particular method is generally a compromise between computational cost and training performance (Parisi et al., 1996).

There are numerous approaches for network training. The most popular algorithms are based on supervised training, which has also been used in the following case study. Paradigms of supervised learning include error-correction learning, reinforcement learning and stochastic learning. An important issue concerning supervised learning is the problem of error convergence, i.e., the minimization of the objective function. Here, the aim is to determine a set of weights which minimises the mean square error. One well-known method, which is common to many learning paradigms, is the least mean square convergence, which was also adopted in the present study.

After a careful and thorough investigation of different training algorithms, the Levenberg-Marquardt back-propagation algorithm (Hagan and Menhaj, 1994), was chosen to optimise the weight and bias parameters during the training process. It is a quick and stable second order non-linear least square technique (Toth et al., 2000).

The Levenberg-Marquardt algorithm is a modification of the classic Newton algorithm for finding an optimal solution to a minimisation problem. It is designed to approach second-order training speed and accuracy without the need to compute the Hessian matrix. The Hessian matrix contains second derivatives of the network errors with respect to the network weights, while the Jacobian matrix contains first derivatives of the network error matrix with respect to weights. If the performance function has the form of a sum of squares (as is typical in training feed-forward networks), then the Hessian matrix can be approximated as:

$$\mathbf{H} = \mathbf{J}^{\mathbf{T}}\mathbf{J} \tag{5.5}$$

with: $\mathbf{J}$ = the Jacobian matrix
$\mathbf{H}$ = the Hessian matrix

and the gradient can be computed as:

$$\mathbf{g} = \mathbf{J}^{\mathbf{T}}\mathbf{e} \tag{5.6}$$

where $\mathbf{e}$ is a matrix of network errors.

5.2.2 Polynomial Neural Nets

Shin and Gosh (1992), Foka (1999) and Ma and Khorasani (2005) applied Polynomial Neural Networks (PoNN) for time series predictions and achieved promising results. Their analyses describe the options of using polynomial nets as a Taylor approximation of the rainfall-runoff function. Consequently, polynomial nets and their potential applicability in the flood forecasting context are discussed in this chapter. Hereafter, "characteristic feature" describes an input vector of a net. It contains important information about the rainfall-runoff process. A general description of polynomial neural networks is given in the following.

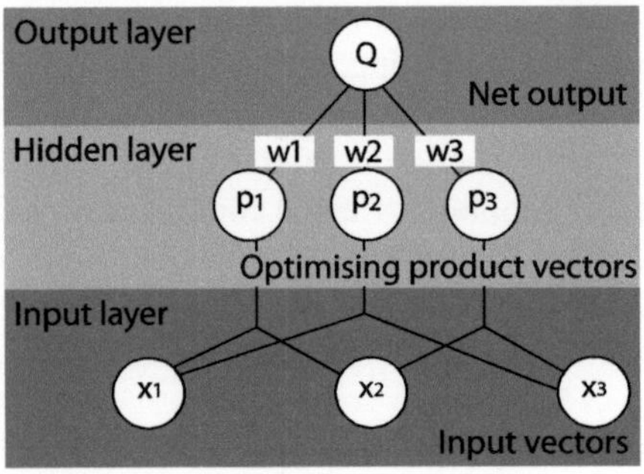

Fig. 5.3 Scheme of a polynomial net with $(x_1 \ldots x_3)$ = input vectors, $(p_1 \ldots .p_3)$ = hidden layer, $(w_1 \ldots .w_3)$ = hidden layer weights, Q = output

5.2.2.1 Basics of Polynomial Neural Networks

Here the overall polynomial approximation of the predicted discharge Q at the outlet of a catchment is described. A polynomial network is a feed-forward network with a single hidden layer. The output of the "hidden" layer is the product of the input terms while the output of the network is the weighted sum of these products (Fig. 5.3). Polynomial nets have only one layer of adaptive weights. This result in a very effective training with short CPU times compared to the classical multi layer nets.

The output of the network is given by:

$$Q = \sum_{i=1}^{N} w_i p_i \tag{5.7}$$

with: Q = output of the PoNN (discharge)
w_i = linear weights of the PoNN (determined in the training)
p_i = product vectors
N = number of vectors (3 in Fig. 5.3)

The PoNN product vectors result from a permutation of the inputs **x** (Fig. 5.3). They are exemplarily derived for a third degree PoNN according to:

$$p = x_k^a x_l^b x_m^c \tag{5.8}$$

Where **x** are input vectors with $k = (1 \ldots N)$, $l = (1 \ldots N)$, $m = (1 \ldots N)$. The exponents a, b, c, are realizations of the integers [0;3]. The power a, b, c of Eq. (5.8) satisfies the criterion:

$$a + b + c \leq g \tag{5.9}$$

where g is the degree of the polynomial applied.

In general the use of second or third degree polynomials is sufficient to guarantee a satisfactory forecast ability of the PoNN. In this case $a + b + c \leq 3$ sets the frame for the definition of product vectors, i.e. three features are multiplied or one squared feature is multiplied with another feature, or just one feature to the power 3 is used. The PoNN transforms the input vectors directly into the flood forecast for a pre-defined river gauge. However, the total number of product vectors rises rapidly when high degree polynomials or a large number of input vectors are used. This can be understood easily from the relation expressed in Eq. (5.10). Here N rises rapidly with increasing g or n:

$$N = \binom{n+g}{g} \tag{5.10}$$

5.2.2.2 Training of Polynomial Nets

A new algorithm which has been developed for the training of polynomial nets in the rainfall runoff context is stepwise serial regression (SSR) by Görner et al. (2006). It consists of a combination of regression methods using both Efroymson`s algorithm as described in Miller (1999) and stepwise regression presented in Meyer-Brötz and Schürmann (1970). In the training process of the PoNN (Fig. 5.4), first all possible product vectors **p** are calculated by means of permutation of the characteristic features (Eq. 5.8). These vectors are stored afterwards in the vector matrix **P.** The training process of the PoNN consists mainly in optimising the net structure and weights. This is achieved by the transformation of **P,** into a new – much smaller – optimised matrix **O**, which represents the structure of the trained net. The corresponding linear weights w_i of the optimised vectors **p** of **O** are determined by SSR. Herein, the general estimator for the target value $\Delta(x)$ is defined as follows:

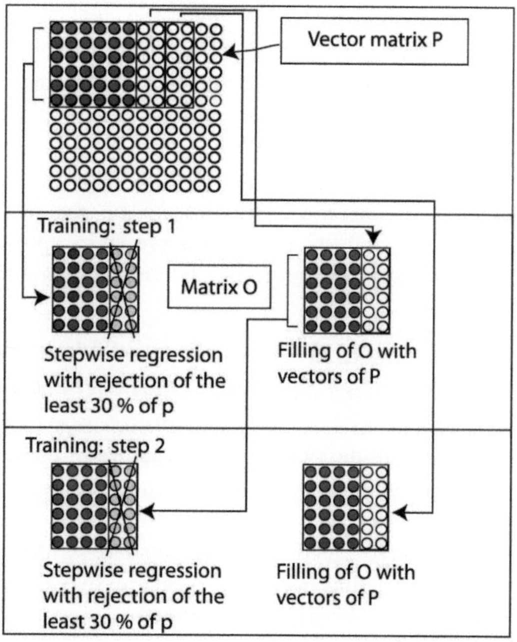

Fig. 5.4 The first two training steps of serial regression polynomial net training

$$\Delta(x) = \mathbf{wO} + \sum_{i=1}^{n} w_i x_i \tag{5.11}$$

where: w = weight
x = regressor

In the selection process the objective function (deviation of the target value from the dependent variable (Δx)) has to be minimised:

$$E\{[z - \Delta(x)]^2\} \Rightarrow Min \tag{5.12}$$

The training process itself starts with the selection of the first n of N vectors of the vector matrix **P** (top part of Fig. 5.4). These vectors represent the initial matrix **O**, which has to be optimised. Now the iterative optimisation (training) can be carried out: The stepwise regression arranges the n vectors in **O** with the goal to reduce the total error (MSE of observed and modelled flow), i.e. the first vector is the most important for the description of Q, the second vector is the second most important and so on. This process is also known as maximum scatter minimisation. The criterion for the scatter minimisation (ΔR) is calculated from the nth Element in **O** according to:

$$\Delta R = \frac{E(x_i z_i)}{E(diagonal \in \mathbf{O})} \tag{5.13}$$

with: $E(x_i z_i)$ = element xz
$E(diagonal \in \mathbf{O})$ = diagonal element in **O**

Based on this ranking, the worst 30% of the vectors are rejected and not considered further (grey vectors in Fig. 5.4). Now vectors from **P** are drawn to fill the places of the rejected vectors in **O**. Again, the n vectors are ranked and reduced by rejection of the least 30%. This procedure is repeated until the n optimal vectors of **O** are selected from all vectors **p** in **P**. In this way the training process runs consecutively through the total amount of all the product vectors **p**. The key advantage of this training method is the information content of the features, which, in contrast to the classical MLFN, can be interpreted in a physically meaningful way. The optimal size of **O** is a function of the catchment characteristics.

5.2.3 Comparative Analysis of Multi Layer Net and Polynomial Network Structures with Regard to Hydrological Problems

To determine the optimal net structure in the context of flood forecasting, preliminary testing involves a comparison of the two methods considered.

- Multi layer feed forward nets are widely applied. Their adoption to flood forecasting has been promoted e.g. by Cullmann et al. (2006)
- Polynomial neural nets have been proposed for river discharge modelling e.g. by Foka (1999)

Both types of ANN were applied and compared in a case study at the Zschopau River in the Ore Mountains in Germany, to provide forecasts for the Kriebstein gauging station. The catchment area at this gauge is 1,757 km^2. A detailed description of the area can be found in Görner et al. (2006). The tested three layer net was generated according to principles described in Cullmann et al. (2006). It was trained with the strategy presented above. The three layer net was fed with 30 input vectors containing information about rainfall, temperature and pre-event moisture state of the watershed. The polynomial network was used with the same input vectors. It was generated and trained according to the standard procedures which were described above. The test was carried out with a lead-time of 24 h for both net types. The input vectors were derived from precipitation, air humidity, wind speed, global radiation and temperature data which were provided as 1 km^2 grids for 47 years in hourly time steps. First of all these data were transformed into runoff by the calibrated conceptual hydrological model WaSiM (Schulla, 1997; Cullmann et al., 2008). The ability of ANN's to represent the simulated runoff was used as the evaluation criterion for the net performance. All 47 years of input data have been used to discriminate between the two methods concerning training speed, operational speed, stability, forecast reliability, and easiness of model set up. The approaches were trained on 80% of the available data – 20% of the database was reserved for validation.

The scheme in Fig. 5.5 shows the principles of the two approaches. The architecture is generally different, but some common features exist and simplify a direct

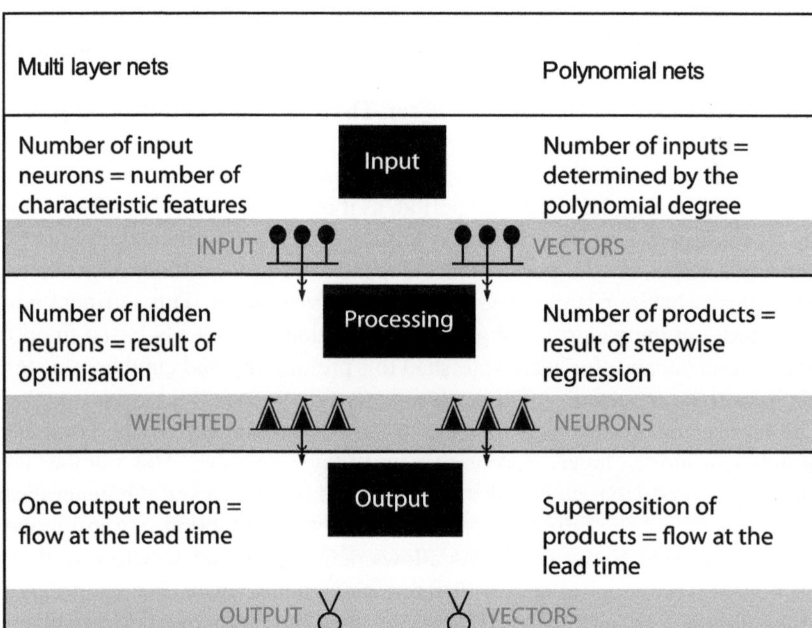

Fig. 5.5 Comparison of architectures multi layer net – polynomial net

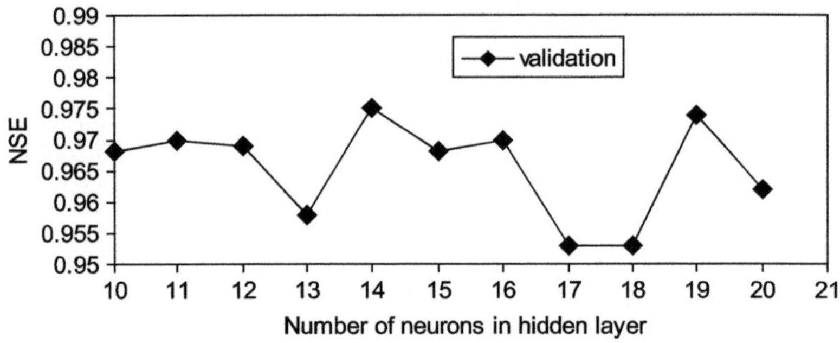

Fig. 5.6 Nash-Sutcliffe efficiency for the test data set versus number of neurons in hidden layer for a 24 h lead-time prediction with the three layer net

comparison of the results. The number of input neurons of the multi layer net and the number of characteristic features used to represent the rainfall runoff process for the considered watershed were equal. For the polynomial net, the input information consists of the same characteristic features, permutated according to the polynomial degree. The processing of the input information was solved in the hidden layer of the three layers net. This layer consists of a number of hidden neurons. Extensive testing is necessary to determine the optimal number of hidden layer nodes. The number of hidden neurons of a multi layer net can be compared to the number of products used for the forecast in the polynomial net. This value is determined by optimisation. The output of both polynomial and multi layer nets was one-dimensional; it represents the runoff at the considered lead-time.

As a first result of this comparison, Fig. 5.6 shows that the Nash-Sutcliffe efficiency (NSE) of the validation period. This criterion varies depending on the number of hidden layer nodes in the multi layer net. The result indicates that the predictive power of the MLFN is unstable for varying net architectures. This is not only the case for the Nash-Sutcliffe efficiency, but also proves for the mean square error of the total runoff during the validation period, as it is shown in Fig. 5.7. The behaviour of the forecast performance criterion is unforeseeable for an increasing number of hidden layer nodes. As the performance depends strongly on the input vectors and weights, this behaviour is characteristic for each watershed. This attribute poses a serious hurdle in the way of easy general application of a MLFN based flood forecasting system because for each watershed this preliminary and cumbersome testing would have to be repeated.

The number of product vectors used in the polynomial net can be compared to the number of hidden layer neurons of a multi layer network. This number has an impact on the predictive power of the polynomial net. However, the generalisation performance of a polynomial net is a clearly defined function of the number of product vectors which are used. This allows defining the net structure in the general set up of a polynomial net for different catchments easily. For the polynomial net, the validation graph (Fig. 5.8) illustrates how the predictive performance rises with an increasing number of product vectors. It reaches an optimum, which is a

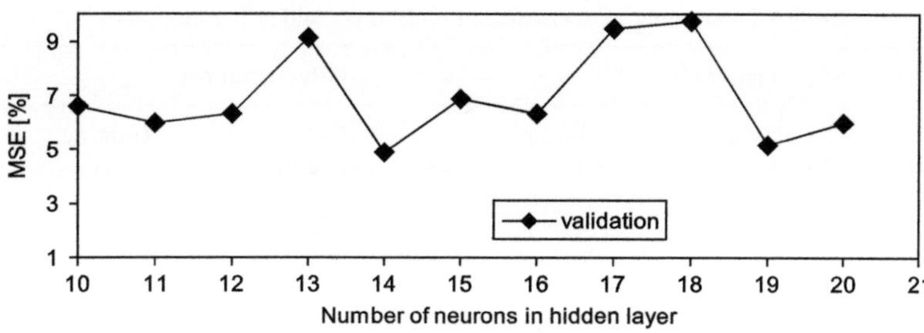

Fig. 5.7 Mean square error for the test data set versus number of neurons in hidden layer for a 24 h lead-time prediction with the three layer net

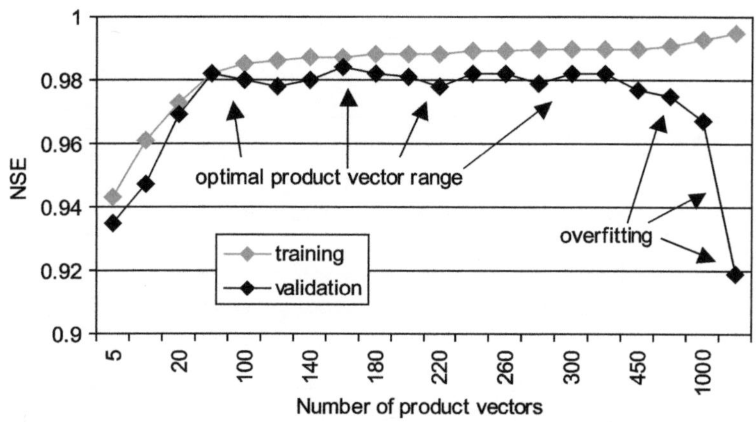

Fig. 5.8 Nash-Sutcliffe efficiency for training and test data versus number of product vectors for a 24 h lead-time prediction with the polynomial net

plateau in the test case (vector numbers ranging from 120 to about 450). Generally the optimal configuration is the one which fully describes the dynamics with the least number of product vectors. The optimum is a function of effort (increasing numbers of vectors lead to more computational efforts) and predictive reliability. The optimal setup may be readapted for each new forecasting task according to the requirements of the users and available resources. Generally, increasing the number of employed product vectors improves the training results of nets. Obviously, with increasing degrees of freedom the ability to generalise decreases. This is visualized in Fig. 5.8, where product vector numbers > 550 show significant reductions in the Nash-Sutcliffe efficiency for the validation period (predictive ability). Again, this is confirmed by a similar behaviour of the mean square error MSE (Fig. 5.9).

In a second test, optimal three layer net architectures were compared to a 3rd grade PoNN with 220 product vectors. The number of product vectors guarantees both good training results and a stable predictive performance (Fig. 5.8). The architecture of the multi layer nets was separately optimised for each lead-time. This results in varying numbers of hidden layer neurons listed in column 2 of Table 5.1.

Table 5.1 Training performance of multi layer and polynomial nets

	Multi layer net			Polynomial net		
Lead time	No. of hidden layer nodes	Training (min)	NSE	Grade/No.of features	Training (min)	NSE
6	14	111	0.97	3/220	45	0.97
12	14	26	0.93	3/220	45	0.98
18	16	111	0.86	3/220	45	0.98
24	14	237	0.85	3/220	45	0.97
36	15	28	0.86	3/220	45	0.97
48	16	34	0.84	3/220	45	0.97
Mean	15	90	0.88	3/220	45	0.97

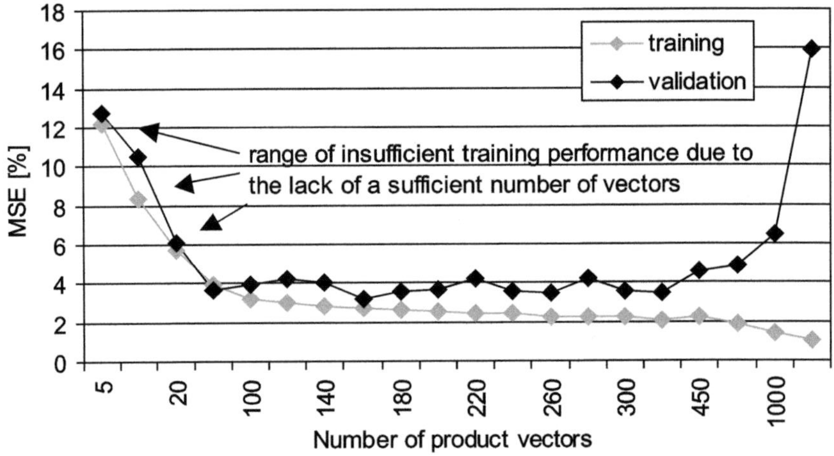

Fig. 5.9 Mean square error for the test data set versus number of neurons in hidden layer for a 24 h lead-time prediction with the three layer net

The optimal number of hidden neurons of the multi layer net varies between 14 and 16 (Fig. 5.7). For the optimal multi layer architectures, the Nash-Sutcliffe efficiency decreases with increasing forecast lead-time, i.e. the predictive power of the multi layer net decreased if longer lead times were considered (Table 5.1). Contrary to this finding, polynomial nets are characterized by a stable forecast performance (NSE) with increasing lead-times. Obviously, this is an essential aspect. It suggests preferring polynomial nets to multi layer nets in the flood forecasting context. This second test also reveals that the training time is constant for polynomial nets. In contrast, multi layer nets exhibit very inhomogeneous training times (Table 5.1). The mean multi layer training time is doubled compared with the time needed to train polynomial nets. The CPU requirement for flood forecasts, which is most important for operational applicability, is the same for the two methods.

The general results of the test are: The predictive power of polynomial nets is superior to the one of classical multi layer feed forward neural networks. Polynomial nets can be configured easily as there is a clear relationship between structure and performance (Fig. 5.8). Their fast and stable training, together with moderate operational CPU requirements (Table 5.1) allow for a general and easy application of polynomial net based forecast tools. The comparison identifies polynomial nets as the method of choice for operational flood forecasting applications. The convincing reason for this statement is the stable performance criterion of the polynomial net based system with increasing lead times. This characteristic is of paramount importance for a reliable forecast and will become more and more important in the future as rainfall forecasts will become more and more accurate.

5.2.4 Optimal Polynomial Network Forecast Strategy

In this section three tests provide means for defining the optimal net structure in terms of the power of the polynomial, the best forecast strategy and the training data requirements. All tests are based on the 585 km^2 catchment of the Freiberger Mulde River (Nossen gauging station) in Germany. In the catchment, the agriculturally used area dominates; around 30% of the catchment is covered with forest. The time of runoff concentration within the catchment is approximately 18 h. The tests have been carried out in this catchment to demonstrate a good predictive performance even for smaller catchments together with the general nature of the approach proposed. The meteorological and runoff database is analogue to the one used in the preceding tests in the Kriebstein basin.

The first test evaluates the dimensionality of the polynomial approach. The net was used with second and third degree approaches, i.e. it was tested for $a + b + c = 2$ and $a + b + c = 3$ (Eq. 5.8). The results of these tests are displayed in Fig. 5.10. The black dots mark the performance of the second degree net. The grey third degree approach (Fig. 5.10) clearly outperforms the second degree approach for all lead times. Based on this finding, polynomial nets were used in the third degree for further testing and development.

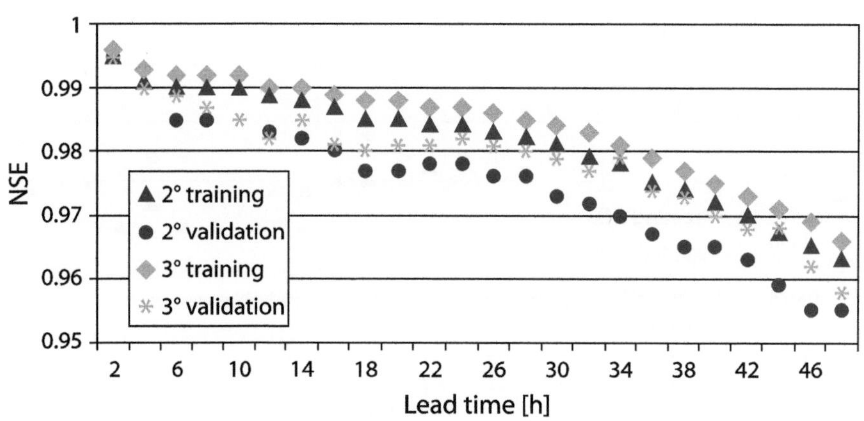

Fig. 5.10 Testing for the optimal polynomial degree (Nash-Sutcliffe efficiency)

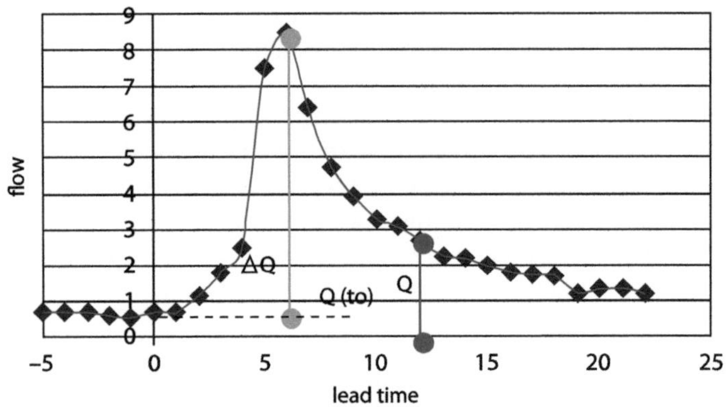

Fig. 5.11 Prediction strategy for Q and ΔQ respectively

In principle it is possible to forecast either the absolute flow at the desired lead time or the flow increment that distinguishes the actual discharge from the flow at the lead time (Fig. 5.11). The forecasting approach developed in this section follows the strategy of incremental predictions for lead-times less than 12 h. The calculation of the forecast at time t for time $t+\Delta t$ (lead time increment) yields the flow increments ΔQ with respect to the actual discharge at time t. However, for lead times of more than 12 h, the prediction does not depend on differences to the actual flow any more. Therefore it is favourable to predict the flow Q for lead times of more than 12 h directly. The underlying principle is shown in Fig. 5.11. In this figure, the grey line refers to the prognosis of differences for up to 12 h while the black line is the absolute value prediction used for longer lead times.

The strategy described was derived from a sensitivity analysis of the Nash-Sutcliffe criteria for both (Q and ΔQ) predictions for lead times of up to 48 h. The results of these analyses are shown in Fig. 5.12, where the grey symbols represent the results for ΔQ, while black symbols denote results for Q.

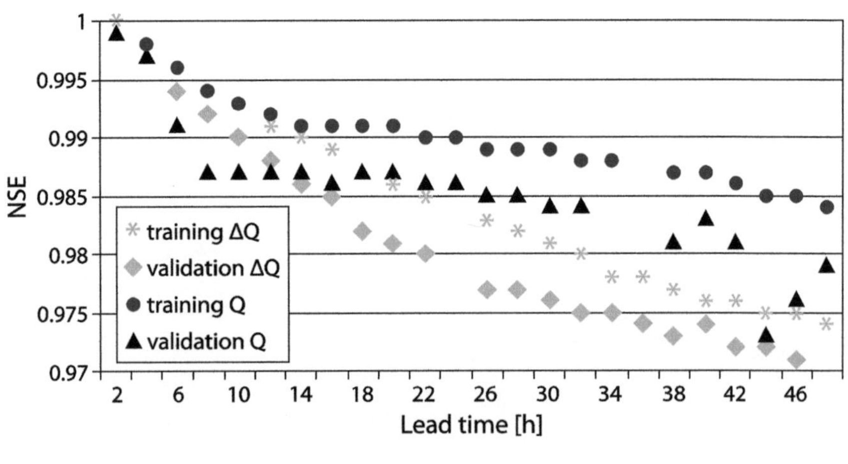

Fig. 5.12 Prediction strategy: Q versus ΔQ (Nash-Stutcliffe-Efficiency)

It can be noticed easily that for longer lead times, the validation performance for the direct flow prediction stays around 0.98, while the grey squares of the ΔQ validation performance criterion steadily decrease – down to values below 0.97. This difference is small, but nevertheless significant, as the Nash-Sutcliffe efficiency was calculated over a period of 47 years. For smaller lead times, the grey validation stars of the delta method yield better results than the direct prediction of Q.

In Fig. 5.13, a zoom on the first seven lead time increments of the test confirms the statements made earlier. For short lead times < 12 h the mean square error of the ΔQ prediction strategy clearly outperforms the direct prediction of Q.

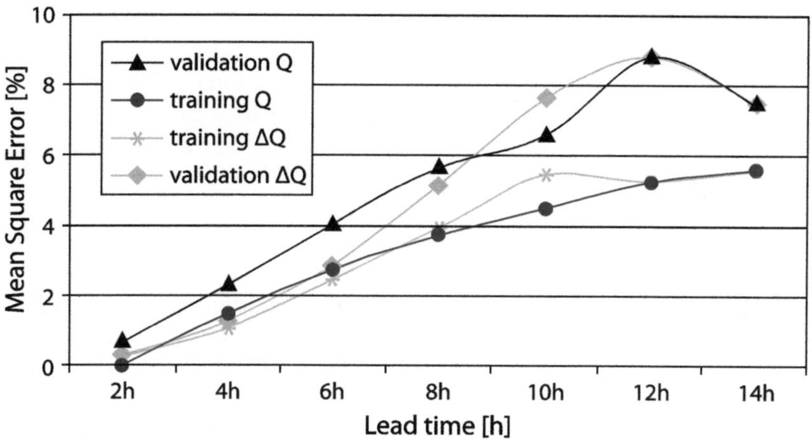

Fig. 5.13 Prediction strategy: Q versus ΔQ (Mean square error)

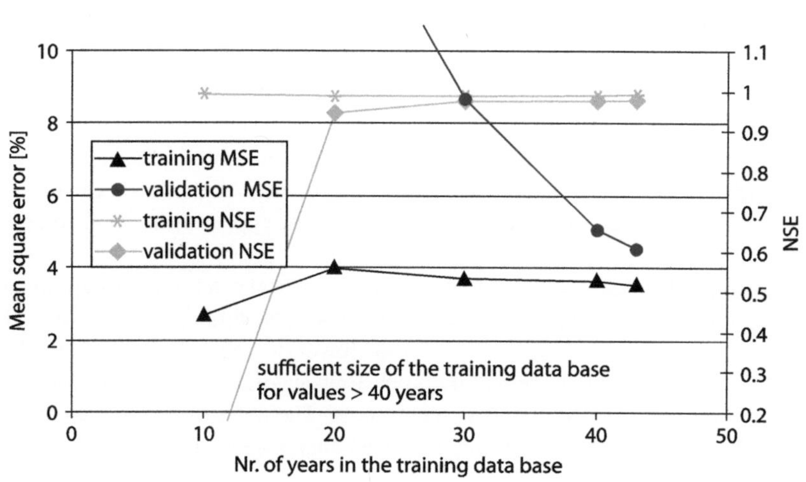

Fig. 5.14 Effect of the size of the training database on training and validation criteria (Nash-Sutcliffe efficiency and mean square error)

A further important characteristic of the setup of the polynomial net is the size of the training database. In a third test, this size had been varied from 10 to 47 years. 20% of the training database is always reserved for testing the predictive power. One year of test data comprises a rough average of 20 flood events. The Nash-Sutcliffe efficiency (Fig. 5.14) yields excellent values for training databases with 20 years of input data. But if the mean square error is additionally considered, it becomes clearly visible, that the predictive power best if more than 40 years of data (600 events) were used in the training of the polynomial network.

5.3 Conclusions

A comprehensive overview of the current status of flood forecasting with ANN was given. At the beginning of this chapter, the approaches used for addressing the lead time issue of forecasting models as well as the implications arising from the need for a detailed operational uncertainty analysis were discussed.

In the following section a detailed description and analysis of possible structures of artificial neural net models were given. This implies an introduction of the newly developed training algorithm for the polynomial neural network. This training approach offers a way for evaluating the importance of single input vectors and provides means to control the physical soundness of the information used in the flood forecasting system. That, in turn, is an essential prerequisite for an automatic detection of suitable input vectors in the setup procedure of a modelling system for a new river basin which is to be investigated. After a complete description of the principles, structures and training procedures, multi layer feed forward models were compared to polynomial network based models. This comparison indicates that an optimal polynomial neural network structure is a comprehensive function of the number of its input vectors. The time needed for training is constant and thus it is easy to calculate the expenditure required for setting up a model. In contrast to these characteristics, it is difficult to define optimal structures for multi layer feed forward neural networks. The training time required for multi layer feed forward based models can not be predicted from the structure. This is a drawback for planning of flood forecasting systems. On the basis of these findings it is recommended to use a polynomial network approach in flood forecasting models.

This contribution closes with recommendations for the setup of net architectures based on several data sets from test catchments. The findings here reveal that a 3rd degree polynomial approach outperforms lower degree polynomials. At the end of this chapter the implications that forecasting absolute flows bring about were compared to forecasting partial flow increments. The last test which was discussed here addressed the size of the training data base that is needed for good performance of the net in our case study.

Generally, polynomial neural networks are well suited in forecasting applications because of their predictable behaviour and the simple structure.

References

ASCE (2000a) Artificial neural networks in hydrology. I: preliminary concepts. By the ASCE Task Committee on Application of Artificial Neural Networks in Hydrology. J Hydrologic Eng 5(2):115–123

ASCE (2000b) Artificial neural networks in hydrology. II: hydrologic applications. By the ASCE Task Committee on Application of Artificial Neural Networks in Hydrology. J Hydrologic Eng 5(2):124–137

Battiti R (1989) Accelerated back-propagation learning: two optimization methods. Complex Syst 3:331–342

Cullmann J, Krausse T, Philipp A (2008) Enhancing flood forecasting with the help of processed based calibration. Phys Chem Earth 33(17–18):1111–1116

Cullmann J, Schmitz GH, Mishra V (2006) Eine neue Strategie zur Hochwasservorhersage in schnellreagierenden Einzugsgebieten – Modellierungsaspekte (A new strategy for flood forecasting in fast reacting watersheds- aspects of modelling) BFG-Veranstaltungen 2006/3, Koblenz

Dawson CW, Wilby R (1998) An artificial neural network approach to rainfall-runoff modelling. Hydrological Sci J 43:47–66

Dawson CW, Wilby R (2001) Hydrological modelling using artificial neural networks. Prog Phys Geogr 25(1):80–108

Foka A (1999) Time series prediction using evolving polynomial neural networks dissertation, University of Manchester, Manchester

Görner W, Cullmann J, Peters R, Schmitz GH (2006) Nutzung künstlicher neuronaler Netze zur Bereitstellung von Entscheidungsgrundlagen für operative und planerische wasser-wirt¬schaftliche Maßnahmen, TU-Dresden (The use of artificial neural networks for operational purposes in water management) Publication of the University of Technology Dresden, Dresden

Hagan MT, Menhaj M (1994) Training feedforward networks with the Marquardt algorithm. IEEE Trans Neural Netw 5:989–993

Hjelmfelt AT, Wang M (1993) Artificial neural networks as unit hydrograph applications. Proceedings of the Engineering Hydrology, ASCE, New York, NY, pp 754–759

Hornik K, Stinchcombe M, White H (1989) Multilayer feedforward networks are universal approximators. Neural Netw 2(5):359–366

Hsu KL, Gupta HV, Sorooshian S. (1995) Artificial neural network modelling of the rainfall-runoff process. Water Resour Res 31(10):2517–2530

Hsu KL, Gupta HV, Sorooshian S (1997) Application of a recurrent network to rainfall-runoff modelling. Proceedings of Aesthetics in the Constructed Environment, ASCE, New York, NY, pp 68–73

Hu TS, Lam KC, Ng ST (2001) River flow time series prediction with a range-dependent neural network. Hydrol Sci J 46(5):729–745

Imrie CE, Durucan S, Korre A (2000) River flow prediction using artificial neural networks: generalizing beyond the calibration range. J Hydrol 233:138–153

Ma L, Khorasani K (2005) Constructive feedforward neural networks using hermite polynomial activation functions. IEEE Trans Neural Netw 16(4):821–823

Maren A, Harston C, Pap R (1990) Handbook of neural computing applications. Academic, San Diego, CA

Meyer-Brötz G, Schürmann J (1970) Methoden der automatischen Zeichenerkennung (methods for automatic character recognition). Akademie Verlag, Berlin, 100–105

Miller A J (1999) Subset selection in regression. Chapman and Hall, London

NWS/NOAA (2010) Glossary. http://www.weather.gov/glossary/ Accessed 18.2.2010

Parisi R, Claudio D, Orlandi G, Rao BD (1996) A generalised learning paradigm exploiting the structure of neural networks. IEEE Trans Neural Netw 7(6):1451–1460

Rojas R (1996) Neural networks: a systematic introduction. Springer, Berlin

Schulla J (1997) Hydrologische Modellierung von Flussgebieten zur Abschätzung der Folgen von Klimaänderungen, Diss., ETH Zürich, CH, 161 pp
Shin Y, Gosh J (1992) Approximation of multivariate functions using ridge polynomial networks. Proceedings of the International Joint Conference on Neural Networks, vol 11, Como, Italy, pp 380–385
Smith J, Eli RN (1995). Neural network models of rainfall-runoff process. J Water Res Manage Plann ASCE 121(6):499–508
Toth E, Brath A, Montanari A (2000) Comparison of short-term rainfall prediction models for real-time flood forecasting. J Hydrol 239:132–147
White H (1989) Learning in artificial neural networks: a statistical perspective. Neural Comput 1:425–464
Zealand CM, Burn DH, Simonovic SP (1999) Short term stream-flow forecasting using artificial neural networks. J Hydrol 214:32–48

Chapter 6
Advances in Regionalising Flood Probabilities

Ralf Merz

Abstract Flood regionalisation methods are used to estimate floods of a given excedance probability, such as the 100-year flood, in ungauged catchments, i.e. catchments, where no local streamflow data are available. They can also be used to improve flood estimates from local data in gauged catchments. The main idea of flood regionalisation methods is to transfer flood information from hydrologically similar gauged catchments (i.e. donor catchments) to catchments, where flood statistics have to be estimated (i.e. target catchments). Different methods for transferring flood information from donor to target catchments have been proposed in the literature. The first type consists of pooling hydrologically similar catchments into a homogeneous group and subsequently using averages within each pooling group to estimate flood probabilities in target catchments. Examples of this type are the Index Flood Method and the Region of Influence approach (ROI). The second type of methods is the application of multiple regressions between flood statistics and catchment attributes. The rationale of this approach is that catchments with similar attributes are also likely to exhibit similar flood generation processes and hence may also behave similarly in terms of their flood frequency response. Catchment attributes include catchment size, land use, geology, elevation, soil characteristics as well as climate variables such as mean annual precipitation. The third type is the application of geostatistical methods, which use spatial proximity as a measure of hydrological similarity. This chapter summarizes the most important methods and recent findings from the literature with a focus on the ungauged catchment case for which flood frequency regionalisation is particularly difficult. For any practical application, the interest resides in how well flood statistics, such as the 100-year flood, can be estimated for a given catchment. Results from some intercomparisons of regionalisation methods are given.

R. Merz (✉)
Institute for Hydraulic and Water Resources Engineering, Vienna University of Technology, Vienna, Austria
e-mail: merz@hydro.tuwien.ac.at

A.H. Schumann (ed.), *Flood Risk Assessment and Management*,
DOI 10.1007/978-90-481-9917-4_6,

Contents

6.1 Introduction

Flood peak estimates associated with a given exceedance probability or return period are needed for many engineering problems. As flood estimates are often required for catchments without streamflow measurements, they are usually obtained by some sort of regional transposition from gauged catchments to the site of interest (Cunnane, 1988; Bobée and Rasmussen, 1995; Hosking and Wallis, 1997). The process of transferring flood information from hydrologically similar catchments (i.e. donor sites) to a catchment of interest (i.e. target site) is generally referred to as regionalisation (Blöschl and Sivapalan, 1995). The attempt to estimate hydrological quantities in catchments where no stream gauge data are available is also known as the Problem of Ungauged Basins. Developing a better understanding of how to transfer hydrological information from gauged to ungauged catchments is a major scientific question in hydrology. This has been recently recognised by the international hydrological community. The International Association of Hydrological Sciences (IAHS) has singled out the problem of ungauged basins as one of their major science questions and has recently initiated an IAHS Decade on Prediction in Ungauged Basins (PUB) (Sivapalan et al., 2003).

There are three main issues in regionalisation: what information is best transferred, from what catchments (i.e. donor catchments) is the information to be taken and, what is the method to be used for calculating the estimates at the site of interest (i.e. the target site). Methods for predicting hydrological quantities in ungauged catchments can be grouped into three main types. The first type of methods is based on the assumption that flood information can be transferred from catchments with similar hydrological behaviour to the site of interest. As usually more than one hydrologically similar catchment can be found, the flood information of several hydrologically similar catchments are grouped or pooled before transferring them to the catchment of interest. These regionalisation schemes are hence termed pooling schemes and usually consist of two steps: pooling the catchments into homogeneous regions or pooling groups according to some sort of similarity measure; and subsequently using averaged flood statistics within each regions. The second type is based on the assumption that a functional relationship between flood statistics and catchment attributes in gauged catchments exists and can be used to estimate flood statistics in ungauged catchments. The most common method of exploiting the functional relationships between flood statistics and catchment attributes are multiple regressions. The third type of methods are geostatistical methods, which are based on the assumption that flood statistics are spatially correlated and hence use the spatial correlation for the spatial interpolation of flood statistics.

This chapter summarizes the most important methods and recent findings from the literature with a focus on the ungauged catchment case for which flood frequency regionalisation is particularly difficult. A comparison of the performance of some of the discussed methods is presented for a comprehensive data set in Austria.

6.2 Regionalisation Approaches

6.2.1 Pooling Schemes

Pooling schemes are based on the assumption that flood information of catchments with similar hydrological behaviour can be grouped together and transferred to the site of interest. The choice of catchments from which information is to be transferred is usually based on some sort of similarity measure, i.e. one attempt to choose those catchments that are most similar to the site of interest. The traditional measure of similarity is spatial proximity with the rationale that catchments that are close to each other will also behave similarly in terms of their flood frequency response as climate and catchment conditions will only vary smoothly in space. A traditional engineering approach which is based on this rationale are regional discharge-area diagrams, where specific discharges of a given probability are plotted against catchment area for catchments within a region or along a river reach. Within the region, the flood behaviour is then assumed to be similar, except of the effect for catchment area, and hence regional flood behaviour is estimated by fitting a line to the observed flood statistics.

In the hydrological literature, pooling schemes commonly follow the idea of the index flood procedure (Dalrymple, 1960). The index flood methodology is based on the identification of homogeneous groups of catchments for which floods with a return period of T years (T-year floods) can be expressed as the product of a scale factor, which is called the index flood, and a growth factor, which describes the relationship between the dimensionless flood and the recurrence interval, T (the so-called growth curve). The index flood is site-dependent while the growth factor is assumed to be valid for the entire homogeneous group of catchments. The idea of the index flood procedure is to estimate the scale of the flood frequency curve (the index flood) locally to account for the local differences between catchments, while the shape of the flood frequency curve exploits information from the entire homogenous group.

The crucial point of the index flood procedure is the identification of fixed homogenous regions. The aim is to form groups of catchments that approximately satisfy the homogeneity assumption, which implies that the site's frequency distributions are identical apart from the site-specific scale factor. Different grouping methods have been proposed in the literature. Traditionally, geographically coherent regions have been applied in regional flood analyses. In most practical applications (e.g. Stedinger et al., 1992, p. 18.35), the regions are found by expert judgement, i.e. by a subjective assessment of which catchments one would expect to behave similar in terms of their flood characteristics. These subjective considerations are usually based on a personal knowledge of the analyst of the catchment characteristics, climatic inputs and runoff response of the catchments. There have also been a number of attempts at identifying homogeneous regions by multivariate statistical methods that are used for grouping catchments according to their similarity in catchment attributes. Cluster analysis is one of the popular statistical methods for combining catchments into groups (e.g. Acreman and Sinclair, 1986; Burn, 1990). The idea of cluster analysis is to identify groups (regions) in such a way that the similarity of catchments within one region is maximized while similarity between regions is minimized. Other methods used to form regions in regional flood frequency analysis are factor analysis, principal component analysis, artificial neural networks, fuzzy sets and canonical correlation analyses. A promising technique for delineating homogenous regions are Classification And Regression Tree (CART) models (Breiman et al., 1984) which, to our knowledge, have not yet been used in regional flood frequency analysis. However, interesting applications to other fields of hydrology can be found in the literature (e.g. Laaha and Blöschl, 2005).

Use of geographically contiguous regions may have a number of drawbacks. First, analysis of observed flood frequency behaviour often reveals small scale variability but catchments that are far apart may still be hydrologically similar (e.g. Pilgrim, 1983). Second, contiguous regions can be highly sensitive to the occurrence of widespread extreme events (Reed and Stewart, 1991). Third, it may be difficult to identify hydrologically meaningful regions. As an alternative it has hence been suggested to pool flood data according to catchment types rather than according to geographical position (Acreman and Sinclair, 1986). Using this idea, Burn (1990) proposed the so-called region of influence (ROI) approach. In the ROI approach, the fixed and contiguous regions are replaced by flexible and overlapping groups that are not necessarily geographically contiguous. Each site of interest (i.e. catchment for which flood statistics are to be estimated) has its own pooling group. The starting point of the ROI approach is the selection of a distance measure defining the hydrological similarity of catchments. The distance measure D_{i0} between station i and station 0 usually contains catchment attributes and has been defined by Burn (1990) as

$$D_{i0} = \left[\sum_{l=1}^{L} W_l (Y_l(\mathbf{x}_i) - Y_l(\mathbf{x}_0))^2 \right]^{1/2} \tag{6.1}$$

where W_l is the weight of the relative importance of attribute l out of L attributes and $Y_l(\mathbf{x}_l)$ is the value of attribute l for station i. Those catchments that are most similar to the site of interest in terms of D_{i0} are included in the group. Reed et al. (1999) introduced the concept "focused pooling" to the grouping of catchment according to catchment attributes, rather than geographical regions. Similarly, they used "pooled growth curves" instead of "regional growth curves".

Pooling of catchments requires the a-priori definition of similarity measures. In the case of spatially contiguous regions, the similarity measures may be based on the differences of at-site flood data, catchment attributes and geographical location. Grouping catchments according to their observed at-site flood statistics may be the most appealing way of fulfilling the homogeneity condition, but it is not always preferable. First, it can not be used in the case of ungauged target sites. Second, the procedure tends to group all catchments with large outliers. The pooled flood frequency curve then tends to overestimate the flood probabilities, while in the remaining regions where, by chance, no extreme floods have been observed, the pooled flood frequency curves tend to underestimate flood probabilities. Catchment attributes usually used for delineating pooling groups include catchment area, climatic indicators, such as mean annual rainfall, and information on topography, geology, soils, land use and catchment morphology. Zrinji and Burn (1994), for example, used catchment area, a catchment shape factor, the percentage of area controlled by swamps and lakes, percentage of barren area, and the mean annual runoff coefficient to group catchments in Newfoundland, Canada using the ROI approach. In the UK Flood Estimation Handbook (IH 1999) the pooling group is initially constructed by seeking those catchments that are nearest to the target catchment in size-wetness-soils space, i.e. those catchments, that are most similar according to catchment size, standardised mean annual rainfall and a baseflow index derived from HOST soil data (Boorman et al., 1995). A similarity measure that has recently gained increased popularity is flood seasonality. In these seasonal approaches, pooling groups are delineated based on the timing of the flood peaks within the year. Piock-Ellena et al. (1999), for example, used cluster analysis based on circular statistics of flood occurrence within the year to identify homogeneous regions. They found that the seasonality of the hydrological flood characteristics can assist in identifying meaningful regions. Castellarin et al. (2001) assessed the effectiveness of seasonality measures for pooling 36 catchments in Northern Italy. They compared the pooling based on the seasonality of maximum annual floods and maximum annual daily rainfall with the pooling based on rainfall and catchment permeability. The results suggested that the seasonality measures provide an efficient description of the catchment hydrological behaviour. In a somewhat different analysis, Burn (1997) applied flood seasonalities to test the homogeneity of regions in the Canadian praries.

The choice of method of forming pooling groups and the assessment of the plausibility of the obtained pooling, involve testing whether the proposed regions may be considered homogeneous or not. Many authors have proposed homogeneity tests in the hydrological literature, including Dalrymple (1960), Wiltshire (1986), Lu and Stedinger (1992), Fill and Stedinger (1995) and Hosking and Wallis (1993, 1997). The L-moment based test of Hosking and Wallis (1993, 1997) compares the between-site variations in sample L-moments for the pooling group with what would be expected for a homogeneous pooling group. Simulations are used to estimate "what would be expected". By repeated simulation of a group of catchments with a common distribution and the same record lengths as observed data, the mean and standard deviation of the dispersion measure are obtained. The homogeneity test of Hosking and Wallis (1993, 1997) is now commonly used, although its superiority over other methods has not been generally demonstrated. Only a few comparisons of the merits and drawbacks of the methods exist in literature. One of them is Viglione et al. (2007), who compared the L-moment based test and two rank tests that do not rely on particular assumptions regarding the parent distribution. They found that L-moment based tests are more powerful when the samples are slightly skewed, while the rank tests have better performance in the case of high skewness. They proposed a simple method to guide the choice of the homogeneity test in a particular application.

An important consideration in the application of a pooling scheme is the number of stations to be included in the pooling group. IH (1999) recommends the *5T rule* as a rule of thumb, which stipulates that the sum of the observed years of all stations in the pooling group should be at least five times the return period of interest. For example, in the case of the 100 year flood being the target return period, the pooling group should contain at least 500 station years.

Once the pooling group is found, the site specific index flood and the regional growth curve have to be calculated. The index flood can be estimated from at-site flood data in the case of gauged catchments. In the case of ungauged catchments which are the focus of this chapter, the index flood is usually estimated from regressions with catchment attributes or data transfer from analogue catchments. In the latter case, the index flood is transferred from a gauged catchment, which is assumed to behave hydrologically similar. IH (1999) define an ideal analogue donor catchment as a catchmnet with a gauge situated just upstream or downstream of the target site, draining a similar catchment area. A catchment of similar sized on an adjacent tributary is also considered as a good analogue catchment. In IH (1999) it is argued that finding useful analogues requires hydrological understanding and experience. An overview of other methods for estimating the index flood, including hydrological simulation, is given in Bochiola et al. (2003)

The pooled growth curve is obtained by selecting a pooled frequency distribution and estimating the distribution parameters by fitting the distribution function to the pooled data. The choice of one parent distribution for all catchments within one pooling group is usually based on a goodness-of-fit measure to the observed data (e.g. Hosking and Wallis, 1993, 1997; Zrinji and Burn, 1994; Castellarin et al., 2001). Several methods exists for fitting the distribution function to the pooled data. The station-year method combines the rescaled flood data from all catchments into a single sample and fits the distribution by treating the combined sample as a single random sample. The method is rarely used in regional flood frequency analysis, because the different accuracies due to varying records lengths of different sites are not reflected in the combined sample. An alternative method is maximum likelihood estimation, which maximises the likelihood to observe the n rescaled samples of the n catchments in the pooling group. The most common approach is to estimate summary statistics such as coefficient of variation and skewness for each catchment individually and combine them to obtain the regional estimates. Similar to the homogeneity test, the method of Hosking and Wallis (1993, 1997) based on L-moments seems to be widely used to estimate the regional growth curves. In this method the pooled L-CV, for example, is calculated as:

$$t_2^P = \frac{\sum_{i=1}^{M} w_i \cdot t_2^{(i)}}{\sum_{i=1}^{M} w_i} \tag{6.2}$$

where $t_2^{(i)}$ is the sample L-CV for the ith most similar catchment i out of catchments and w_i is the weight of catchment i. A standard choice for w_i is to weight by record length (w_i = record length). This approach gives more emphasis to long records. IH (1999) recommends weights according to catchment similarity and record length. In this method, the most similar catchment, in terms of the similarity measure used to form the pooling group, obtains the similarity weight one, and decreasing weigths are given to subsequent sites. The total weight of each site is then calculated by multiplying the similarity weight and the record length.

6.2.2 Functional Relationships to Catchment Attributes

The rationale of functional relationships between flood characteristics and catchment attributes is that catchment attributes are thought of as surrogates of the hydrological processes within a catchment. Catchments with similar attributes are hence likely to exhibit similar runoff generation processes and flood characteristics (Acreman and Sinclair, 1986). Catchment attributes include catchment size, land-use, geology, topographic elevation, soil characteristics as well as climate variables such as mean annual precipitation. Stream flow is usually not considered as a catchment attribute in order to make this group of methods applicable to the ungauged catchment case. The availability of various digital catchment attributes in Geographical Information Systems (GIS) has boosted the use of functional relationships in regional flood analyses. For many regions of the world, (e.g. UK (IH, 1999) and Baden-Württemberg, Germany (LfU, 1999)) GIS data have been used for detailed analyses of the regional flood behaviour for operational use.

The functional relationships are traditionally exploited by multiple regressions between flood statistics and catchment attributes (e.g. Tasker, 1987; Merz and Blöschl, 2005). In multiple regressions, the estimate of the flood statistics at the site of interest is expressed as

$$\hat{Y}(\mathbf{x}_0) = c_0 + c_1 \cdot Y_1(\mathbf{x}_0) + c_2 \cdot Y_2(\mathbf{x}_0) + \ldots + c_n \cdot Y_n(\mathbf{x}_0) + \varepsilon \tag{6.3}$$

where $\hat{Y}(\mathbf{x}_0)$ is the flood sample moment or flood quantile (or a transformation thereof) to be estimated, $Y_1(\mathbf{x}_0)$, $Y_2(\mathbf{x}_0)$, $Y_3(\mathbf{x}_0)$ are the catchment attributes at locations x_i, $c_0, c_1, c_2, \ldots, c_n$ are the regression coefficients and ε is the model error. Many catchment attributes used in flood regionalisation are highly skewed which introduces biases. A transformation to standard normally distributed catchment attributes reduces possible biases. As an approximation logarithmic transformations are sometimes used.

In multiple regressions, one of the most serious problems is multicollinearity, i.e. when at least one of the catchment attributes is highly correlated with another catchment attribute or with some linear combination of other catchment attributes. If multicollinearity is present, the regression coefficient can be highly unstable and unreliable. The variance inflation factor (VIF) (Hirsch et al., 1992) has been proposed for diagnosing multicollinearity and guiding the reduction of the number of catchment attributes to be used in the regression.

An alternative way of approaching the problem of multicollinearity is stepwise regression (e.g. Brath et al., 2001). Instead of selecting the most important variables a-priori, in stepwise regression the choice of predictive variables is carried out by an iterative procedure. The stepwise selection of predictor variables can be forward, backwards or a combination thereof. In a forward mode, the model starts with no variables, trying out the variables one by one and including them if they are "statistically significant". Backward elimination involves starting with all candidate variables and testing them one by one for statistical significance, deleting those that are not significant. The statistical significance of a candidate variable is usually tested by the F-Test, but other techniques such as Marlow's Cp (e.g. Laaha and Blöschl, 2007) exist.

The choice of catchment attributes is critically important for obtaining reliable regionalisation models. Information on geology, soils, topography, land use, river network density and long term rainfall behaviour, such as mean annual precipitation, is widely used because it is commonly available in ungauged catchments. In the Flood Estimation Handbook (IH, 1999), for example, the flood median in catchments in the UK is estimated as a function of catchment area, standardised average annual rainfall, an index to account for the retention of reservoirs and lakes and two parameters representing soil properties. In the regression model of the federal state of Baden-Württemberg, Germany (LfU, 1999) the flood quantiles are functions of catchment area, the percent of urban area and forests, a slope parameter of the river section, the maximum length of the flow path, the flow path length from the catchment centre to the outlet, mean annual precipitation and an empirical landscape factor to capture the variability of the flood processes related to the geological units. Schumann and Pfützner (2002) showed, that most of the flood variability of an mountainous region in Germany can be explained by a multi regression using two predictor variables, i.e. elevation and drainage density. Pfaundler (2001) developed a stepwise multiple regression model for 231 Swiss catchments, in which catchment area, mean annual maximum daily rainfall, river network density and mean permeability are used as predictive attributes for mean annual floods. Uhlenbrook et al. (2002) found a good correlation of flood quantiles of 29 catchments in Southern Germany with catchment precipitation, catchment area, a slope parameter, the portion of forests and the portion of farmland. In Merz and Blöschl (2008c), the predictive power of various catchment attributes is analysed in terms of the correlation of the flood moments (i.e. mean annual flood (MAF), coefficient of Variation (CV) and skewness (CS)) to a range of catchment indicators for 459 Austrian catchments. The results indicate that MAF and CV are strongly correlated with indicators characterising the hydro-climatic conditions of the catchments, such as mean annual precipitation, long term evaporation and the base flow index. For the catchments analysed, the flood moments were not significantly correlated with attributes usually used in regression analyses to represent runoff generation, such as geology, soil types, land use and the SCS curve number. In Fig. 6.1 MAF and CV are plotted against mean annual precipitation (MAP) (Fig. 6.1, left) and the SCS curve number (US SCS, 1972) (Fig. 6.1, middle) derived from land use and soil data. A relative strong correlation is found between flood characteristics and MAP. In catchments with low MAP rates, floods discharges tend to be moderate, with a large variability between the years, while in regions with higher rainfall rates, flood discharges tend to be high in every year. Hence MAP is positively correlated to the mean annual flood discharges and negatively correlated to CV. In contrast the SCS curve numbers have very little predictive power. The tables of the SCS curve number method that are used to estimate the CN from land use and soil type do not seem to apply to the catchments of the study area even though they are widely used in many countries. This discrepancy is related due the spatial patterns of forest and agricultural land in Austria. In the SCS method forest is associated with small curve numbers, because of the large infiltration into forest soils. Forest is the dominant land use class at the northern rim of the high Alps, but this is a region of persistent rainfall events caused by orographic enhancement, which produces large flood discharges. Indicators of runoff generation that do have significant predictive power for flood moments are dynamic catchment attributes derived from observed

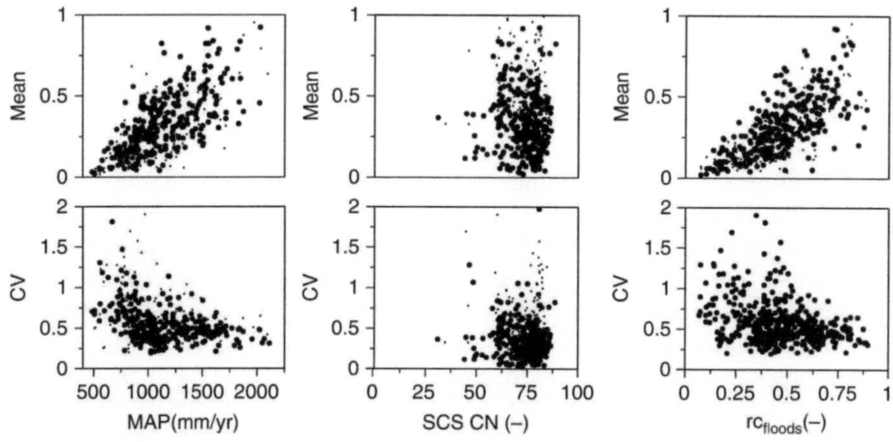

Fig. 6.1 Statistical flood moments plotted against long term mean annual precipitation (*MAP*) (*left*), the SCS curve number (*middle*) and the mean runoff coefficient of annual flood events (*right*). Catchments with a flood record longer than 25 years are plotted as *large dots*, those with less than 25 years are plotted as *small dots* (from Merz and Blöschl, 2008)

rainfall-runoff. In Fig. 6.1 (right) MAF and CV are plotted against mean runoff coefficients of annual flood events. The runoff coefficients are positively correlated to mean annual flood discharges and negatively correlated to CV. This means that in catchments with higher runoff coefficients, flood runoff tends to be large, but there is a lower variability of the flood runoff between the years. In contrast, in catchments with lower runoff coefficients, flood runoff tends to be lower, but larger floods can also occur, even rarely. The caveat of using event runoff coefficients for predicting flood statistics is that the event runoff coefficients have been estimated from event runoff data (Merz et al., 2006) and are hence not available in ungauged catchments. This underlines the importance of developing new indicators at the regional scale that are more representative of the causative flood processes than current indicators are.

Observed data may suggest that useful multiple regressions between catchment attributes and flood characteristics can only be found for some parts of the study region, while for other parts different regressions are more efficient. In this case, one would use regressions to estimate floods in smaller, hydrologically similar sub-regions. Delineating these sub-regions poses a similar problem as homogenous pooling groups. Hence the methods for delineating such pooling groups can also be used for identifying sub-regions.

Fitting regression models to observed data is often complicated by single extreme values or outliers. Extreme values may act as leverage points forcing the regression line close to them, particularly if the least squares are used for parameter estimation. Eliminating the outliers may increase the goodness-of-fit, but may not necessarily result in an increase in the predictive power of the model. Statistical diagnostics that can be used to assess the regression model are leverage statistics, prediction residuals and PRESS statistics (Hirsch et al., 1992). However, a hydrological assessment of the plausibility of the selected regression model, including an analysis of the underlying data and hydrological processes in the region, is deemed very important to assure a model free of spurious correlations and data artefacts.

6.2.3 Geostatistical Methods

An alternative to the use of pooling schemes and functional relationships are geostatistical methods. The main strength of geostatistical methods is that they are best linear unbiased estimators (BLUE); best meaning that the mean squared error is a minimum, linear meaning that the estimate is a weighted mean of the data in the area, and unbiased meaning that the mean expected error is zero (Journel and Huijbregts, 1978). Geostatistical methods are widely used in hydrology for various purposes (e.g. Blöschl, 2006). A geostatistical method used for predicting flood quantiles is Ordinary Kriging (Merz and Blöschl, 2005). In Ordinary Kriging the estimated value $\hat{Z}(\mathbf{x}_0)$ of a flood characteristic is a weighted linear combination of the observed flood characteristics of n gauged catchments Z_i

$$\hat{Z}(\mathbf{x}_0) = \sum_{i=1}^{n} \lambda_i Z_i \tag{6.4}$$

The weights λ_i are determined such as to minimize the estimation variance, while ensuring the unbiasedness of the estimator which leads to a system of linear equations called the kriging system. The only information of hydrological similarity required in the kriging system is the semivariogram values for different lags. One appealing feature of kriging is that the method provides an estimate of the estimation variance of the regionalised variables, which is usually not possible with other regionalisation schemes. This estimation variance is very useful for assessing the credibility of the regional estimate. Another advantage of kriging is the ability to account for uncertainty of flood characteristics at gauged catchments resulting from measurement errors and short samples. The uncertainty (i.e. the estimation error) of the flood statistics decreases with number of years of the flood record and increases with the order of the moment and the probability of the flood quantile. For example, the first moment (the mean) can be estimated from a flood peak record with relatively little error while the third moment (the skewness) is always associated with substantial estimation error. If the errors associated with each flood characteristic Z_i are non-systematic, uncorrelated with each other, uncorrelated with the moment Z and have a known variance σ_i^2, the kriging system can be extended to account for these errors (De Marsily, 1986, p. 300):

$$\sum_{k=1}^{n} \lambda_k \gamma(\mathbf{x}_i - \mathbf{x}_k) - \lambda_i \sigma_i^2 + \mu = \gamma(\mathbf{x}_i - \mathbf{x}_0) \quad i = 1, \ldots, n \tag{6.5}$$

$$\sum_{t=1}^{n} \lambda_i = 1$$

where μ is the Lagrange parameter, the $\mathbf{x}_i$ and $\mathbf{x}_k$ are the coordinates of the catchment centroids, and $\gamma(\mathbf{x}_i - \mathbf{x}_k)$ is the semivariogram (without representing errors) of the flood characteristic Z for the lag of catchment centroids $\mathbf{x}_i$ and $\mathbf{x}_k$. Note that each catchment i can have a different error. It is therefore possible to account for different record lengths in different catchments. De Marsily (1986) termed the method Kriging with Uncertain Data (KUD). Note that the ordinary kriging (OK) system is identical to Eq. (6.5) with $\sigma_i^2 = 0$. Merz and Blöschl (2005) estimated the error variance due to short record lengths in a Monte Carlo analysis by drawing samples of a given size from a known distribution and estimating the variance of the flood moments between different samples. As a next step they parameterised this error variance as a function of record length and the order of the moment by fitting a power law of the form:

$$\sigma_k^2(m) = a \cdot m^{-b} \tag{6.6}$$

where m is the number of years of record in a catchment and k is the order of the moment.

One of the issues with the use of geostatistical methods in hydrology is that they have evolved in the mining industry where the main problem consisted of estimating the expected ore grade (and its uncertainty) of a block using point samples of the ore grade in the area. To this end, the spatial correlations of pairs of points were plotted versus their Euclidian distance which was then used to estimate the variable at the location of interest for a given block size from the point samples. The problem in catchment hydrology is quite different. Unlike mining blocks, catchments are nested and water follows a stream network. It is therefore clear that upstream and downstream catchments would have to be treated differently from neighbouring catchments that do not share a subcatchment. Most geostatistical regionalisation methods do not take into account the nested nature of catchments (e.g. Daviau et al., 2000; Adamowski and Bocci, 2001; Eaton et al., 2002; Merz and Blöschl, 2005), but there is promising recent work that does account for it. Sauquet et al. (2000) developed a method of analysing correlations along the stream network which they then used for mapping annual runoff for ungauged catchments using water balance constraints in the estimation procedure. Skøien et al. (2006) proposed the Top-Kriging approach which is based on similar concepts. They showed that it indeed outperforms the traditional Euclidian framework. Top-Kriging takes both catchment area and distance along the stream network into account (Skøien et al., 2006) and is the most natural way of statistically interpolating along stream networks as no additional assumptions beyond the standard geostatistical assumptions are needed

In Fig. 6.2 a comparison of the estimates of the normalised specific 100 year flood from Top-kriging and Ordinary Kriging for the Mur River in Austria is shown. Figure 6.2a, shows the catchment areas of the Mur and tributaries as well as the stream gauges. Figure 6.2b presents the estimates of the normalised specific 100 year flood $Q_{100\,N}$ from Top-kriging (top) and Ordinary Kriging (bottom) colour coded on the stream network. The results derived from measurements are shown as circles in both figures with the same colour coding. To compare smaller and large catchments, the specific 100 year floods were normalised to a catchment area of 100 km^2 before interpolation (Skøien et al., 2006). Hence, the normalised specific 100 year discharges shown in Fig. 6.2b represent floods from hypothetical catchments with an area of 100 km^2.

The Top-kriging estimates on the main river are similar to the results derived from measurements on the main river (gauges 1–3) and they do not change much along the reach. The estimates on the northern tributaries are much smaller than those on the main stream which is consistent with the measurements on the same tributaries (gauges 5 and 7). This is also reflected in the estimates for the other northern tributaries. On the southern side, the results from measurements are larger, so the Top-kriging estimates are generally much larger than those on the northern side of the Mur. The Ordinary Kriging estimates differ substantially from the Top-kriging estimates. The main difference is that the estimates are not similar along the stream network as is the case of Top-kriging but similar along Euclidian distance in space. Although gauge 7 gave a value of 0.4 (red colour) from measurements, most of the Ordinary Kriging estimates along this tributary are around 0.6 (yellow to green colours). This is because the estimates along this tributary are too much influenced by the measurements along the main river while they should be mainly influenced by the downstream gauge as is the case in Top-kriging. On the other hand, the estimates on the main stream are somewhat underestimated by Ordinary Kriging as they are too much affected by the measurements on the tributaries.

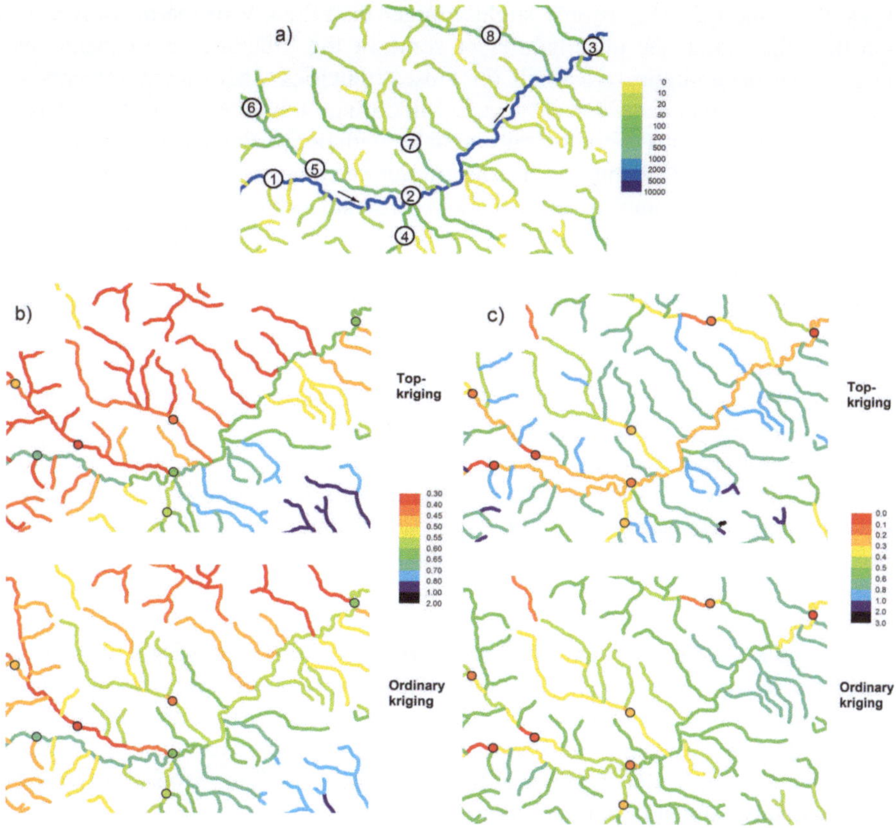

Fig. 6.2 Comparison of Topkriging and Ordinary Kriging flood estimate and their uncertainty. (a) Upstream catchment area of the Mur river and tributaries. Numbers refer to stream gauges. (b) Estimates of the specific 100 year flood Q_{100N} ($m^3/s/km^2$) standardised to a catchment area of 100 km^2 from Topkriging (*top*) and Ordinary Kriging (*bottom*). (c) Uncertainties of the Q_{100N} estimates, expressed as the coefficient of variation (from Skøien et al., 2006)

The uncertainties of the estimates in the Mur region are shown in Fig. 6.2c colour coded on the stream network. The kriging variances ${\sigma_R}^2$ are scaled by the estimates Q_{100N}, and the coefficient of variation (CV) of the estimate is shown:

$$\mathrm{CV} = \frac{\sigma_R}{Q_{100\,N}} \tag{6.7}$$

Both procedures estimate the lowest uncertainties close to the measurements, equal to or larger than the CV of the measurements. Note that the uncertainties on the stream network have been plotted for stream reaches of finite lengths as for these the catchment boundaries were available. The uncertainties estimated by Top-kriging and Ordinary Kriging are very different for most of the stream network. Top-kriging (Fig. 6.2c top) gives relatively small uncertainties on the main river with

CVs of around 0.2. This is only slightly larger than the CV of the measurements. On the other hand, the uncertainties of some of the tributaries are considerably larger. The uncertainties are small for those tributaries where measurements are available, but rather large for tributaries without any measurements. It is interesting that the uncertainty increases substantially with decreasing catchment area. For some of the smallest catchments, i.e. headwater catchments, CVs of more than one are estimated. This points to very uncertain estimates, which is not surprising as no measurements are near. The uncertainties estimated by Ordinary Kriging (Fig. 6.2c bottom) contradict what one would intuitively expect. Most disturbing is that some of the smallest catchments have uncertainties equal to or smaller than the uncertainty of the main river. This is of course a result of the uncertainty being a function of Euclidian distance between catchment centre and measurements only, and not a function of the size and nesting of the catchments. Topkriging is hence a much better regionalisation method than Ordinary Kriging.

One could argue that spatial proximity and similarity of catchment attributes are not mutually exclusive pieces of information, so a combination of kriging and catchment attributes may provide more information than any of the individual regionalisation approaches alone. One method that combines kriging and catchment attributes is external drift kriging. In external drift kriging, local regression with an auxiliary variable is directly represented in the kriging system and the auxiliary variable is assumed to be error free and perfectly related to the flood characteristics (Deutsch and Journel, 1997, p. 67). An alternative is georegression, which is a two step procedure. In a first step, flood statistics are related to catchment attributes by a multiple regression. In a second step, the corresponding residuals are spatially interpolated using ordinary kriging. In georegression, kriging is used to spatially distribute the residual to minimise biases.

To benefit from the features of kriging but assuming that catchment attributes are a better indicator of similarity than spatial proximity, Chokmani and Ouarda (2004) developed a kriging procedure in the physiographical space, a multidimensional space defined by catchment attributes. They used canonical correlation analysis or principal component analysis to construct the physiographical space representing the distance of catchments according to their similarity in terms of catchment attributes. They then used kriging to interpolate flood statistics in the physiographical space.

6.3 Performance of Regionalisation Approaches

In the previous sections, different types of regionalisation methods were presented. For any practical application, the interest resides in which method should be used to estimate flood probabilities for a particular site of interest. Each of the methods has its merits and drawbacks, but comprehensive intercomparisons of regional flood frequency methods are rare in the literature. Pandey and Nguyen (1999) tested the performance of nine parameter estimation methods for multiple regressions using data from 71 stations in Quebec, Canada. For that data set, nonlinear models outperformed the linear models for parameter estimation. However, the differences were small. Most of the linear models had slightly higher estimation errors and they tended to under-predict floods from large catchments. Ouarda et al. (2008) tested the performance of four methods to delineate homogeneous pooling groups using 29 catchments from Mexico. They analysed hierarchical cluster analysis, canonical correlation analysis, a revised version of the canonical correlation analysis and canonical kriging. While all methods performed quite adequately, results pointed to some advantages of the canonical correlation analysis based methods over the other methods.

A comprehensive comparison of different types of flood regionalisation methods has been published by Merz and Blöschl (2005). They examined the predictive performance of various methods for the ungauged catchment case based on a jack-knifing comparison of locally estimated and regionalised flood quantiles for 575 Austrian catchments. Different variants of the ROI approach, as the most flexible type of pooling schemes, different variants of multiple regressions and different variants of kriging and georegression were compared. For each catchment, the first three flood moments (i.e. mean, coefficient of variation (CV) and skewness (CS)) were estimated by the regionalisation schemes and the flood probabilities were calculated by the GEV distribution. If the record is sufficiently long, the difference between these regional estimates and estimates based on at-site data mainly represent the error that is introduced by the regionalisation. For 122 catchments with records of 40 years or more they estimated the normalised mean error (nme) as a measure of the bias and the normalised standard deviation error (nsdve) as a measure of the random error. In Fig. 6.3 the biases and the random error are plotted against return period for those variants of each type of regionalisation that gave the best performance. Figure 6.3 shows that the errors of the two geostatistical approaches

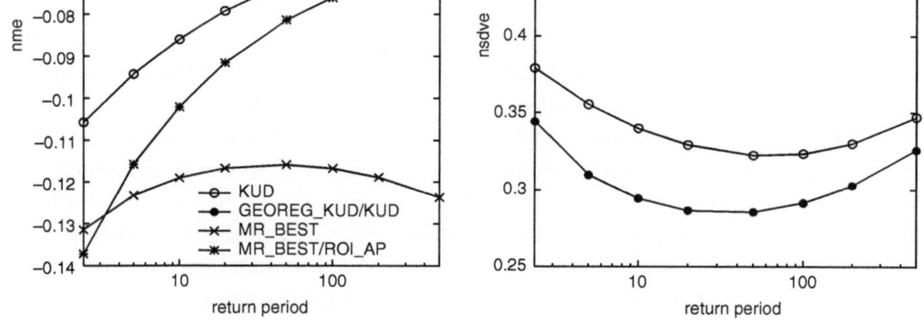

Fig. 6.3 Comparison of bias (*left*) and random error (*right*) of flood quantile regionalisation for four regionalisation types. *Open circles* (KUD): kriging; *Full circles* (GEOREG_KUD/KUD): georegression; Crosses (MR_BEST): multiple regression; Asterisks (MR_BEST/ROI_AP): Region Of Influence Approach. The GEV distribution and product moments are used. Data from 575 Austrian catchments. (from Merz and Blöschl, 2005)

(open circles – Kriging; full circles – Georegression) are significantly smaller than those of the approaches based on catchment attributes. For example, for a 100-year flood, the random errors of the geostatistical approaches are 0.29 and 0.33 while they are 0.45 and 0.47 for the multiple regressions and the Region Of Influence (ROI) approach, respectively. These differences very clearly indicate that, for the Austrian data set, spatial proximity is a significantly better predictor of flood frequency than are catchment attributes. Apparently, the catchment attributes are not very representative of the real physical controls of the flood frequency processes. The georegression gives slightly smaller errors than the kriging (KUD) method. This means that information on catchment attributes will improve the geostatistical regionalisation for all return periods but the improvement is not large. The random errors of the multiple regression approach are slightly smaller than those of the combined ROI-regression approach, but the biases are larger. It is also interesting that the biases of the geostatistical methods are significantly smaller than those of the attribute-based approaches. Kriging is an unbiased estimator, and the biases are indeed small, while, apparently, both the multiple regression and ROI approaches yield larger biases.

In the engineering practice, flood frequency regionalisation is often supported by expert judgement (e.g. IH, 1999). In the study of Merz and Blöschl (2005) a reproducible comparison that is solely based on the hard data available is represented. One would expect that local expert knowledge will improve the predictive performance of all methods, particularly for the ROI approach, but this is difficult to quantify in an objective way.

6.4 Discussion

In recent years the development of a number of different flood regionalisation approaches has substantially advanced the estimation of flood statistics in ungauged catchments. This chapter has reviewed numerous methods that differ in the way information of hydrological similar gauged catchments are transferred to ungauged catchments. Floods in ungauged catchments can be estimated by pooling hydrological similar gauged catchments into homogeneous groups and subsequently using averages within each pooling group or by exploiting functional relationships between flood characteristics and catchment attributes. A third avenue is the spatial interpolation of flood characteristics.

There is no single best method that gives the best estimates in all applications. The performance of the regionalisation methods depends on the hydrological situation in the region under study and, probably even more, on the user's knowledge of the hydrological processes and the available data. Each of the methods has some merits and drawbacks, which should be considered in the choice of methods. For example, using spatially contiguous regions account for the smooth variation of climate and catchment attributes in space and the large hydrological similarity between upstream and downstream neighbours. However, the sensitivity of spatially contiguous regions to widespread extreme events and the observed small scale variability in combination with the physical rationale of hydrological similarity resulting from similarity in catchment attributes makes functional relationships or pooling based on similarity in catchment attributes more appealing. The selection of the most suitable regionalisation method is not straightforward. An analysis of similar, gauged catchments in the same region can provide some assistance. Also applying different regionalisation approaches may provide a deeper insight in the estimation process and hence increase the confidence in the regionalised flood characteristics.

Regionalisation methods tend to reflect the regional trend, but some catchments in the study region may exhibit local particularities. Locationspecific behaviour may not always be visible in the available data used for regionalisation. Merz and Blöschl (2008a) give an example, where flood discharges in St. Aegyd in Lower Austria are only about 20% of those of the neighbouring catchments that are similar in size and catchment attributes. This difference resulted from a small portion of highly permeable areas in St. Aegyd which is highly efficient in decreasing flood runoff due to the near stream location. This would be difficult to quantify in a regionalisation method, but such local particularities may be detected by hydrological reasoning, based on additional information, such field surveys or observed data of other hydrological quantities. The process understanding may complement the statistical estimation methods and hence, give a more complete representation of flood processes at a given site.

The combination of statistical regionalization methods and process understanding based on additional information to give more reliable estimates of flood characteristics in ungauged catchments is a promising new research direction. Examples of hydrological reasoning based on expanded information and the combination with statistical estimation methods for the case flood estimation in gauged catchments is given in Merz and Blöschl (2008a, b).

Acknowledgements Financial support of the Austrian Academy of Sciences (APART [Austrian Programme for Advanced Research and Technology] –fellowship) is thankfully acknowledged. I am grateful to Günter Blöschl for providing valuable comments on an early draft of this manuscript.

References

Acreman MC, Sinclair CD (1986) Classification of drainage basins according to their physical characteristics; an application for flood frequency analysis in Scotland. J Hydrol 84: 365–380

Adamowski K, Bocci C (2001) Geostatistical regional trend detection in river flow data. Hydrol Processes 15:3331–3341

Blöschl G (2006) Geostatistische Methoden bei der hydrologischen Regionalisierung. In: Wiener Mitteilungen, Wasser-Abwasser-Gewässer, Band 197. Technische Universität Wien, pp 21–40

Blöschl G, Sivapalan M (1995) Scale issues in hydrological modelling – a review. Hydrol Processes 9:251–290

Bobée B, Rasmussen P (1995) Recent advances in flood frequency analysis, U.S National. Report to International Union of Geodesy and Geophysics 1991–1994. Rev Geophys 33(Suppl): 1111–1116

Bochiola D, De Michele C, Rosso R (2003) Review of recent advances in index flood estimation. HESS 7(3):283–296

Boorman DB, Hollis JM, Lilly A (1995) Hydrology of soil types: a hydrologically based classification of the soils of the United Kingdom, IH Report No. 126, Institute of Hydrology, Wallingford

Brath A, Castellarin A, Franchini M, Galeati G (2001) Estimating the index flood using indirect methods. Hydrol Sci J 46(3):399–418

Breiman L, Friedman JH, Olshen R, Stone CJ (1984) Classification and regression trees. Wadsworth International Group, Belmont

Burn DH (1990) Evaluation of regional flood frequency analysis with a region of influence approach. Water Resour Res 26(19):2257–2265

Burn DH (1997). Catchment similarity for regional flood frequency analysis using seasonality measures. J Hydrol 202:212–230

Castellarin A, Burn DH, Brath A (2001) Assessing the effectiveness of hydrological similarity measures for flood frequency analysis. J Hydrol 241:270–285

Chokmani K, Ouarda TBMJ (2004) Physiographical space-based kriging for regional flood frequency estimation at ungauged sites. Water Resour Res 40:W12514. doi:10.1029/2003WR002983

Cunnane C (1988) Methods and merits of regional flood frequency analysis. J Hydrol 100: 269–290

Dalrymple T (1960) Flood frequency methods. US Geological Survey Water-Supply Paper, 1543-A, 11–51

Daviau J-L, Adamowski K, Patry GG (2000) Regional flood frequency analysis using GIS, L-moment and geostatistical methods. Hydrol Processes 14:2731–2753

De Marsily G (1986) Quantitative hydrogeology. Academic, San Diego, CA, 440 pp

Deutsch CV, Journel AG (1997) GSLIB, Geostatistical software library and user's guide. Oxford University Press, New York, Oxford, 384 pp

Eaton B, Church M, Ham D (2002) Scaling and regionalization of flood flows in British Columbia, Canada. Hydrol Processes 16:3245–3263

Fill HD, Stedinger JR (1995) Homogeneity tests based upon Gumbel distribution and a critical appraisal of Darlymple's test. J Hydrol 166:81–105

Hirsch RM, Helsel DR, Cohn TA, Gilroy EJ (1992) Statistical analysis of hydrological data. In: Maidment R (ed) Handbook of hydrology. McGraw-Hill, New York, NY, pp 17.1–17.55

Hosking JRM, Wallis JR (1993) Some statistics useful in regional frequency analysis. Water Resour Res 29:271–281

Hosking JRM, Wallis JR (1997) Regional frequency analysis: an approach based on L-moments. Cambridge University Press, New York, NY

Institute of Hydrology (IH) (1999) Flood estimation handbook. Institute of Hydrology, Wallingford

Journel AG, Huijbregts CJ (1978) Mining geostatistics. Academic, London

Laaha G, Blöschl G (2005) A comparison of low flow regionalisation methods – catchment grouping. J Hydrol 323:193–214. doi: 10.1016/j.jhydrol.2005.09.001

Laaha G, Blöschl G (2007) A national lowestimation procedure for Austria. Hydrol Sci J 52(4):625–644

LfU Baden-Württemberg (Hrsg.) (1999) Hochwasserabfluss-Wahrscheinlichkeiten in Baden-Württemberg; Reihe: Oberirdische Gewässer/Gewässerökologie; Karlsruhe

Lu L, Stedinger JR (1992) Sampling variance of normalized GEV/ PWM quantile estimators and a regional homogeneity test. J Hydrol 138:223–245

Merz R, Blöschl G (2005) Flood frequency regionalisation – spatial proximity vs. catchment attributes. J Hydrol 302(1–4):283–306

Merz R, Blöschl G, Parajka J (2006) Spatio-temporal variability of event runoff coefficients. J Hydrol 331:591–604. doi:10.1016/j.jhydrol.2006.06.008

Merz R, Blöschl G (2008a) Flood frequency hydrology 1: temporal, spatial and causal expansion of information. Water Resour Res 44:W08432. doi:10.1029/2007WR006744

Merz R, Blöschl G (2008b) Flood frequency hydrology 2: combining data evidence. Water Resour Res 44:W08433. doi:10.1029/2007WR006745

Merz R, Blöschl G (2008c) Process controls on the statistical flood moments a data based analysis. Hydrological Processes. doi:10.1002/hyp.7168

Ouarda TBMJ, Ba KM, Diaz-Delgado C et al (2008) Intercomparison of regional flood frequency estimation methods at ungauged sites for a Mexican case study. J Hydrol 348:40–58. doi: 10.1016/j.jhydrol.2007.09.031

Pandey GR, Nguyen VTV (1999) A comparative study of regression based methods in regional flood frequency analysis. J Hydrol 225:92–101

Pfaundler M (2001) Adapting, analysing and evaluating a flexible Index Flood regionalisation approach for heterogeneous geographical environments. Schriftenreihe des Institutes für Hydromechanik und Wasserwirtschaft, ETH Zürich, Zürich, 189p

Pilgrim DH (1983) Some problems in transferring hydrological relationships between small and large drainage basins and between regions. J Hydrol 65:49–72

Piock-Ellena U, Merz R, Blöschl G, Gutknecht D (1999) On the regionalisation of flood frequencies – Catchment similarity based on seasonality measures. In: Proceedings of the 28th Biennial IAHR Congress in Graz, Austria, Aug 22–27. Technical University of Graz, Institute for Hydraulics and Hydrology, CD Rom: 434.htm

Reed DW, Stewart EJ (1991) Discussion on dam safety: an evaluation of some procesures for design flood estimation. Hydrol Sci J 36(5):499–502

Reed DW, Jakob D, Robson AJ, Faulkner DS, Stewart EJ (1999) nRegional frequency analysis: a new vocabulary. In: Gottschalk L, Olivry JC, Reed D, Rosbjerg D (eds) Hydrological extremes: understanding, predicting, mitigating. Proc. Birmingham Symp. July 1990, Wallingford. IAHS Publ. no. 255, pp 237–243

Sauquet E, Gottschalk L, Leblois E (2000) Mapping average annual runoff: a hierarchical approach applying a stochastic interpolation scheme. Hydrol Sci J 45:799–815

Schumann A, Pfützner B (2002) Regionalized flood estimation at ungauged sites supported by GIS, Proceedings of an International Conference on Flood Estimation, Berne, März 6–8, International Commission for the Hydrology of the Rhine basin, CHR Report II-17, Swiss Federal Office for Water and Geology, Bern-Ittigen, S. 669–677

Sivapalan M, Takeuchi K, Franks SW, et al (2003) IAHS decade on predictions in ungauged basins (PUB), 2003–2012: shaping an exciting future for the hydrological sciences. Hydrol Sci J 48:857–880

Skøien J, Merz R, Blöschl G (2006) Top-Kriging – geostatistics on stream networks. HESS 10:277–287

Stedinger JR, Vogel RM, Foufoula-Georgiou E (1992) Frequency analysis of extreme events, Ch. 18. In: Maidment DR (ed) Handbook of hydrology. McGraw-Hill, New York, NY

Tasker GD (1987) Regional analysis of flood frequencies. In: Singh VP (edr) Regional flood frequency analysis. Reidel, Dordrecht, pp 1–9

Uhlenbrook S, Steinbrich A, Tetzlaff D, Leibundgut Ch (2002) Regional analysis of the generation of extreme floods. Proceedings of the 4th international conference on FRIEND, Bridging the gap between research and practice, 18–22 March, Cape town, South Africa, IAHS Publ. No. 274, 243–250

US-SCS (1972) National engineering handbook Section 4 Hydrology. US Department of Agriculture, Washington, DC

Viglione A, Laio F, Claps P (2007) A comparison of homogeneity tests for regional frequency analysis Water Resour Res 43:W03428. doi:10.1029/2006WR005095

Wiltshire SE (1986) Identification of homogeneous regions for flood frequency analysis. J Hydrol 84:287–302

Zrinji Z, Burn DH (1994) Flood frequency analysis for ungauged sites using a region of influence approach. J Hydrol 153(1–4):1–21

Chapter 7
Rainfall Generators for Application in Flood Studies

Uwe Haberlandt, Yeshewatesfa Hundecha, Markus Pahlow, and Andreas H. Schumann

Abstract This chapter discusses various approaches for stochastic rainfall synthesis focusing on methods for generation of short time step precipitation as required for flood studies. A brief introduction motivates the utilisation of rainfall generators for flood modelling. Then special characteristics of rainfall as stochastic process are discussed. The rainfall models presented in the following are classified in alternating renewal models, time series models, point process models, disaggregation and resampling approaches. They are usually applied for continuous unconditional simulation of rainfall series in time and/or in space. Two case studies at the end of the chapter illustrate the application of daily and hourly space-time precipitation models for flood studies.

Contents

U. Haberlandt (✉)
Institute for Water Resources Management, Hydrology and Agricultural Hydraulic Engineering, Leibniz Universität Hannover, Hannover, Germany
e-mail: haberlandt@iww.uni-hannover.de

A.H. Schumann (ed.), *Flood Risk Assessment and Management*,
DOI 10.1007/978-90-481-9917-4_7,

7.1 Introduction

Reliable flood risk assessment and the development of effective flood protection measures require thorough knowledge about flood frequencies at different points in a catchment. The classical approach to obtain design flows is to carry out local or regional flood frequency analysis using long records of observed discharge data (e.g. Hosking and Wallis, 1997; Stedinger et al., 1993). If flow data are not available or if impacts of climate or land use change are to be investigated, rainfall-runoff modelling is a good alternative, either using event based or continuous simulation. A disadvantage of the event based simulation is the required assumption about equal return periods for the design storm and the resulting design flood. This is usually not given considering e.g. the initial soil moisture conditions in the catchment, which may lead to different floods for the same storm event. This problem can be avoided with continuous rainfall-runoff simulation and the design flood is derived by flood frequency analysis of long series of simulated flows. However, such kind of hydrological modelling requires long continuous rainfall series with high temporal and sufficient spatial resolution. Given the restricted availability of those observed data, synthetic precipitation is used more commonly for this purpose (Aronica and Candela, 2007; Blazkova and Beven, 2004; Cameron et al., 1999; Moretti and Montanari, 2008).

Over recent years, several stochastic precipitation models for short time step rainfall have been proposed. To the early approaches belong the alternating renewal models, which are based on event series of wet-dry spells (Acreman, 1990; Grace and Eagleson, 1966; Haberlandt, 1998; Pegram and Clothier, 2001). Those models have a simple structure, the estimation of parameters from point observations is straightforward and the models can easily be applied to rainfall synthesis at single locations. However, they are usually not able to simulate space-time rainfall for several stations in a catchment.

The classical approach of time series models is more suitable for daily or longer time step precipitation due to difficulties with modelling the high intermittence of short time step data and the large number of required parameters (Haan et al., 1976; Wilks, 1998). Bárdossy and Plate (1992) used a multivariate AR(1) model for space-time rainfall synthesis with special data transformation to consider intermittency.

Advanced approaches for rainfall modelling with sub-daily time steps are the point process models such as Neyman-Scott or Bartlett-Lewis rectangular pulse models (Cowpertwait, 2006; Onof et al., 2000; Rodríguez-Iturbe et al., 1987b), which can also be extended to simulate space-time rainfall. They are based on the physical structure of the rainfall process and describe probabilistically arrival times of storms and cells within storms as well as cell intensities and durations.

Other models for synthetic rainfall generation, which can be applied for flood studies include different disaggregation approaches (Koutsoyiannis et al., 2003; Lu and Yamamoto, 2008; Olsson, 1998) and various resampling methods (Bárdossy, 1998; Lall and Sharma, 1996). A somewhat dated but good overview of precipitation models is given by Foufoula-Georgiou and Georgakakos (1991). Other reviews about precipitation models can be found e.g. in Onof et al. (2000) or in Wheater et al. (2005).

This chapter is organised as follows. In Section 7.2 the basics about precipitation as stochastic process are introduced. In Sections 7.3–7.7 five classes of rainfall models are discussed. At the end of the chapter in Sections 7.8 and 7.9 two examples regarding daily and hourly rainfall synthesis are presented, respectively.

7.2 Precipitation as Stochastic Process

A stochastic process (Y_{t}) is a sequence of (usually correlated) random variables $Y_1, Y_2, \ldots, Y_n$. A time series represents one finite realisation of such a process. Important characteristics of stochastic processes are their *dimensionality*, *persistence* and *stationarity*. One-dimensional (univariate) processes Y_{t} consider one variable as a function of time. Multi-dimensional (multivariate) processes Y_{1t},Y_{2t} or Y_{tu} consider more than one variable and/or depend on time t and location u. Hydrological processes are usually persistent, i.e. they have a certain memory. Subsequent values are dependent.

A process has *Markov-character*, if all required information for Y_t can be obtained solely from the q previous time steps:

$$F(Y_t|Y_{t-1}, Y_{t-2}, \ldots, Y_1) = F(Y_t|Y_{t-1}, Y_{t-2}, \ldots, Y_{t-q}). \tag{7.1}$$

A stochastic process is called *stationary*, if the n-dimensional probability distribution describing the process is independent of time (i.e. *time invariant*):

$$F(Y_1, Y_2, Y_3, \ldots, Y_n) = F(Y_1 + \tau, Y_2 + \tau, Y_3 + \tau, \ldots, Y_n + \tau). \tag{7.2}$$

Usually it is sufficient to assume *second order stationarity*, which requires constant expected value and variance as well as a covariance, which depends solely on the time lag. Time series are often separated into a stationary and a non-stationary part. Then the stationary part can be modelled as a stochastic process. In hydrology time series are often separated into a trend T_t, a periodic component S_t and a random part Z_t:

$$Y_t = T_t + S_t + Z_t. \tag{7.3}$$

Three different types of stochastic processes can be distinguished: *normal type*, *point process* and *alternating process*. The normal type process shows constant or steady varying characteristics without sudden change (e.g. river flow). The point process describes events with short duration occurring at random points in time (e.g. flood spillway discharge). For the alternating process two mutually exclusive normal processes take turns. A special case is the *intermittent process* for which one of the alternating processes is zero.

The rainfall process is a complex intermittent (with wet-dry periods) multidimensional process (considering space and time). It shows spatial and temporal persistence and characteristic pattern. Usually precipitation occurrence and precipitation intensity are considered as different processes.

Figure 7.1 shows an observed rainfall time series in different temporal resolution. It becomes clear, that with decreasing time step the rainfall intermittency and variability increases. Also, a certain clustering of rainfall events can be seen, which in addition complicates the modelling of the rainfall process.

The classification of precipitation models is difficult, because many approaches have a number of common characteristics. Here, the models are classified as follows and will be presented thereafter accordingly:

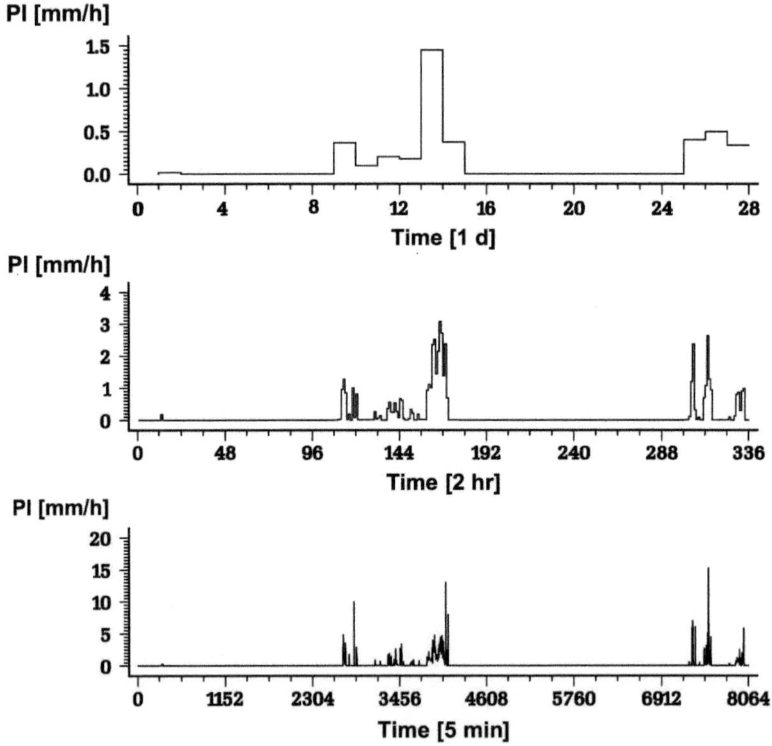

Fig 7.1 Observed rainfall times series plotted in different time resolutions (Station: Stuttgart-Vaihingen-University, 1.02.1990–28.02.1990)

1. Alternating renewal models
2. Time series models
3. Point process models
4. Disaggregation models
5. Resampling models

7.3 Alternating Renewal Models

Alternating renewal models (ARM) describe the precipitation process by dividing the time series into dry and wet spells. The entire precipitation process is separated into an external and an internal structure. The external structure characterises the occurrence and the amount of the precipitation events, as explained by the four random variables dry spell duration (*D*), wet spell duration (*W*) and wet spell amount (*V*) or wet spell intensity (*I*) with $I = V/W$. The internal structure describes the precipitation distribution within the wet spells. Figure 7.2 shows a scheme of the rainfall event process.

The precipitation occurrence process can be treated as an alternating renewal process if the durations of the spells are independent and if the unique spell states are identically distributed. This process is completely determined by establishing probability distribution functions (pdf) for the variables *D* and *W*. Different distribution functions such as Weibull, Exponential, Gamma, Lognormal etc. have been employed for this purpose (see e.g. Acreman, 1990; Grace and Eagleson, 1966; Haberlandt, 1998).

The precipitation amount or the intensity can also be modelled using probability distributions. However, *V* or *I* cannot be treated as independent of *W*. To consider these dependence conditional probabilities, bivariate distribution functions or copulas (Nelsen, 2006) can be applied. The latter have the advantage that arbitrary marginal distributions can be fitted independently for *W* and *I*. To generate the two

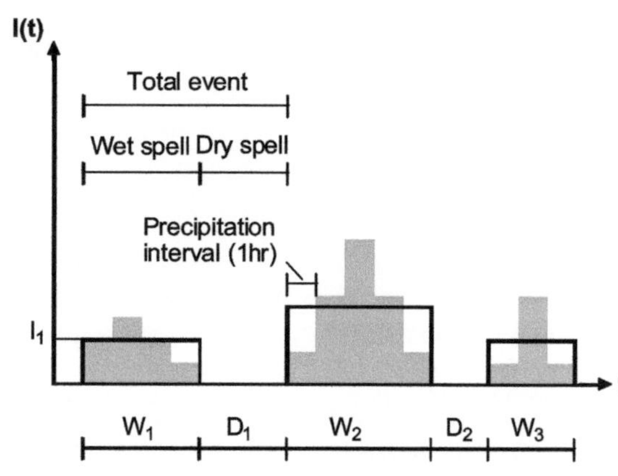

Fig. 7.2 Scheme of the precipitation event process (from Haberlandt et al., 2008)

dependent random variables W and I using a 2-copula the procedure outlined by De Michele and Salvadori (2003) can be followed.

The identification and parameter estimation of the alternating renewal model depends on the definition of a rainfall event. Rainfall events can be defined by a minimum rainfall amount V_{min} and a minimum separation time between events D_{min}. The minimum rainfall amount V_{min} excludes negligible events, which would complicate the fitting of probability distributions. The second criterion D_{min} should theoretically guarantee the statistical independence of subsequent events. In the literature values of D_{min} range from 1 h (Acreman, 1990) to 12 h (De Michele and Salvadori, 2003). If D_{min} is greater than the target time step, then the generated values underestimate the true precipitation intensity if the model is not able to consider clustering within the wet spells (i.e. zero rainfall amounts for certain time steps within a wet spell). To avoid complicated model formulations and underestimation of intensities it should be tried to keep D_{min} as small as possible.

The internal rainfall structure for an ARM can be modelled e.g. using time series models, empirical urn models, or special rainfall profiles for the disaggregation of the wet spell intensity I into equidistant time step rainfall intensities. As an example for a simple profile model a mixture of two exponential functions is shown in Fig. 7.3. This internal rainfall model is completely defined by the four random variables W, V, I_P and T_P. Since W and V are already given by the external model, only the wet spell peak I_P and the wet spell peak time T_P have to be specified here. The wet spell peak can be estimated by a simple regression to the mean wet spell intensity of the event using all stations in the study region. Realisations of the wet spell peak time t_P are generated from a uniform distribution. The main advantage of this simple profile model is that it needs no station specific parameters. This approach is obviously a strong simplification of the internal rainfall process. However, from the internal structure only a sufficient approximation of the peak intensity is expected. The internal rainfall variability is assumed to be of minor importance, compared to the external one for the generation of extreme flow events (see also Haberlandt, 1998).

Alternating renewal models are simple in structure and allow easy parameter estimation. They are especially suited for the simulation of the event series of short time step rainfall. In their original structure they are not appropriate for space-time simulation of rainfall. However, they can be coupled with other rainfall models to

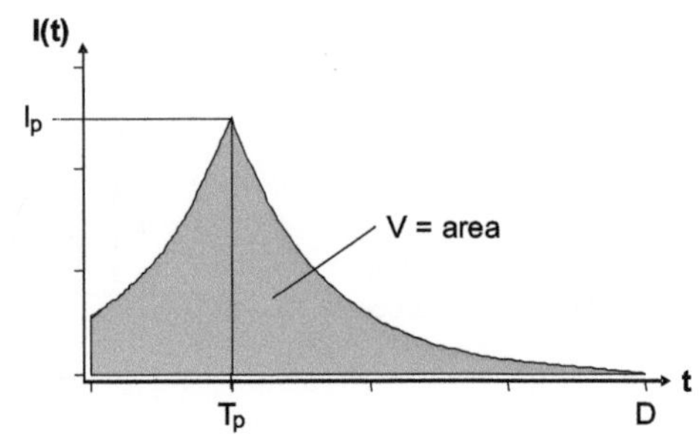

Fig. 7.3 Scheme of a simple profile model, which can be used to disaggregate wet spells into precipitation intensities (from Haberlandt et al., 2008)

allow multivariate synthesis (see Section 7.9). ARM parameters can be related to climate and catchment properties allowing regionalisation (e.g. Haberlandt, 1998) and stochastic downscaling.

7.4 Time Series Models

For modelling purposes, an observed precipitation series is regarded as a realisation of a stochastic process. Therefore, a mathematical expression that is able to describe the probability structure of the time evolution of the observations in which the temporal dependence between the elements of the series is reflected would be suitable to model the process. Such an expression is referred to as a time series model. Time series models are derived from the statistical information contained in the observations and, therefore, they are used to simulate synthetic series, which have the same statistical properties as the observed series.

There are different classes of time series models and the selection of a suitable type to model a given time series depends on the nature of the data. Models that are commonly used in the modelling of precipitation time series are briefly discussed in the following subsections.

7.4.1 Markov Chains

A precipitation time series at fine time scales, such as daily and hourly, generally displays a temporal intermittence. Therefore, in addition to establishing a suitable model for the amount of precipitation at each time step whenever precipitation occurs, there is a need to find a model that represents its occurrence. For this reason, precipitation at daily and sub-daily time scales is usually modelled using two separate models – one for the occurrence and one for the amount of precipitation.

The occurrence of precipitation is usually modelled using a class of time series models known as the Markov Chain, which is a suitable model to represent time series of discrete values. The discrete values of the time series are regarded as the different states of the Markov Chain and the succession of these states in the chain is described using transition probabilities. Generally, these transition probabilities are conditional probabilities that depend only on a few previous time steps of the chain (see Eq. 7.1). Accordingly, the transition probability governing the state of the next time step depends only on the states of the q time steps preceding it. Such a model is referred to as a Markov Chain of order q. The occurrence of precipitation is commonly modelled using a first order Markov Chain, in which the transition probability for the next state depends only on the present state (Haan et al., 1976; Richardson, 1982; Roldan and Woolhiser, 1982; Stern and Coe, 1984; Wilks, 1998). Designating the occurrence of precipitation or a wet state at a given time step by one and a dry state by zero, the occurrence process can be described using a two-state (wet and dry), first-order Markov Chain model. The model is described using four transition probabilities corresponding to: a wet state following a dry state (p_{01}), a dry state following a dry state (p_{00}), a wet state following a wet state (p_{11}), and a dry state following a wet state (p_{01}). Note that p_{01} and p_{00} sum up to 1 and therefore it suffices to know either one of them. The same holds for p_{11} and p_{10}. Therefore, a two-state, first-order Markov process can be fully described using two transition probabilities.

The transition probabilities are estimated from the observed time series. Once these probabilities are known, simulation of a time series is performed using a uniform [0,1] random number generator. The generated random number is compared with the conditional probability of a wet state following the current state. If the number is less than the transition probability, the next state will be wet and dry otherwise. For each wet state, the amount of precipitation can be estimated by randomly drawing a number from an appropriate distribution for the amounts of precipitation, e.g. a Gamma distribution for daily precipitation amounts (Katz, 1977; Stern and Coe, 1984), using an assumption that the amounts on wet time steps are serially independent. Higher order Markov-Chains and chains with more than two states are also used. In particular, they can be used for direct generation of rainfall amounts, dividing the wet states into several classes of precipitation amounts (Haan et al., 1976). However, the number of required parameters might become quite large then.

7.4.2 ARMA Models

Autoregressive moving-average (ARMA) models belong to a class of time series models that is commonly applied for the modelling of continuous time series. It is composed of an autoregressive component that describes how an observation at a given time depends directly on one or more previous observations and a moving average component, which describes how an observation depends on random shocks of the generating stochastic process at the current and previous time steps. An ARMA(p,q) model, which contains an autoregressive component of order p and a moving average component of order q is expressed as:

$$x_{t+1} - \mu = \sum_{i=1}^{p} \phi_i (x_{t-i+1} - \mu) + \varepsilon_{t+1} + \sum_{j=1}^{q} \theta_j \varepsilon_{t-j+1}, \tag{7.4}$$

where ϕ_i and θ_i are the autoregressive and moving average parameters, respectively, and ε_t is a random shock or white noise term at time t that is independent and identically distributed with a mean of 0 and a variance of σ_ε^2. A Gaussian distribution is often assumed for the white noise. The parameter μ is the mean of the time series. Eq. (7.4) applies to a stationary time series and therefore μ is the same for each time step. If the data show an apparent non-stationarity, such as a seasonal variation displayed by precipitation, the data can first be transformed into a stationary series by subtracting the variable mean and dividing the resulting anomaly by the standard deviation, which may also be variable. This transformation reduces the data into a stationary time series of zero mean and unit variance. After prediction is made by the model, the actual values of the time series are then obtained using a back transformation.

An important element in the application of the ARMA model is the selection of the orders p and q. It is always desirable to have a parsimonious model which can model the data adequately. However, choosing the orders of the ARMA model is not straightforward and usually a trial and error procedure is followed by studying the autocorrelation structure of the time series (see e.g. Hipel and McLeod, 1994). Objective criteria, such as the Akaike information criteria (Akaike, 1974) can also be employed. Once the orders of the model are selected, estimation of the model parameters can be performed using the method of maximum likelihood. A special class of ARMA models is a pure autoregressive model of order AR(p). For such a model, both the order and the model parameters can easily be estimated from the autocorrelation structure.

Considering precipitation, ARMA models are mainly applied to the modelling of precipitation at coarse time scales, such as monthly and annual, where there is no temporal intermittence (Delleur and Kavvas, 1978; Kavvas and Delleur, 1975). However, it is also possible to use such models for precipitation series at finer time scales, where the data show temporal intermittence with data transformation. The data can be treated as coming from a continuous distribution, usually a normal distribution, with the dry states regarded as coming from the non-positive portion of the distribution (see e.g. Bárdossy and Plate, 1992; Hutchinson, 1995). An ARMA model can then be applied to the transformed data and simulated negative values are set to zero (no precipitation). Daily precipitation amounts generally show a skewed distribution and to account for this behaviour, a power transformation is often applied to the values simulated by the model to obtain the precipitation amounts.

Transformation of the precipitation time series into a normally distributed variable can also be exploited for a multi-site modelling of precipitation, since a multi-normal distribution is amenable to the classical multivariate stochastic analysis. For example, Bárdossy and Plate (1992) used a multivariate autoregressive AR(1) model using a power transformed multi-normal distribution to model daily series of precipitation at multiple locations. A similar approach is implemented in the case study example in Section 7.8.

7.4.3 DARMA Models

As discussed in Section 7.4.1, Markov Chains are used to model discrete time series. Whenever longer term persistence is displayed by the data, a higher order Markov Chain is needed to model the time series. However, this leads to a large number of parameters. One would face a similar problem if one would prefer to model both the amounts and the occurrence of a precipitation series using a multi-state Markov Chain by discretising the precipitation amounts into different classes. Discrete ARMA models (Jacobs and Lewis, 1978a, b) constitute a more general class of discrete time series models that can be used more parsimoniously than Markov Chains to handle multi-state discrete time series or discrete time series with longer persistence structure. Their structure is based on the ARMA model.

A DARMA($p, q + 1$) model is a sequence X_n formed by a probabilistic combination of elements of a sequence Y_n, which are independent and identically distributed (i.i.d.) with a probability distribution given by π.

$$\begin{aligned} X_t &= U_t T_{t-S_t} + (1 - U_t) Z_{t-q-1} \\ Z_t &= V_t Z_{t-A_t} + (1 - V_t) Y_t \end{aligned} , \tag{7.5}$$

where U_t and V_t are i.i.d. binary random variables with $P(V_t = 1) = \alpha$ and $P(U_t = 1) = \beta$; A_t is an i.i.d. random variable defined on $\{1, 2, \ldots, p\}$ with $P(A_t = k) = \varphi_k$; S_t is an i.i.d. random variable defined on $\{1, 2, \ldots, q\}$ with $P(S_t = k) = \theta_k$. DARMA models were employed for daily precipitation modelling by Chang et al. (1984a, b).

7.4.4 Advantages and Disadvantages

Markov Chain models are attractive in their non-parametric nature, ease of applicability and interpretability. Their disadvantage is their lack of parsimony when used to account for long term persistence. DARMA models have an advantage of being alternative solutions to circumvent this problem. Their disadvantage is, however, they lack physical interpretability. In addition, they display a discontinuous memory. The temporal dependence is realised only through runs of constant values in the generated series. Similar to the Markov Chain, ARMA models are interpretable and easy to apply for continuous variables which do not show temporal discontinuity. Their disadvantages are that they are not readily applicable if there is a discontinuity in the data and that they make restrictive assumptions about the distribution of the data.

7.5 Point Process Models

Advanced approaches for rainfall modelling with sub-daily time steps are the point process models (Cowpertwait, 2006; Onof et al., 2000; Rodríguez-Iturbe et al., 1987a, b). They are based on the physical structure of the rainfall process. A point process PP describes the occurrence or position of events in the model space. The model space of PPs can be considered as time domain or as space-time domain. The process is called a *marked point process* if not only occurrence but also rainfall amount or intensity is simulated. Discrete and continuous PP can be distinguished. Common for shorter time steps are the continuous PP's, which are discussed in the following.

The simplest PP is the *Poisson process*, simulating only the rainfall occurrence, where impulses with durations towards zero and heights of one occur at random points T_i in time. The number of impulses and the inter-arrival time are Poisson and exponentially distributed, respectively. The *Poisson-White-Noise* (PWN) model is the simplest marked PP with an attached rainfall amount U_i to the random impulses

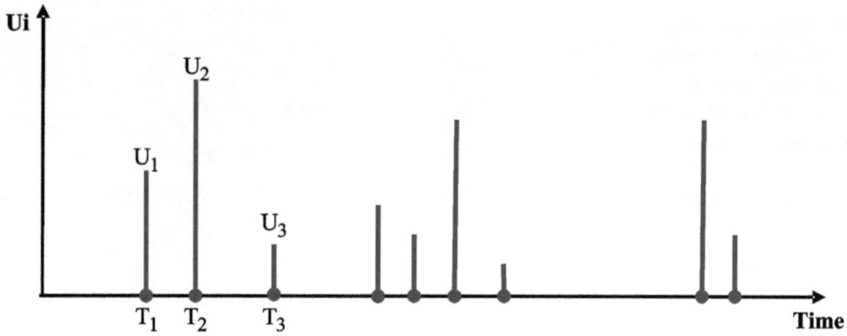

Fig. 7.4 Scheme of the Poisson-White-Noise (PWN) process

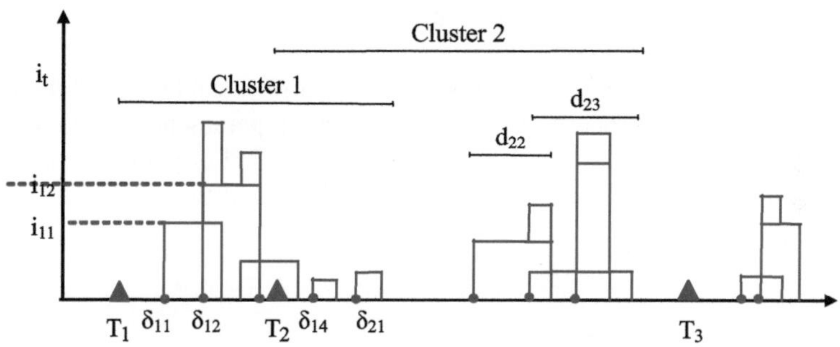

Fig. 7.5 Scheme of the Newman-Scott-Rectangular-Pulse (NSRP) process with T_i arrival time of clusters and δ_{ij} arrival time of events

(see Fig. 7.4). The rainfall amounts are independent from each other and from the occurrence process as well as identically distributed.

The *Poisson-Rectangular-Pulse* (PRP) model extends the PWN by simulating rectangular impulses with durations d_i and intensities i_i. The rectangular impulses are assumed to be i.i.d.. Often also duration and intensity are assumed to be independent. The *Newman-Scott-Rectangular-Pulse* (NSRP) model considers rainfall occurrence at two levels, clusters and rectangular impulses (see Fig. 7.5).

Clusters arrive according to a PP and impulses are modelled by probability distribution functions. All random variables are i.i.d.. Clusters and events can overlap. A very similar model is the *Bartlett-Lewis-Rectangular-Pulse* (BLRP) model.

A *spatial-temporal point process* model can be constructed in analogy to the temporal models. These models again use a hierarchical representation of rainfall. The smallest element is a spatial rain cell. The rain cells cluster within larger-scale structures that can be called storms. These storms themselves cluster again, to form rain events (see Fig. 7.6). Storm centers originate in a Poisson process in space and time, and they move with a constant velocity (see e.g. Wheater et al., 2005).

The theory of point processes is based on the physical structure of the continuous rainfall process. Since observations are often possible only at discrete points or as

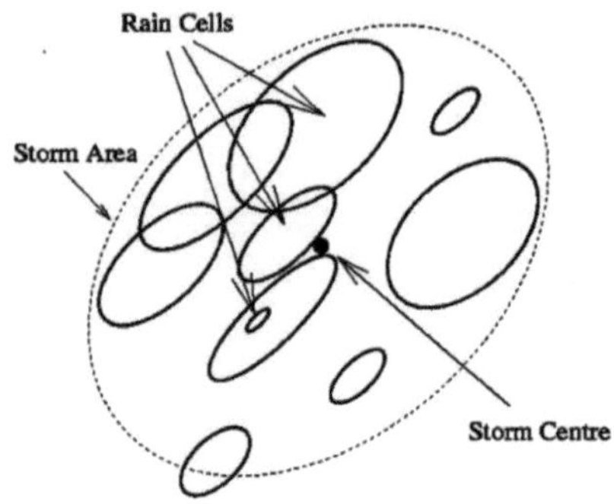

Fig. 7.6 Scheme for the spatial structure of storms within the spatial-temporal rainfall model based on the Poisson cluster process (reproduced with kind permission from Wheater et al., 2005)

integral over time, the parameter estimation for PP is quite complicated. It requires usually observed radar data or relies on optimisation.

Advantages of the point process models are their physical basis and flexibility in describing the rainfall process. A disadvantage is the parameter estimation in case of insufficient observations, which may lead to some kind of parameter equifinality (Beven and Freer, 2001). The latter makes conditioning of the model parameters on climate difficult.

7.6 Disaggregation Models

Disaggregation models use observed rainfall on larger time steps (e.g. days) and partition those values into smaller ones (e.g. hours). In principle several rainfall models presented in this chapter can be operated as disaggregation models, conditioning the synthesis on the larger time step rainfall sums (Bárdossy, 1998; Koutsoyiannis et al., 2003). Disaggregation allows simulation in stages for different time steps using each a suitable approach, e.g. modelling daily rainfall with a Markov Chain and then disaggregating it with a random cascade. The *random cascade models* are based on scaling properties of rainfall and are designed especially for the disaggregation of rainfall in time or space (Gaume et al., 2007; Lovejoy and Schertzer, 2006; Olsson, 1998).

One such approach, a multiplicative random cascade model with exact mass conservation (Güntner et al., 2001), is discussed exemplarily in the following for disaggregating daily rainfall into hourly values. The model divides the observed 24 h series subsequently into two equal size non-overlapping time intervals or boxes (see Fig. 7.7). If the total rainfall volume in a box is V, $V_1 = W_1 \times V$ is assigned to the first half and $V_2 = W_2 \times V$ to the last with:

$$W_1, W_2 = \begin{cases} 0 \text{ and } 1 & \text{with probability } P(0/1) \\ 1 \text{ and } 0 & \text{with probability } P(1/0) \\ x \text{ and } 1 - x & \text{with probability } P(x/1 - x) \end{cases}, \tag{7.6}$$

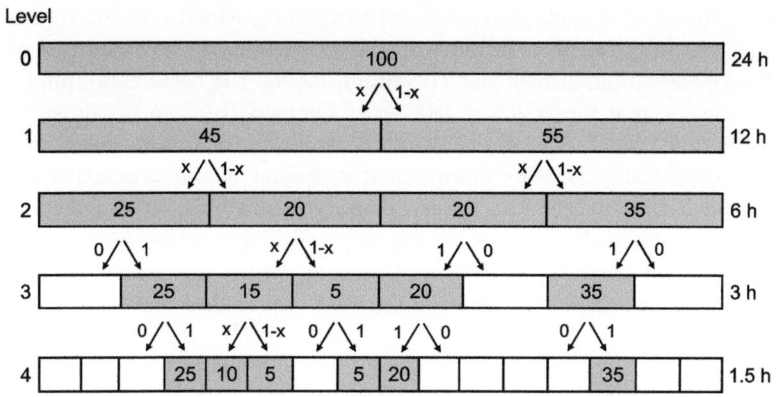

Fig. 7.7 Scheme of a multiplicative random cascade model (adapted from Olsson, 1998)

where $0 < x < 1$ and $P(0/1) + P(1/0) + P(x/1 - x) = 1$. The values x are usually calculated using probability distribution functions.

Usually the random cascade assumes scale invariance, meaning that the parameters are constant over all disaggregation levels. In this case the parameters can be calculated from aggregation of the observed (daily) data to larger time steps. If scale invariance cannot be granted, then the parameters can also be regionalised, e.g. transferred from the nearest high resolution (hourly) rainfall station.

An important advantage of disaggregation models is to make maximum use of the denser rainfall networks for daily data to generate short time rainfall with a conservation of the observed rainfall volume in space and time on the daily time step. A disadvantage is the restriction to the length of the observation period for rainfall synthesis. A currently not satisfyingly resolved problem is the spatial-temporal disaggregation using random cascade models.

7.7 Resampling Models

A common feature shared by precipitation modelling approaches presented in the foregoing sections is that they make an assumption about the probability distribution governing the stochastic process. This assumption may not always hold and sometimes it might even be difficult to determine well defined distributions for a given dataset. In addition, they make an assumption about the temporal and spatial dependence structures in a parametric way. Due to this, the models are often limited to reproducing only a certain number of the desired properties of the observed precipitation series. In order to circumvent such problems, models that make little assumption about the distributional property of the observations or the parametric form of the model have been introduced. These models fall into the category of resampling models.

Resampling methods are based on randomly selecting data samples from the historical observation to produce another data set that has the same properties as the observations. One such method is *bootstrapping*, which consists of randomly sampling data from the observation with replacement using the empirical distribution of the observations (Efron and Tibshirani, 1998). The basic assumption behind bootstrapping is that the original data were sampled from an unknown generating distribution probabilistically with replacement and a similar sampling design on the original data using its empirical distribution would produce another set that will have similar features. *Permutation* is another resampling technique, in which the resampled data consist of the original data with the ordering swapped randomly (i.e. resampling without replacement). Two approaches of modelling precipitation series that make use of the two approaches are briefly discussed below.

7.7.1 k-Nearest Neighbourhood Bootstrapping

This approach involves generation of precipitation series based on bootstrap sampling from the historical observations in such a way that a number of desirable features of the observed series are preserved. The basic idea behind this approach is that elements of a time series that have similar features will be followed by elements that are similar with one another. The similarity of features of the elements is inferred using a distance metric defined on a multi-dimensional space, whose coordinates are the different features that are considered important defining attributes of the time series. For instance, these features can be amounts of precipitation in the previous time steps, other auxiliary variables such as circulation indices or air temperature. The variables are often standardised to avoid scale effects and a weighted form the Euclidean distance can be employed as a distance metric.

If the precipitation at a given time t is known, generation at the next time step $t + 1$ is performed as follows. First k observations are identified that are closest to the precipitation at time t, based on the defined distance metric. Then one of these observations is selected randomly and the observation at the subsequent time step is adopted as precipitation for time $t + 1$. Unlike the standard bootstrap method, sampling is not performed from the empirical distribution of the neighbouring samples. Instead, a non-parametric smooth k-nearest neighbourhood density function that assigns weights to the neighbourhood observations that monotonically decreases with their distance is employed. The approach is discussed in detail by Lall and Sharma (1996) and its application in the modelling of daily precipitation series can be found in Rajagopalan and Lall (1999) and Brandsma and Buishand (1998).

The k-nearest neighbourhood approach has an attractive feature in that it can easily be extended to a multi-site precipitation generation model in a non-parametric way (Buishand and Brandsma, 2001). Here, the coordinates of the feature space constitute the precipitation at the individual sites. The dimensionality of the space can also be reduced using summary statistics over a few regions or multivariate techniques such as principal component analysis.

7.7.2 Simulated Annealing

Another resampling approach for modelling precipitation series that reproduces a range of characteristics of the observed series is simulated annealing. This approach is based on the Markov Chain Monte Carlo simulation technique (Gilks et al., 1996; Hastings, 1970). A synthetic series of precipitation of a desired length is first generated from a probability distribution of the precipitation amounts. The distribution can be either an empirical distribution of the observations or a fitted parametric one. Subsequently, a number of elements are abstracted randomly from this synthetic series based on the distribution of the total number of wet states for all time steps. The precipitation at the remaining time steps is set to zero. The series thus generated may not reproduce the autocorrelation structure and many other properties of the observed series. Simulated annealing can be effectively employed then to render the series the required properties by repeatedly reshuffling its elements.

Whether the generated time series has the required properties is judged using an objective function, which are measures of the closeness of the properties of the simulated series to that of the observations. Each time two elements of the generated series are reshuffled the objective function is recomputed and compared with the objective function before reshuffling was done. If an improvement is achieved, the reshuffling is accepted. If not, it is conditionally accepted with a probability that is a function of the value of the objective function and a global time varying parameter known as the annealing temperature. This procedure is repeated several times until convergence of the objective function is achieved. Interested readers are referred to Bárdossy (1998) for details of this approach. An example for an application to obtain the spatial rainfall dependence structure using simulated annealing is given in Section 7.9.

7.7.3 Advantages and Disadvantages

One obvious advantage of resampling methods is that they make little assumption about the distributional properties of the random process and the functional form of the dependence structure. This feature enables the approaches to capture a range of desirable properties of the observed precipitation series since they can directly be used in the model design in a non-parametric way. However, resampling methods are computationally expensive. In addition, since resampling is performed only on the observed series, the methods pose limitations in the extrapolation of the extremes. Therefore, their applicability in problems where extremes at the temporal scale are to be generated is not apparent as no values beyond the extremes that were historically observed can be generated.

7.8 Example for Daily Rainfall Synthesis

An example of the application of some of the modelling principles discussed in the foregoing sections is presented in this section. It is shown how daily precipitation

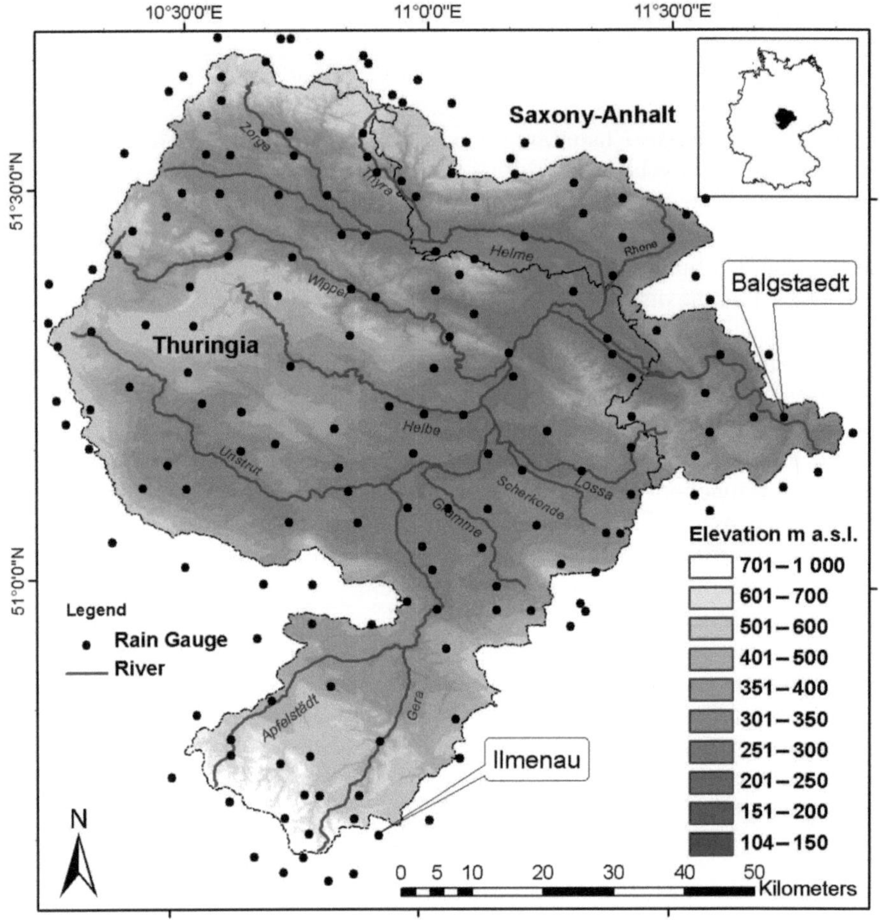

Fig. 7.8 Unstrut catchment with 122 daily precipitation stations used in this study. The highlighted stations Balgstaedt and Ilemnau will be considered in detail

series at multiple locations can be modelled with the objective of reproducing the statistical properties of the observations that are relevant to flood risk analysis in a watershed (Hundecha et al., 2009). Modelling was carried out on 122 daily precipitation stations within and around the Unstrut catchment, which is located in Central Eastern Germany. The catchment has a drainage area of ~6,340 km^2 and variable topography as shown in Fig. 7.8.

Recently, the catchment has seen recurrent flooding episodes and it was required to carry out a flood risk assessment in a bid to provide a flood protection strategy in the future. The assessment considered a range of flood events with recurrence intervals of up to 1,000 years. The available precipitation time series at the observation stations are too short to enable generation of runoff data of sufficient length from which statistical inferences could be made. Therefore, there was a need to generate sufficiently long synthetic precipitation time series to drive a hydrological model that generates the runoff. Since the spatial distribution of the precipitation at a given time is one of the determining factors of the resulting runoff, the inter-site correlation structure of the precipitation displayed by the observed series should be reproduced by the generated synthetic series. The implemented model should therefore be a multi-site model that preserves the spatial structure. In addition, the distribution of the precipitation amounts should be such that the extremes are well captured and extrapolation into lower frequency events is possible.

7.8.1 The Modelling Steps

The implemented modelling approach is the multivariate version of the autoregressive model applied on the precipitation series transformed into a normally distributed variable, which was discussed in Section 7.4.2. Transformation into a normally distributed variable is performed for two reasons. First, the inter-site correlation structure can easily be accounted for using the classical multivariate stochastic methods if normal distributions are used. Second, both the occurrence process and the amounts of precipitation can be modelled using a single model setup without the need for separate models. This is achieved by censoring the generated values such that values less than a certain cut-off point are set to represent dry days.

The first step of the modelling process is to identify the appropriate parametric distribution for the daily amounts of precipitation at each station. Since the distribution of daily amounts of precipitation is generally skewed with a bias towards low amounts, the Gamma distribution is often used as an appropriate model, although other distributions such as mixed exponential and the Kappa distributions have been used in some cases (Mielke, 1973; Wilks, 1998; Woolhiser and Roldan, 1982). For the time series in the present study area, however, a single distribution was found to be inadequate to represent the entire range of the observations. The data show an apparent non-homogeneity in their distribution, which could be a result of the difference in the mechanisms that produce the different regimes of precipitation.

One possibility to circumvent the above problem would be to implement two different distributions to characterise the low to medium amounts and high amounts, respectively. The low to medium amounts, which constitute the bulk of the data, can be modelled using the commonly used Gamma distribution, while the high amounts are characterised by a distribution that is used to model the extreme values of a random variable. In line with the statistical theory of extreme values, a Generalised Pareto Distribution (GPD) was proposed to characterise the high amounts of precipitation. A GPD is a model that is used to model the extremes of a random variable beyond a sufficiently high threshold (Coles, 2001). However, in addition to identifying a suitable threshold above which a GPD is used, the resulting density function of the Gamma-GPD model becomes discontinuous at the threshold. To avoid this, the Gamma distribution and GPD are mixed over the entire domain of precipitation amounts using a dynamic mixing weight that gives more emphasis for higher precipitation amounts. This approach was proposed by Frigessi et al. (2003) and later applied to modelling of precipitation by Vrac and Naveau (2007). The mixture model is given by:

$$f(z,u) = \frac{[1 - p(z,u)]h(z,u) + p(z,u)g(z,u)}{K(u)}, \tag{7.7}$$

where $f(z,u)$is the probability density function of the daily precipitation z at location u; $p(z,u)$ is the mixing weight; $K(u)$ is a normalizing constant that ensures the area under $f(z,u)$ is unity. $h(z,u)$ and $g(z,u)$ are the probability density functions of the Gamma and GPD respectively:

$$h(z,u) = \frac{(z/\beta(u))^{\gamma(u)-1}\exp(-z/\beta(u))}{\beta(u)\Gamma(\gamma(u))}, \tag{7.8}$$

where $\gamma(u) > 0$ and $\beta(u) > 0$ are the shape and scale parameters, respectively, corresponding to location u; $\Gamma(.)$ is the Gamma function.

$$g(z,u) = \frac{1}{\alpha(u)}\left(1 + \frac{\xi(u)\,z}{\alpha(u)}\right)^{-1/\xi(u)^{-1}}, \tag{7.9}$$

where $\xi(u)$ and $\alpha(u) > 0$ are the shape and scale parameters, respectively. When $\xi(u) > 0$, the distribution displays a heavy tail behaviour. When $\xi(u) < 0$, it is bounded and when $\xi(u) = 0$, the distribution reduces to an exponential distribution $g(z,u) = 1/\alpha(u)\exp(-z/\alpha(u))$ with a light tail. These different features of the GPD make it a suitable flexible distribution to model the different extreme behaviours of the daily precipitation.

The mixing weight $p(z,u)$ is a function that gives more emphasis to the Gamma distribution at low values of precipitation and more weight to the GPD at high values, while keeping the continuity of the distribution. A mixing weight proposed by Frigessi et al. (2003) is used here:

$$p(z,u) = \frac{1}{2} + \frac{1}{\pi}\arctan\left(\frac{z - \nu(u)}{\tau(u)}\right), \tag{7.10}$$

where $\nu(u) > 0$ and $\tau(u) > 0$ are the location and steepness parameters respectively. Note that the limit of $p(z,u)$ for large z values is unity, thereby making the tail of the mixture distribution governed by the GPD for large z. The normalizing constant $K(u)$ is then given by:

$$K(u) = 1\frac{1}{\pi}\int_0^\infty [g(z,u) - h(z-u)]\arctan\left(\frac{z - \nu(u)}{\tau(u)}\right)dz, \tag{7.11}$$

In order to account for the seasonal variability of daily precipitation, different models are fitted to each month of the year.

The stochastic process generating the precipitation series at multiple sites is described using a multivariate first order autoregressive model using a multivariate standard normal distribution $\mathbf{W}(t) = (W(t, u_1), \ldots, W(t, u_n))$:

$$\mathbf{W}(t) = r(\mathbf{W}(t-1) - \mathbf{w}) + \mathbf{C}\boldsymbol{\Psi}(t) + \mathbf{w}, \tag{7.12}$$

where $\boldsymbol{\Psi}(t) = (\psi(t, u_1), \ldots, \psi(t, u_{\rm n}))$ is a random vector of independent normally distributed random variables with zero mean and unit variance. $\mathbf{w} = (w(u_1), \ldots, w(u_{\rm n}))$ is the mean vector of $W(t, u)$, whose elements are all 0 since a standard normal variate is used here, r is the lag-1 day autocorrelation function, and $\mathbf{C}$ a $n \times n$ matrix that takes the spatial variability of the process into account. Some simplifying assumptions are made to reduce the complexity of Eq. (7.11) as discussed in Stehlik and Bárdossy (2002). The lag-1 autocorrelation r is assumed to be spatially invariable at the spatial scale of the study area. This assumption is based on the fact that precipitation at a daily timescale is caused by a large scale circulation feature. With a related further assumption that the lag-0 correlation matrix $\mathbf{M_0}$ and the lag-1 correlation matrix $\mathbf{M_1}$ are related as $r\mathbf{M_0} = \mathbf{M_1}$, the autoregressive parameter matrix can be reduced to the spatially invariant lag-1 autocorrelation r. Furthermore, the matrix $\mathbf{C}$ is related to $\mathbf{M_0}$ and $\mathbf{M_1}$ through a relation derived by Bras and Rodríguez-Iturbe (1994):

$$\mathbf{C}\mathbf{C}^T = \mathbf{M_0} - \mathbf{M}_1\mathbf{M}_0^{-1}\mathbf{M}_1^T. \tag{7.12}$$

The parameters r and $\mathbf{C}$ have to be estimated before simulation can be performed using Eq. (7.12). They should be estimated from the observations. Therefore, the observations should be transformed into standard normal variates. Note that dry days are regarded as sampled from the portion of the normally distributed variable to the left of a certain threshold value. This value is estimated as the value of a standard normal variate corresponding to the cumulative distribution function (cdf) value equal to the probability of dry days. Therefore, transformation of the observations to a standard normal variate is performed using the relationship:

$$F(z, u) = \frac{\Phi(x) - (1 - p_w(u))}{p_w(u)} \tag{7.14}$$

where $f(z, u)$ is the cdf of the daily precipitation at location u, and $\Phi(\cdot)$ is the cdf of a standard normal variate.

The required parameters of the generating process can then be estimated from the transformed data. Details of the estimation procedure are presented in Bárdossy and Plate (1992). Simulation produces normally distributed variables. The actual precipitation amounts are then estimated through back transformation using Eq. (7.14), so that the simulated values are regarded as coming from the actual distribution of the precipitation.

7.8.2 Simulation of Daily Precipitation

The parameters of the generating model were estimated from daily precipitation data at the 122 stations for the period 1969–2001. 100 realisations of daily precipitation series, each 100 years in length were generated. Two stations were selected to study the model performance: Ilmenau, where the extremes of daily precipitation are characterised by a heavy tail, and Balgstaedt, where the tail of the daily precipitation distribution is lighter.

Figure 7.9 shows the monthly mean precipitation amounts and the standard deviations of the daily precipitation on the monthly basis at the two stations and the corresponding values for the areal mean daily precipitation. The solid and dotted lines show respectively the median values and the 90% confidence intervals of the simulated values from the 100 realisations. From the figure, it can be seen that the observed values are in most cases contained within the confidence intervals of the simulated values. In addition, the model reproduces the seasonal variations of both the monthly mean and the standard deviations. The monthly standard deviations of the areal average precipitation show less seasonal variation and they are estimated with less uncertainty as can be seen from the narrower confidence interval.

Figure 7.10 shows evaluation of the model's ability to reproduce features related to the occurrence of precipitation. The figure shows comparison of the observed and simulated monthly probabilities of wet days, as well as spells of dry and wet days. The model generally reproduces both the probability of wet days and the spell

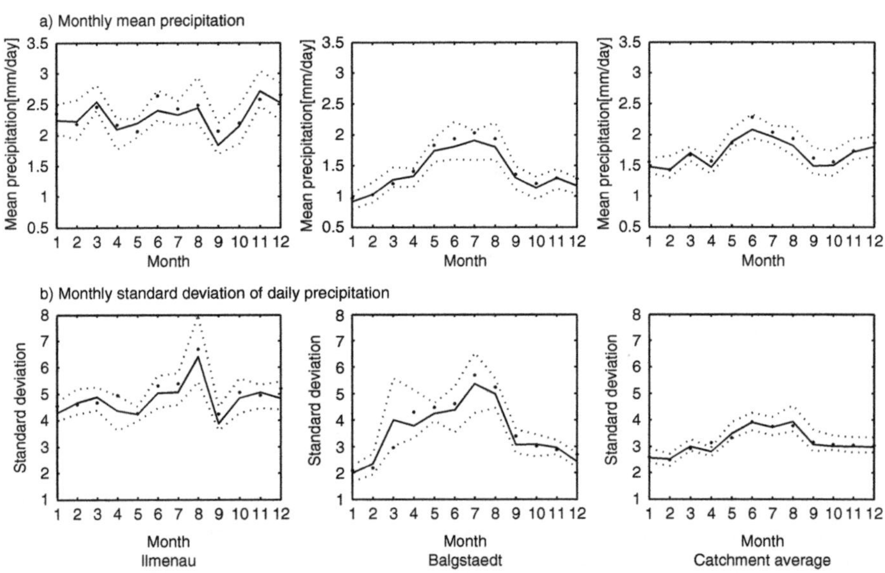

Fig. 7.9 Observed and simulated monthly mean and standard deviations of daily precipitation amounts (*solid circles*: observations; *solid lines*: simulated median values; dotted lines: simulated 90% confidence interval)

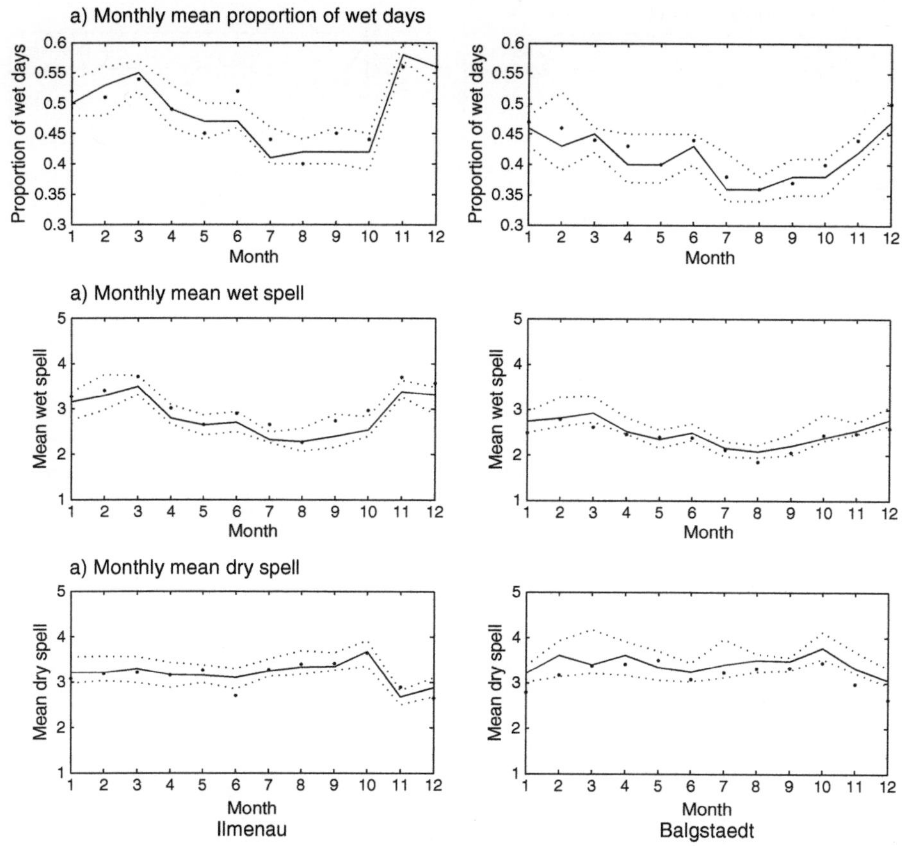

Fig. 7.10 Observed and simulated monthly mean proportion of wet days and spell lengths. *Solid circles* indicate observed values, *solid lines* indicate median of the simulated, and *dotted lines* indicate the 90% confidence interval of the simulated

lengths well. But one can see from the figure that the spell lengths are fairly constant through out the year and therefore, it would be worthwhile to investigate whether the model reproduces the frequency distributions of the spell lengths. Figure 7.11 shows the comparison between the frequencies of wet and dry spells of the observed series and the corresponding median values of the simulated series. Like the mean spell lengths, the frequencies are reproduced well, suggesting that the model is able to capture the occurrence structure of precipitation.

The most important feature the model needs to reproduce in the context of this work is the distribution of the extremes of the daily precipitation. Figure 7.12 shows the improvement achieved in modelling the tail of the distribution of the daily precipitation by implementing the mixture model. The figure shows the Q-Q plot of the observed and modelled daily precipitation for the month of August, when flood associated extreme precipitation occurs, for the period 1969–2001. The heavy tail of the distribution of the daily precipitation at station Ilmenau is now better reproduced

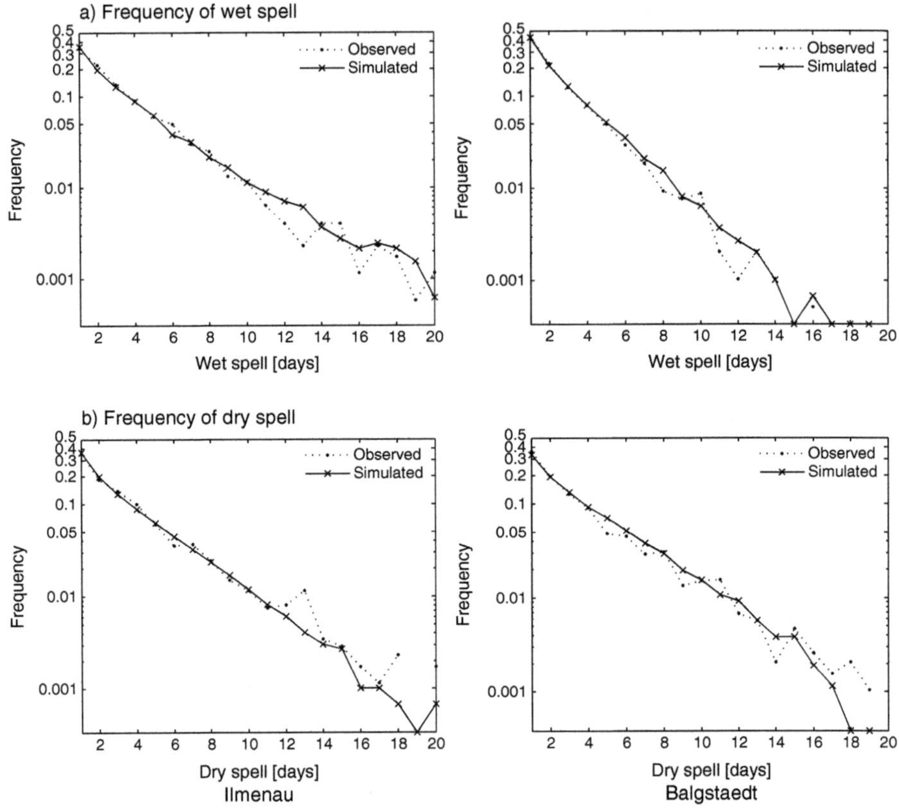

Fig. 7.11 Observed and median of simulated frequencies of durations of wet and dry days over the entire observation and simulation periods

by the mixture model. Although the tail at station Balgstaedt is not as heavy as at Ilmenau, the Gamma model does not capture it well and the mixture model has improved its representation.

Figure 7.13 shows comparisons between the extreme value distributions of the observations and the simulated series, which are estimated by fitting Generalized Extreme Value (GEV) distributions to the annual maximum daily precipitations. The confidence intervals for the simulations were estimated from the distributions fitted to the 100 realisations, while they were estimated from distributions fitted to 1,000 bootstrap samples taken from the GEV distribution fitted to the observed data. The extremes at both stations are reasonably reproduced. At station Balgstaedt, the model tends to slightly overestimate the higher extremes. Nevertheless, the overestimation is not very high and it can be considered acceptable since such quantiles are in the extrapolation region. The extremes of the catchment average daily precipitation are reproduced well by the model as the confidence interval of the model simulated extremes is well contained within the corresponding intervals of the observations. The overall good performance of the model in terms of reproducing

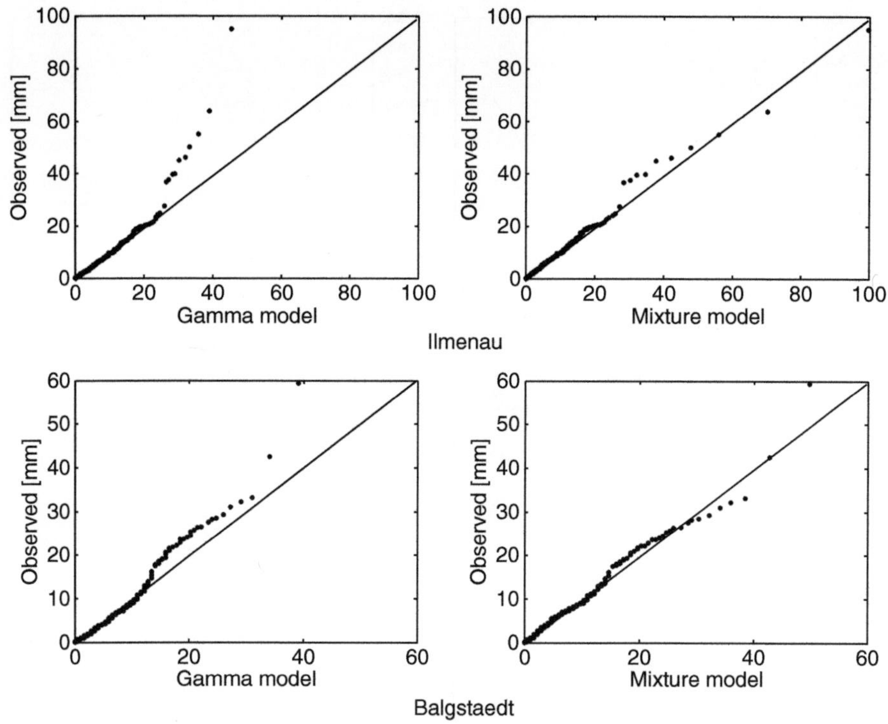

Fig. 7.12 Q-Q plots of the observed and modelled daily precipitation using a Gamma and mixture model for the month of August. The *solid line* indicates the 1:1 relationship

the extremes of daily precipitation both at station and catchment scale suggests the suitability of the model for the intended flood risk study.

7.9 Example for Hourly Rainfall Synthesis

This example shows a two step procedure for stochastic synthesis of continuous hourly space-time rainfall and its application for derived flood frequency analysis. First, a single-site alternating renewal model is presented to simulate independent hourly precipitation time series for several locations. In the second step a multi-site resampling procedure using simulated annealing is applied on the synthetic point rainfall event series to reproduce the spatial dependence structure of rainfall. In a case study synthetic precipitation is generated for some locations with short observation records in two mesoscale catchments of the Bode river basin located in northern Germany. The synthetic rainfall data are then applied for derived flood frequency analysis using the hydrological model HEC-HMS. In the following sections a brief overview of this work is given. Detailed information can be found in Haberlandt et al. (2008).

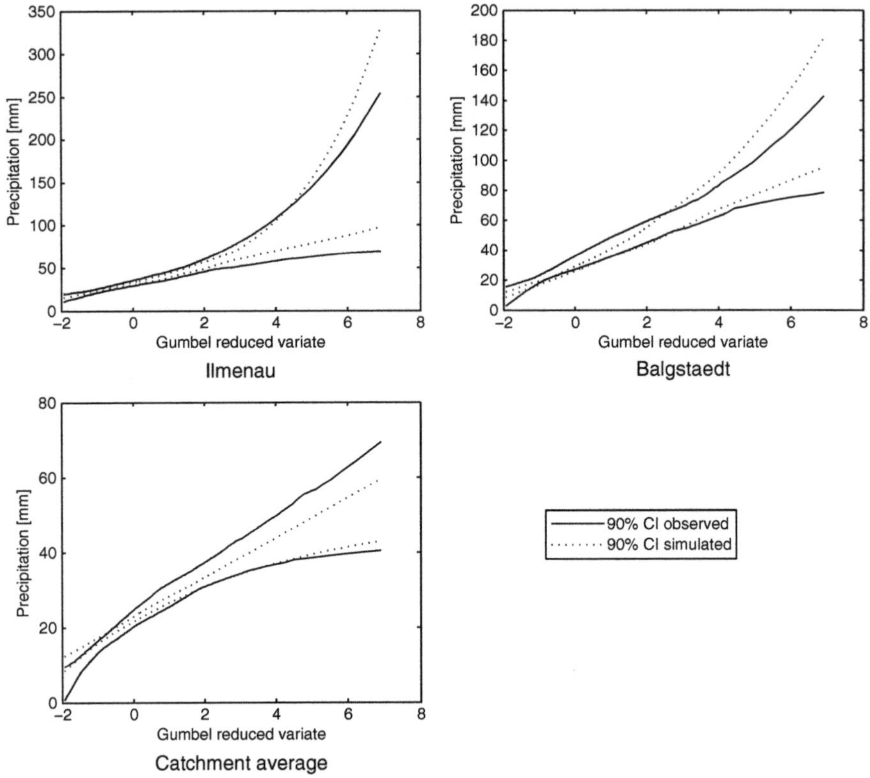

Fig. 7.13 Comparison of the 90% confidence intervals of the GEV distributions fitted to the observed and simulated annual maximum daily precipitation

7.9.1 Methodology of Precipitation Synthesis

In the first step of the rainfall generation process single site precipitation synthesis is carried out using an alternating renewal model (ARM) as described in Section 7.3. For modelling of the wet spell durations W a generalised extreme value distribution is used. Dry spell durations D are modelled by a Weibull distribution function. The wet spell intensity I is modelled using a Kappa distribution function. The dependence between wet spell intensity and duration is described by a 2-copula. Here, after some simple comparisons the Frank copula is chosen, mainly because of practical reasons such as its fully covered dependence range, easy parameter estimation and simulation features. Besides, the Frank copula has been successfully applied for linking rainfall duration and intensity before (De Michele and Salvadori, 2003). The alternating renewal model part describing the external structure of the rainfall process has 11 station specific parameters in total, which are estimated for summer (May to October) and winter seasons (November to April) separately. For the simulation of the internal precipitation structure, the simple profile model shown in Section 7.3 is applied.

In the second step of the rainfall generation process the synthetic precipitation event time series for several locations in the study region are resampled in order to reproduce the spatial dependence structure of the rainfall process. It is important to note, that for resampling not the hourly data are used but the event series. This is in accordance with the basic idea of the alternating renewal process from step one, which assumes the independence of subsequent events. Thus, the resampling on the event time series preserves the temporal rainfall structure for single time series and does not need to consider any autocorrelation. The resampling is carried out using simulated annealing as explained in Section 7.7.2. The objective function, which is used, needs to reflect the spatial dependence structure of the rainfall process. Three bivariate criteria are defined for this purpose: the probability of bivariate rainfall occurrence at two stations, the Pearson's coefficient of correlation and the rainfall continuity measure proposed by Wilks (1998), which estimates the expectation of the rainfall intensity of one station conditioned on the rainfall status dry or wet of the neighbouring station.

7.9.2 Data, Study Region and Hydrological Model

The hybrid rainfall model is tested and applied for derived flood frequency analysis in two mesoscale catchments within the Bode river basin in northern Germany (Fig. 7.14). The two catchments Holtemme and Selke have drainage areas of 168 and 105 km^2, respectively. Floods are generated either by frontal rainfall, frontal

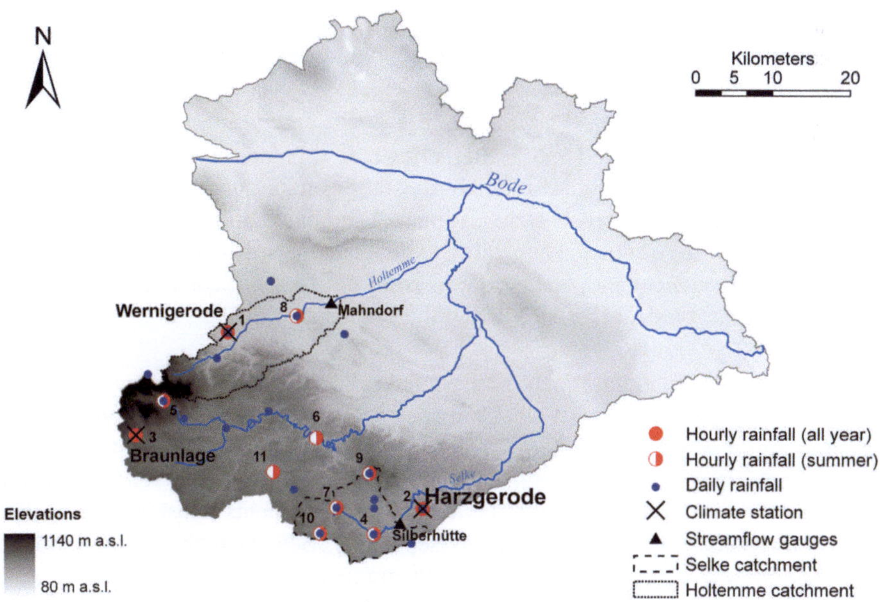

Fig. 7.14 Study region with meteorological stations and streamflow gauges

rainfall on snow smelt or convective storms. Large floods in the Selke catchment occur mainly in the winter season, while floods in the Holtemme take place mostly in the summer time.

A total number of 23 recording rainfall stations with hourly data and 19 non-recording rainfall stations with daily data are employed for the case study. The daily stations are used only as additional information for hydrological modelling. The 23 stations with hourly records are located in an extended study area within and around the Bode river basin and are used in this total set for the robust estimation of the spatial dependence criteria. Only a subset of 11 hourly stations can be used directly for rainfall synthesis and hydrological modelling regarding the two mesoscale catchments. The length of the observation periods varies between 9 and 14 years from 1993 to 2006.

For runoff simulations the hydrological model HEC-HMS (Scharffenberg and Fleming, 2005) is used. HEC-HMS is a conceptual semi-distributed rainfall-runoff model and offers various tools for the description of the hydrological processes. The model is operated here continuously on an hourly time step. It uses the soil moisture accounting (SMA) algorithm for runoff generation, the Clark Unit Hydrograph for the transformation of direct runoff, two linear reservoirs to consider interflow and base flow transformation and a simple river routing where the flows are only lagged in time. Snow melt is calculated externally using the degree-day method. Potential evapotranspiration is computed also externally using the method proposed by Turc-Wendling (Wendling et al., 1991).

7.9.3 Application

The hybrid precipitation model is applied and validated for the study region in the following three step procedure:

1. Parameters are estimated for the alternating renewal model and single site rainfall is generated independently for all hourly stations in the study region.
2. Spatial dependence criteria are estimated and the hourly rainfall series are resampled using simulated annealing to generate the spatial rainfall structure.
3. Derived flood frequency analysis is carried out for both catchments using the rainfall-runoff model HEC-HMS and the synthetic rainfall data.

Probability distribution functions are fitted to the event variables dry spell duration (D), wet spell duration (W) and wet spell intensity (I). For separation of events a minimum dry spell duration D_{min} of 1 and 2 h for summer and winter seasons, respectively, have been selected (see Section 7.3). The parameters for the distribution functions are estimated using the method of L-moments (Hosking and Wallis, 1997) for winter and summer seasons separately. Several realisations of hourly rainfall time series, each 100 years in length are generated for all stations.

Table 7.1 shows a comparison between observed and simulated event characteristics exemplarily for the three all year rainfall stations. The comparison shows

Table 7.1 Comparison of event characteristics considering the wet spell amount *V* for three rainfall stations from 14 years observed and 200 years simulated rainfall data each

Name of rainfall station	Average number of events [–]		Average *V* [mm]		Std. Dev. of *V* [mm]		Skewness of *V* [–]		Rainfall sum [mm]	
	obs	sim	obs	sim	obs	sim	Obs	sim	obs	sim
Summer season (May–October):										
Wernigerode	102	92	3.88	3.72	6.02	6.09	4.93	7.51	396	343
Harzgerode	89	86	3.68	3.56	5.36	5.68	4.81	7.73	328	308
Braunlage	134	128	5.33	5.24	8.42	8.96	5.08	6.19	714	672
Winter season (November–April):										
Wernigerode	102	88	3.77	3.53	5.21	4.47	3.85	5.28	382	312
Harzgerode	78	76	3.39	3.25	4.93	4.15	7.49	6.63	266	246
Braunlage	115	110	7.44	7.02	11.5	10.1	3.91	4.21	854	768

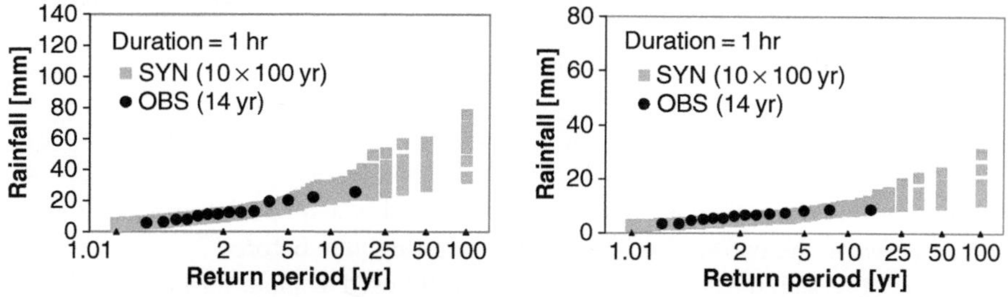

Fig. 7.15 Empirical distribution functions of observed (OBS) and synthetic (SYN) maximum rainfall for the station Harzgerode (*left*: Summer, *right*: Winter)

sufficient agreement between observed and simulated statistics, with a slight underestimation of mean rainfall and a somewhat larger deviation for the higher order moments. Note, that only those features are used here for validations, which do not represent model variables in the precipitation model.

In addition a frequency analysis is carried out on the annual maximum precipitation series for different durations. The results are presented in Fig. 7.15 for the station Harzgerode. It can be seen, that the observed values are plotted mostly within the range of the simulated realisations.

The resampling of the synthetic event series is carried out separately for the two catchments and the two seasons using simulated annealing for all available hourly rainfall gauges. The performance of the simulated annealing algorithm depends on the number of rainfall stations included and on the specific criterion considered. The results improve with fewer stations. So, it is favourable to process the two neighbouring catchments separately. Fig. 7.16 shows a comparison of hourly rainfall time series before and after the resampling process for two neighbouring rainfall stations, which are located about 20 km apart. The effect of the resampling procedure

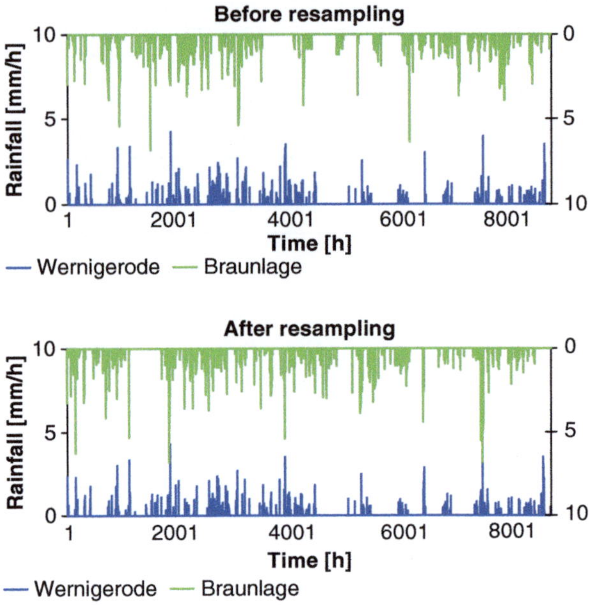

Fig. 7.16 Comparison of hourly precipitation for the stations Wernigerode and Braunlage before and after resampling the events of the latter station using simulated annealing

becomes quite clear. Wet and dry spells as well as intensity peaks correspond much better between the two stations after the resampling than before.

In the third step the stochastic rainfall data are used as input for the hydrological model HEC-HMS. The hydrological model was calibrated using hourly and daily rainfall data, but synthetic precipitation is only generated for the hourly stations. So, the generated hourly data are also transferred to the daily station locations, which were used in calibration, applying nearest neighbour interpolation and a correction factor taking into account the different mean seasonal precipitation amounts at the specific locations. Then, synthetic areal rainfall for subcatchments is calculated by Thiessen interpolation from all daily and hourly station locations. To provide long term hourly data of evapotranspiration for continuous modelling simply repetitions of observed data are used here.

Figure 7.17 compares observed and simulated flood frequencies from annual series for the two catchments Selke and Holtemme at the streamflow gauges Silberhütte and Mahndorf, respectively. Simulated flows are shown based on hourly rainfall data using the short observed rainfall time series and 10 synthetic rainfall realisations each 100 years in length. In addition observed annual peak flows are plotted, which have notable longer records than the simulated ones using observed precipitation.

The observed maximum flows and the simulated ones using observed rainfall data are lying mostly within the range of the simulated flows based on stochastic precipitation data, although located somewhat more in the lower part of the synthetic range. Despite this slight overestimation of flows the overall picture shows

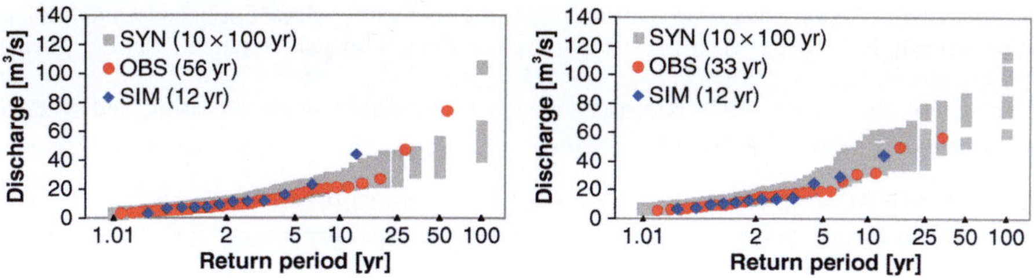

Fig. 7.17 Empirical distribution functions of observed discharge (OBS), simulated discharge using observed rainfall (SIM) and 10 realisations simulated discharge each 100 years in length using synthetic rainfall (SYN) for the Selke (*left*) and Holtemme (*right*) catchments

the ability of the precipitation model to provide suitable input for derived flood frequency analysis.

References

Acreman MC (1990) Simple stochastic model of hourly rainfall for Farnborough, England. Hydrol Sci J 35(2):119–148

Akaike H (1974) A new look at the statistical model identification. IEEE Trans Automatic Control AC-19:716–723

Aronica GT, Candela A (2007) Derivation of flood frequency curves in poorly gauged Mediterranean catchments using a simple stochastic hydrological rainfall-runoff model. J Hydrol 347(1–2):132–142

Bárdossy A (1998) Generating precipitation time series using simulated annealing. Water Resour Res 34(7):1737–1744

Bárdossy A, Plate EJ (1992) Space-time model for daily rainfall using atmospheric circulation patterns. Water Resour Res 28(5):1247–1259

Beven K, Freer J (2001) Equifinality, data assimilation, and uncertainty estimation in mechanistic modelling of complex environmental systems using the GLUE methodology. J Hydrol 249(1–4):11–29

Blazkova S, Beven K (2004) Flood frequency estimation by continuous simulation of subcatchment rainfalls and discharges with the aim of improving dam safety assessment in a large basin in the Czech Republic. J Hydrol 292(1–4):153–172

Brandsma T, Buishand TA (1998) Simulation of extreme precipitation in the Rhine basin by nearest neighbour resampling. Hydrol Earth Syst Sci 2:195–209

Bras RL, Rodriguez-Iturbe I (1994) Random functions and hydrology. Dover, New York, NY, 559 pp

Buishand TA, Brandsma T (2001) Multisite simulation of daily precipitation and temperature in the Rhine basin by nearest-neighbor resampling. Water Resour Res 37(11): 2761–2776

Cameron DS, Beven KJ, Tawn J, Blazkova S, Naden P (1999) Flood frequency estimation by continuous simulation for a gauged upland catchment (with uncertainty). J Hydrol 219(3–4):169–187

Chang TJ, Kavvas ML, Delleur JW (1984a) Daily precipitation modelling by discrete autoregressive moving average processes. Water Resour Res 20:565–580

Chang TJ, Kavvas ML, Delleur JW (1984b) Modelling of sequences of wet and dry days by binary discrete autoregressive moving average processes. J Climate Appl Meteorol 23: 1367–1378

Coles S (2001) An introduction to statistical modeling of extreme values. Springer, London, 208 pp

Cowpertwait PSP (2006) A spatial-temporal point process model of rainfall for the Thames catchment, UK. J Hydrol 330(3–4):586–595

Delleur JW, Kavvas ML (1978) Stochastic models for monthly rainfall forecasting and synthetic generation. J Appl Meteorol 17:1528–1536

De Michele C, Salvadori G (2003) A Generalized Pareto intensity-duration model of storm rainfall exploiting 2-Copulas. J Geophys Res 108 (D2):4067. doi:10.1029/2002JD002534

Efron B, Tibshirani RJ (1998) An introduction to the bootstrap. Chapman and Hall, New York, NY, 436 pp

Foufoula-Georgiou E, Georgakakos KP (1991) Hydrologic advances in space-time precipitation modeling and forcasting. In: Bowles DS, Connel O (eds) Recent advances in the modeling of hydrologic systems. Kluwer, Dordrecht, 47–65 pp

Frigessi A, Haug O, Rue H (2003) A dynamic mixture model for unsupervised tail estimation without threshold selection. Extremes 5:219–235

Gaume E, Mouhous N, Andrieu H (2007) Rainfall stochastic disaggregation models: calibration and validation of a multiplicative cascade model. Adv Water Resour 30(5):1301–1319

Gilks W, Richardson S, Spiegelhalter D (1996) Markov chain Monte Carlo methods in practice. CRC Press, Boca Raton, FL, 486 pp

Grace RA, Eagleson PS (1966) The synthesis of short-time-increment rainfall sequences. Hydrodynamics Laboratory, Massachusetts Institute of Technology, Cambridge, MA

Güntner A, Olsson J, Calver A, Gannon B (2001) Cascade-based disaggregation of continuous rainfall time series: the influence of climate. Hydrol Earth Syst Sci 5:145–164

Haan CT, Allen DM, Street JO (1976) A Markov Chain model of daily rainfall. Water Resour Res 12(3):443–449

Haberlandt U (1998) Stochastic rainfall synthesis using regionalized model parameters. J Hydrol Eng 3(3):160–168

Haberlandt U, Ebner von Eschenbach A-D, Buchwald I (2008) A space-time hybrid hourly rainfall model for derived flood frequency analysis. Hydrol Earth Syst Sci 12:1353–1367

Hastings W (1970) Monte Carlo sampling methods using Markov Chains and their applications. Biometrika 57:97–109

Hipel KW, McLeod AI (1994) Time series modelling of water resources and environmental systems. Elsevier, Amsterdam, 1013 pp.

Hosking JRM, Wallis JR (1997) Regional frequency analysis: an approach based on L-moments. Cambridge University Press, New York, NY, 240 pp.

Hundecha Y, Pahlow M, Schumann A (2009) Modeling of daily precipitation at multiple locations using a mixture of distributions to characterize the extremes. Water Resour Res 45:W12412. doi:10.1029/2008WR007453

Hutchinson MF (1995) Stochastic space-time weather models from ground-based data. Agric Forest Meteorol 73:237–264

Jacobs PA, Lewis PAW (1978a) A discrete time series generated by mixture I: correlation and run properties. J R Stat Soc B 40(1):94–105

Jacobs PA, Lewis PAW (1978b) A discrete time series generated by mixture II: asymptotic properties. J R Stat Soc B 40(2):222–228

Katz RW (1977) Precipitation as a chain-dependent process. J Appl Meteorol 16:671–676

Kavvas ML, Delleur JW (1975) Removal of periodicities by differencing and month subtraction. J Hydrol 26:335–353

Koutsoyiannis D, Onof C, Wheater HS (2003) Multivariate rainfall disaggregation at a fine timescale. Water Resour Res 39(7): 1173. doi:10.1029/2002WR001600

Lall U, Sharma A (1996) A nearest neighbor bootstrap for resampling hydrological time series. Water Resour Res 32:679–693

Lovejoy S, Schertzer D (2006) Multifractals, cloud radiances and rain. J Hydrol 322(1–4):59–88

Lu M, Yamamoto T (2008) Application of a random cascade model to estimation of design flood from rainfall data. J Hydrol Eng 13(5):385–391

Mielke PW, Jr (1973) Another family of distributions for describing and analyzing precipitation data. J Appl Meteorol 10(2):275–280
Moretti G, Montanari A (2008) Inferring the flood frequency distribution for an ungauged basin using a spatially distributed rainfall-runoff model. Hydrol Earth Syst Sci 12:1141–1152
Nelsen RB (2006) An introduction to Copulas. Springer, New York, NY, 270 pp
Olsson J (1998) Evaluation of a scaling cascade model for temporal rainfall disaggregation. Hydrol Earth Syst Sci 2:19–30
Onof C, Chandler RE, Kakou A, Northrop P, Wheater HS, Isham V (2000) Rainfall modelling using Poisson-cluster processes: a review of developments. Stoch Environ Res Risk Assess 14:384–411
Pegram GGS, Clothier AN (2001) High resolution space–time modelling of rainfall: the "string of beads" model. J Hydrol 241(1–2):26–41
Rajagopalan B, Lall U (1999) A k-nearest-neighbor simulator for daily precipitation and other weather variables. Water Resour Res 35:3089–3101
Richardson CW (1982) Stochastic simulation of daily precipitation, temperature, and solar radiation. Water Resour Res 17:182–190
Rodríguez-Iturbe I, Cox DR, Isham V (1987a) Some models for rainfall based on stochastic point processes. Proc R Soc, London A 410:269–288
Rodríguez-Iturbe I, Febres de Power B, Valdés JB (1987b) Rectangular pulses point process models for rainfall: analysis of empirical data. J Geophys Res 92(D8):9645–9656
Roldan J, Woolhiser DA (1982) Stochastic daily precipitation models 1. A comparison of occurrence processes. Water Resour Res 18(5):1451–1459
Scharffenberg WA, Fleming MJ (2005) Hydrologic modelling system, HEC-HMS. User's Manual, 248 pp
Stedinger JR, Vogel RM, Foufoula-Georgiou E (1993) Frequency analysis of extreme events. In: Maidment DR (ed) Handbook of hydrology. MacGraw-Hill, New York, NY, pp 18.1–18.66
Stehlík J, Bárdossy A (2002) Multivariate stochastic downscaling model for generating daily precipitation series based on atmospheric circulation. J Hydrol 256(1–2):120–141
Stern RD, Coe R (1984) A model fitting analysis of daily rainfall data (with discussion). J R Stat Soc A 147:1–34
Vrac M, Naveau P (2007) Stochastic downscaling of precipitation: from dry events to heavy rainfalls. Water Resour Res 43:W07402. doi:10.1029/2006WR005308
Wendling U, Schellin H-G, Thomä M (1991) Bereitstellung von täglichen Informationen zum Wasserhaushalt des Bodens für die Zwecke der agrarmeteorologischen Beratung. Meteorologische Zeitschrift 41:468–474
Wheater H, Chandler R, Onof C, Isham V, Bellone E, Yang C, Lekkas D, Lourmas G, Segond ML (2005) Spatial-temporal rainfall modelling for flood risk estimation. Stoch Environ Res Risk Assess (SERRA) 19(6):403–416
Wilks DS (1998) Multisite generalization of a daily stochastic precipitation generation model. J Hydrol 210(1–4):178–191
Woolhiser DA, Roldan J (1982) Stochastic daily precipitation models. 2. A comparison of distribution amounts. Water Resour Res 18:1461–1468

Chapter 8
Copulas – New Risk Assessment Methodology for Dam Safety

Bastian Klein, Andreas H. Schumann, and Markus Pahlow

Abstract Consideration of a broad range of hydrological loads is essential for risk-based flood protection planning. Furthermore, in the planning process of technical retention facilities it is necessary to use flood events, which are specified by several characteristics (peak, volume and shape). Multivariate statistical methods are required for their probabilistic evaluation. Coupled stochastic-deterministic simulation may be applied to generate a runoff time series, since the required amount of data is generally not available. Even the effect of complex flood protection systems may be evaluated through generation of a data base by means of stochastic-deterministic simulations with subsequent statistical analysis of the individual hydrological load scenarios. Multivariate frequency analyses of correlated random variables are useful to specify these scenarios statistically. Copulas are a very flexible method to estimate multivariate distributions, because the marginal distributions of the random variables can differ. Here a methodology for flood risk assessment is presented which was applied in two case studies in Germany.

Contents

B. Klein (✉)
Institute of Hydrology, Water Resources Management and Environmental Engineering, Ruhr-University Bochum, Bochum, Germany
e-mail: Klein@bafg.de

A.H. Schumann (ed.), *Flood Risk Assessment and Management*,
DOI 10.1007/978-90-481-9917-4_8,

8.1 Introduction

Recent flood events and the related drastic consequences indicate that it is important to consider a broad range of different hydrological loads for the design of flood protection structures such as flood control reservoirs and polders, instead of using the traditional approach of considering only a single design flood of a predefined return period. For single flood protection structures the different hydrological inflows can be generated stochastically (see e.g. De Michele et al., 2005; Klein and Schumann, 2006) or by rainfall-runoff modeling. In large catchments with interacting flood protection structures the efficiency in flood mitigation depends mainly on the spatial distribution of the flood event. To consider these aspects in the design phase spatially distributed precipitation events can be generated by coupling a spatially distributed stochastic rainfall generator with a distributed or semi-distributed rainfall-runoff model to generate hydrological scenarios (see e.g. Blazkova and Beven, 2004; Haberlandt et al., 2008). These hydrological scenarios have to be evaluated probabilistically by frequency analysis to provide a base for risk analyses of flood protection structures or systems.

In most cases, the return period of flood events is defined by a univariate analysis of a random variable such as the flood peak (Hosking and Wallis, 1997; Stedinger et al., 1993). However, technical flood retention depends on several coinciding random variables such as flood peak, volume, shape of the hydrograph and duration. Hence univariate frequency analysis may lead to an over- or underestimation of the risk associated with a given flood (Salvadori and De Michele, 2004). For the design of flood control structures the flood peak plays a fundamental role. However, for flood storages such as reservoirs and polders the flood volume is also very important and should be considered in design and planning. For example an event with a very large peak and a small volume could be stored in the flood control storage of a reservoir, whereas an event with a smaller peak but a larger volume may well lead to flood spillage.

If a dependency exists between the random variables it is advantageous to define the return period of a flood event by using joint probabilities for the flood characteristics that are relevant for the design.

8.2 Copula Theory

One reason for the fact that multivariate statistical models are rarely used for multivariate analysis of random variables in hydrology is that they require considerably more data for the estimation of a reliable model than the univariate case. Owing to the data requirements, also the application of multivariate analysis of correlated random numbers in practice is mainly reduced to the bivariate case.

Several multivariate distributions are available for multivariate frequency analysis. In 1976, Hiemstra derived the bivariate lognormal probability distribution for bivariate frequency analysis (Hiemstra et al. 1976). More recently bivariate distributions such as the bivariate exponential distribution (Singh and Singh, 1991), the bivariate normal distribution (Bergmann and Sackl, 1989; Goel et al., 1998; Yue, 1999), the bivariate gamma distribution (Yue, 2001; Yue et al., 2001) and the bivariate extreme value distribution (Shiau, 2003; Yue et al., 1999) have been applied to bivariate frequency analysis (e.g. for the correlated variables flood peak and flood volume or flood volume and flood duration). Sandoval and Raynal-Villasenor (2008) applied a trivariate extreme value distribution for frequency analysis.

The application of these multivariate distributions to model the dependency between the correlated random variables has a number of drawbacks. First of all, it is a precondition that the marginal distributions have to come from the same family of distributions or the random variables have to be transformed otherwise (e.g. normalized in the case of the bivariate normal distribution). This is not necessarily fulfilled by the dependent variables characterizing a given process. Furthermore, most of the available distributions are limited to the bivariate case and an extension to the multivariate case can not be derived easily. In addition, the dependencies between the random variables of these distributions are generally described by the Pearson's coefficient of linear correlation. However, in reality the dependency structure of the flood variables may not be Gaussian.

These problems can be avoided by applying copulas to model the dependencies between correlated random variables. A joint distribution of correlated random variables can be expressed as a function of the univariate marginal distributions using a copula. Copulas enable one to model the dependency structure independently from the marginal distributions. A large number of different copulas are available to describe this dependency. Therefore it is possible to build multidimensional distributions which differ in their marginal distributions.

Copulas have been extensively used in financial studies (see e.g. Cherubini et al., 2004; Embrechts et al., 2003), but more recently copulas also have been implemented in hydrology for multivariate analysis of hydrological random variables. Salvadori and De Michele (2004) provide a general theoretical framework for exploiting copulas to study the return periods of hydrological events. In several works copulas have been used for the bivariate analysis of the properties of flood events (Karmakar and Simonovic, 2008, 2009; Shiau et al., 2006; Zhang and Singh, 2006), for the bivariate frequency analysis of the properties of precipitation (De Michele and Salvadori, 2003; Evin and Favre, 2008; Zhang and Singh, 2007a) and for the analysis of droughts (Shiau et al., 2007). In De Michele et al. (2005) the dependency between peak and volume is described using the concept of copulas. In this work the adequacy of dam spillways is tested by means of Monte-Carlo simulation for the generation of volume-peak combinations from copula and marginal distributions and their transformation into hydrographs. Favre et al. (2004) presented modeling results of bivariate extreme values using copulas for different case studies. Trivariate analyses of the dependency of random variables using copulas have been applied to model the properties of precipitation events (Grimaldi and Serinaldi, 2006a; Kao and Govindaraju, 2008; Salvadori and De Michele, 2006; Salvadori and De Michele, 2007; Zhang and Singh, 2007b) and for the trivariate flood frequency analysis of flood peak, volume and event duration (Grimaldi and Serinaldi, 2006b; Serinaldi and Grimaldi, 2007; Zhang and Singh, 2007c). In Renard and Lang (2006) the use of copulas for multivariate extreme value analysis is demonstrated with different case studies.

8.2.1 Basic Principles of Copula Theory

Only a short overview of the theory of copulas is presented here. More detailed discussions can be found in Joe (1997), Nelsen (1999), Salvadori and De Michele (2004) and Salvadori et al. (2007). A copula is a function which exactly describes and models the dependency structure between correlated random variables independently of their marginal distributions. It is defined as mapping $C : [0,1]^{\mathrm{n}} \rightarrow [0,1]$, such that for all $u_1, u_2, \ldots, u_{\mathrm{n}} \in [0,1]$ the copula function $C(u_1, u_2, \ldots, u_n) = 0$ if at least one of the arguments is equal to zero and $C(u_1, u_2, \ldots, u_k, \ldots, u_n) = u_k$ if all the arguments are 1 except u_k.

The link between the copula and the multivariate distribution is provided by the theorem of Sklar (1959):

$$F_{X_1,X_2,\ldots,X_n}(x_1, x_2, \ldots, x_n) = C[F_{X_1}(x_1), F_{X_2}(x_2), \ldots, F_{X_n}(x_n)] \tag{8.1}$$

where $F_{X_1,X_2,\ldots,X_n}(x_1, x_2, \ldots, x_n)$ is the joint cumulative distribution function (cdf) with the continuous marginals $F_{X_1}(x_1), F_{X_2}(x_2), \ldots, F_{X_n}(x_n)$ of the random variables.

The content of this chapter is concentrated on bivariate analysis of the random variables X and Y with the marginal distributions $F_X(x)$ and $F_Y(y)$ with a joint cumulative distribution function (cdf):

$$F_{X,Y}(x, y) = C[F_X(x), F_Y(y)]. \tag{8.2}$$

Because copulas are invariant for strictly increasing transformations of X and Y it is possible to use two uniformly distributed random variables U and V on [0,1], defined as $U = F_X(x)$ and $V = F_Y(y)$.

Assuming continuous marginal distributions with the probability density functions $f_X(x)$ und $f_Y(y)$ the joint probability density function (pdf) becomes:

$$f_{X,Y}(x, y) = c[F_X(x), F_Y(y)]\, f_X(x) f_y(y) \tag{8.3}$$

where c is the density function of the copula C:

$$c(u,v) = \frac{\partial^2 C(u,v)}{\partial u \partial v}. \tag{8.4}$$

Three special cases of copulas are of particular importance:

1. The copula of the Fréchet-Hoeffding lower bound W:

$$W\,(u,v) = \max\,\{u + v - 1, 0\} \tag{8.5}$$

2. The copula of the Fréchet-Hoeffding upper bound M:

$$M\,(u,v) = \min\,\{u,v\} \tag{8.6}$$

3. The independence copula Π:

$$\Pi(u,v) = u \cdot v. \tag{8.7}$$

The copulas W and M are the general bounds for any copula C and any pair $(u,v) \in [0,1]^2$:

$$W\,(u,v) = \max\,(u + v - 1, 0) \le C(u,v) \le M(u,v) = \min\,(u,v). \tag{8.8}$$

In Fig. 8.1 the level curves, which connect the points (u, v) where the value of the copula has the same value t for the three special cases, $W(u,v) = t$, $\mathrm{M}(u,v) = t$ and $\Pi(u,v) = t$ are illustrated.

Conditional distributions such as the conditional distribution of U for a given value of $V = v$ can be calculated easily using copulas (other conditional laws see e.g. Salvadori and De Michele, 2004; Salvadori et al., 2007; Zhang and Singh, 2006):

$$P(U \le u | V = v) = \frac{\partial}{\partial v} C(u,v). \tag{8.9}$$

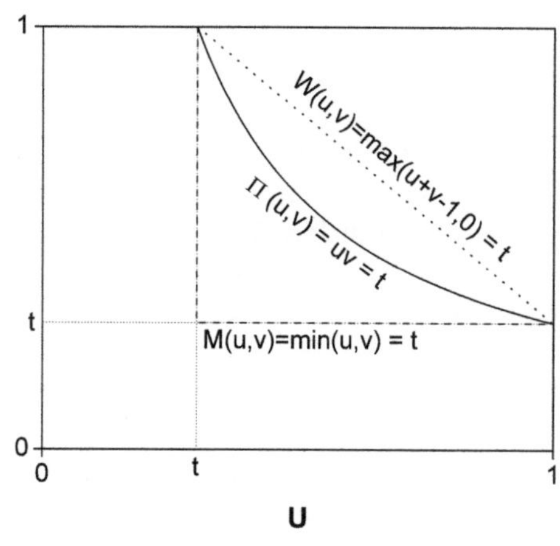

Fig. 8.1 Level curves of the special cases $W(u,v), M(u,v)$ and $\Pi(u,v)$

The empirical copula (Deheuvels, 1979) derived from a random sample $(X_1, Y_1), \ldots, (X_n, Y_n)$ of size n is formally defined by:

$$C_n(u, v) = \frac{1}{n} \sum_{i=1}^{n} 1\left(\frac{R_i}{n+1} \leq u, \frac{S_i}{n+1} \leq v\right) \tag{8.10}$$

where R_i is the rank of $X_i \in \{X_1, \ldots, X_n\}$ and S_i is the rank of $Y_i \in \{Y_1, \ldots, Y_n\}$ of the sample and 1(A) the indicator function of the set A.

A large variety of families of copulas described e.g. in Joe (1997) and Nelsen (1999) are available to model multivariate dependencies. There are generally three families of copulas that are applied in hydrological practice: the Archimedian copulas, the elliptical copulas (including the Gaussian, Student and Cauchy copulas) and the extreme-value copulas (including the Gumbel-Hougaard copula -the only Archimedian extreme-value copula-, the Galambos copula and the Hüsler-Reiss copula). The Archimedian copulas are widely used because they can be constructed easily and a large variety of copulas belongs to this family.

8.2.2 Archimedian Copulas

Archimedian copulas are expressed by the copula generator $\varphi : [0, 1] \rightarrow [0, \infty]$, which is a convex, continuous decreasing function with $\varphi(1) = 0$:

$$C_\theta(u, v) = \varphi^{-1}[\varphi(u) + \varphi(v)]. \tag{8.11}$$

In Table 8.1 four widely used one parameter Archimedian copulas are presented with their corresponding generator φ (t), where θ is the parameter of the generator function.

All four copulas are applicable to positively correlated random variables but only the Ali-Mikhail-Haq and the Frank copula can model negatively correlated random variables as well. The Ali-Mikhail-Haq copula is not appropriate for highly correlated (both positive and negative) variables.

Besides one-parameter Archimedean copulas several two-parameter copulas are described in Nelsen (1999) and Joe (1997). The advantage of these two-parameter copulas is that they may be used to capture more than one type of dependencies. One parameter models the upper tail dependency and the other parameter models the lower tail dependency of the copula. As an example the two-parameter copula BB1 (as described in Joe (1997)), which has also the form of the Archimedian copula family (Eq. 8.11) is presented in Table 8.1.

8.2.3 Parameter Estimation

There are several methods employed to estimate the parameter θ of the copula function. The two standard rank-based nonparametric measures of dependency,

Table 8.1 Summary of various 2-dimensional Archimedean copulas and their generator (Joe, 1997; Nelsen, 1999)

Copula		$\varphi(t)$
Gumbel-Hougaard:		
$\exp\left\{-\left[(-\ln u)^{\theta}+(-\ln v)^{\theta}\right]^{1/\theta}\right\}$	$\theta \geq 1$	$(-\ln t)^{\theta}$
Ali-Mikhail-Haq:		
$\frac{uv}{1-\theta(1-u)(1-v)}$	$-1 \leq \theta < 1$	$\ln\frac{1-\theta(1-t)}{t}$
Frank:		
$-\frac{1}{\theta}\ln\left[1+\frac{(e^{-\theta u}-1)(e^{-\theta v}-1)}{e^{-\theta}-1}\right]$	$\theta \neq 0$	$-\ln\left(\frac{e^{-\theta t}-1}{e^{-\theta}-1}\right)$
Cook-Johnson:		
$\left[u^{-\theta}+v^{-\theta}-1\right]^{-1/\theta}$	$\theta \geq 0$	$t^{-\theta}-1$
BB1		
$C_{\theta}(u,v)=\left\{1+\left[\left(u^{-\theta_1}-1\right)^{\theta_2}+\left(v^{-\theta_1}-1\right)^{\theta_2}\right]^{1/\theta_2}\right\}^{-1/\theta_1}$	$\theta_1 > 0, \theta_2 \geq 1$	$\varphi(t)=\left(t^{-\theta_1}-1\right)^{\theta_2}$

Spearman's ρ and Kendall's τ can be expressed solely in terms of the copula function. Therefore the parameter of the copula can be estimated from the measure of dependency of the bivariate sample. Kendall's τ can be expressed as:

$$\tau = \int_{[0,1]^2} C_{\theta}(u,v)dC_{\theta}(u,v) - 1 \tag{8.12}$$

which, for Archimedian copulas with a generator function φ, is reduced to

$$\tau = 1 + 4\int \frac{\varphi(t)}{\varphi'(t)}dt, \tag{8.13}$$

where φ' is the derivative of φ with respect to t.

Another estimation method is the maximum pseudolikelihood method (Genest and Favre, 2007) which is also called canonical maximum likelihood. It can also be used if θ is multidimensional. The method of maximum pseudolikelihood involves maximizing a rank-based log-likelihood function of the form:

$$l(\theta) = \sum_{i=1}^{n} \log\left[c_{\theta}\left(\frac{R_i}{n+1}, \frac{S_i}{n+1}\right)\right], \tag{8.14}$$

where R_i is the rank of $X_i \in \{X_1 \ldots X_n\}$ and $Y_i \in \{Y_1 \ldots Y_n\}$ of the sample of size n and the copula density distribution $c_{\theta}(u,v)$ after Eq. (8.4)

In Joe (1997) a parametric two-step procedure called the "inference from margins" or IFM method is recommended to estimate the parameter of the model. An estimate of θ is obtained by maximizing the log-likelihood function

$$l(\theta) = \sum_{i=1}^{n} \log \{c_\theta[F_X(x_i), F_y(y_i)] f_X(x_i) f_y(y_i)\}. \tag{8.15}$$

In general estimation methods are based on a measure of dependency. The maximum pseudolikelihood estimation method estimates the parameters independently from the marginal distributions. They only model the rank-based dependency between the two random variables. In contrast the IFM depends on the estimation of the margins and therefore the parameters of the copula are unduly affected if the marginal distributions are wrong (see e.g. Kim et al., 2007).

8.2.4 Identification of the Appropriate Copula Model

Methods to identify the copula which provides the best fit to the observed data are required, since there is a large variety of different copula families available to model the dependency between random variables. Several methods exist to select the appropriate model.

8.2.4.1 Graphical Diagnostics

Possibly the most natural way for the identification of the appropriate copula is the graphical comparison between the scatter plot of the pairs $[R_i/(n+1), S_i/(n+1)]$ (representing the empirical copula C_n) from the bivariate data with pairs (U_j, V_j) from an artificial data set, generated from the copula $C_{\theta n}$, or by comparing the pairs of the bivariate data with the pairs (X_j, Y_j) generated from the copula $C_{\theta n}$. and transformed back into the original units using the marginal distributions $F_X(x)$ and $F_Y(y)$. Algorithms to generate random pairs (U, V) from a copula are described e.g. in Whelan (2004). A simple simulation algorithm to generate random pairs (u,v) from the copula is given in Genest and Favre (2007):

1. Generate U from a uniform distribution on the interval (0,1).
2. Given $U = u$, get $V = Q_u^{-1}(U^*)$ from the inverse function of the conditional distribution according to Eq. (8.16), where U^* is another random number generated from the uniform distribution on the interval (0,1).

$$Q_u(v) = P(V \le v | U = u) = \frac{\partial}{\partial u} C(u, v). \tag{8.16}$$

Another option is the comparison between the parametric and nonparametric estimate of the probability $K_C(t)$ (Genest and Rivest, 1993) that the value of the copula $C(u, v)$ is smaller than or equal to $t (0 < t \le 1)$:

$$K_C(t) = P[C(u, v) \le t], 0 < t \le 1. \tag{8.17}$$

Let

$$B_C(t) = \{(u, v) \in [0, 1]^2 : C(u, v) \leq t\}, 0 < t \leq 1 \tag{8.18}$$

where B_C (t) is the region within $[0,1]^2$ lying on, below and to the left of the level curve L_t with

$$L_t = \{(u, v) \in [0, 1]^2 : C(u, v) = t\}, 0 < t \leq 1 \tag{8.19}$$

of the joint distribution. The level curve is a function of the two variables u and v that connects points where the copula has the same value t. The probability distribution K_C (t) provides a unique probability measure of the set B_C (t) (see Salvadori and De Michele, 2004). For Archimedian copulas the parametric estimate of $K_{C_\theta}(t)$ can be calculated easily using

$$K_{C\theta}(t) = t - \frac{\varphi(t)}{\varphi'(t)} \tag{8.20}$$

and the level curve L_t as a function of u:

$$L_t(u) = \varphi^{-1}[\varphi(t) - \varphi(u)]. \tag{8.21}$$

The nonparametric estimate of $K_{C_n}(t)$ can be calculated after Genest et al. (2006):

$$K_{C_n}(t) = \frac{1}{n}\sum_{i=1}^{n} 1(V_{in} \leq t); t \in [0, 1] \tag{8.22}$$

where

$$V_{in} = \frac{1}{n}\sum_{j=1}^{n} 1(x_j \leq x_i, y_j \leq y_i) = \frac{1}{n}\sum_{j=1}^{n} 1(R_j \leq R_i, S_j \leq S_i). \tag{8.23}$$

8.2.4.2 Goodness-of-Fit Statistics

There are different possibilities to compare and evaluate the goodness of fit of the different possible copula functions to model the dependency of the random variables. The "Root Mean Square Error" (RMSE) goodness-of-fit measure is the mean squared difference between the empirical and the estimated distribution functions:

$$\text{RMSE} = \sqrt{E(x_c - x_0)^2} = \sqrt{\frac{1}{n-k}\sum_{i=1}^{n}[x_c(i) - x_0(i)]^2}, \tag{8.24}$$

where x_c (i) is the i-th computed value (here: value of the fitted theoretical distribution function), x_0 (i) the i-th observed value (here: value of the empirical distribution function), k the number of model parameters used to obtain the calculated value and n the number of observations.

The Akaike's information criterion AIC (Akaike, 1974) and the Bayesian information criterion BIC (Schwarz, 1978) are measures of the goodness of fit of an estimated statistical model and can also be used for model selection. Both criteria depend on the maximized value of the likelihood function L for the estimated model. The AIC is defined as

$$\text{AIC} = (-2) \cdot \ln(L) + 2 \cdot k \tag{8.25}$$

where k is the number of model parameters and the BIC is defined as:

$$\text{BIC} = (-2) \cdot \ln(L) + k \cdot \ln(n) \tag{8.26}$$

where n is the sample size. Both criteria differ from each other with increasing sample size. According to these criteria, the model with the lowest value of the AIC and BIC is the best model.

The multidimensional Kolmogorov-Smirnov (KS) goodness-of-fit test uses (see Saunders and Laud, 1980), in a similar way as for the univariate KS-Test, the maximal distance between the empirical and the theoretical distribution function as a goodness-of-fit measure. Standard tables or bootstrap resampling can be applied to find the critical values for a significance level α or the p-values for the goodness-of-fit test.

In Genest et al. (2006, 2009) and Genest and Favre (2007) several methods for formal goodness-of-fit testing of copulas are presented. Two of these methods use the difference between the empirical and theoretical distribution of $K_C\,(t)$ (see Eq. 8.17) as test measure. The first methodology is based on the Cramér-von-Mises statistics and the second one on the Kolmogorov-Smirnov statistics. To estimate the critical values for a significance level α and the p-value a bootstrap methodology is presented for the goodness-of-fit test of copulas. Another bootstrap methodology based on the Cramér-von-Mises statistics as goodness-of-fit test for copulas is described in Genest and Favre (2007) and Genest and Remillard (2008). There the difference between the empirical and theoretical copula is used as measure.

8.2.5 Bivariate Frequency Analysis

The joint distribution of Eq. (8.2) is the probability $P(X \leq x, Y \leq y) = F_{X,Y}(x, y)$ that X and Y are less than or equal to the specific thresholds x and y, respectively. Given the assumption of total independence between the two random variables, the joint probability $F_{X,Y}(x, y)$ becomes the product of the two individual probabilities $F_X\,(x)$ and $F_Y\,(y)$:

$$P(X \leq x, Y \leq y) = F_{X,Y}(x, y) = F_X(x)F_Y(y). \tag{8.27}$$

The joint probability $F_{X,Y}(x, y)$ is reduced to the univariate probability $F_X\,(x)$ or $F_Y\,(y)$ if the two random variables are completely dependent. In this case the random variables X and Y are functions of each other:

$$P(X \leq x, Y \leq y) = F_{X,Y}(x, y) = F_X(x) \text{ or } F_{X,Y}(x, y) = F_Y(y). \tag{8.28}$$

If there are dependencies, the probability that X and Y both exceed x and y is defined as

$$\begin{aligned} P(X > x \wedge Y > y) &= 1 - F_X(x) - F_Y(y) + F_{X,Y}(x,y) \\ &= 1 - F_X(x) - F_Y(y) + C[F_X(x), F_Y(y)]. \end{aligned} \tag{8.29}$$

In the case of independence between the two random variables the joint return period is the product of the two univariate return periods:

$$T_{X,Y}^{\wedge} = \frac{\mu_T}{1 - F_X(x) - F_Y(y) + F_X(x)F_Y(y)} = \frac{\mu_T}{(1 - F_X(x))(1 - F_Y(y))} = T_X T_Y, \tag{8.30}$$

where μ_T is the mean interarrival time between two successive events ($\mu_T = 1$ year, considering annual values).

Under consideration of dependency the corresponding joint return period (labeled with the logical AND operator "$\wedge$") is expressed as

$$T_{X,Y}^{\wedge} = \frac{\mu_T}{P(X > x \wedge Y > y)} = \frac{\mu_T}{1 - F_X(x) - F_Y(y) + C[F_X(x), F_Y(y)]}. \tag{8.31}$$

This joint return period is always larger than the maximum of the univariate return periods $T_X = \mu_T/(1 - F_X(x))$ and $T_Y = \mu_T/(1 - F_Y(y))$ of the individual random variables.

The probability with which either X or Y exceed their respective thresholds x or y is defined as

$$P(X > x \vee Y > y) = 1 - F_{X,Y}(x,y) = 1 - C[F_X(x), F_Y(y)]. \tag{8.32}$$

Here the joint return period (labeled with the logical OR operator "$\vee$") is expressed as

$$T_{X,Y}^{\vee} = \frac{\mu_T}{P(X > x \vee Y > y)} = \frac{\mu_T}{1 - C[F_X(x), F_Y(y)]}. \tag{8.33}$$

This joint return period is always smaller than the minimum of the two univariate return periods T_X and T_Y of the individual random variables:

$$T_{X,Y}^{\vee} \leq \min[T_X, T_Y] \leq \max[T_X, T_Y] \leq T_{X,Y}^{\wedge}. \tag{8.34}$$

In the case of independence of the two random variables this joint return period becomes:

$$T_{X,Y}^{\vee} = \frac{\mu_T}{1 - F_X(x)F_Y(y)} = \frac{T_X T_Y}{T_X + T_Y - \mu_T}. \tag{8.35}$$

In Fig. 8.2 the regions of $[0,1]^2$ embodying the probability masses for the two cases: exceedance probability $P(X > x \vee Y > y)$ -"OR" case- and $P(X > x \wedge Y > y)$ -"AND" case- are illustrated as shaded areas.

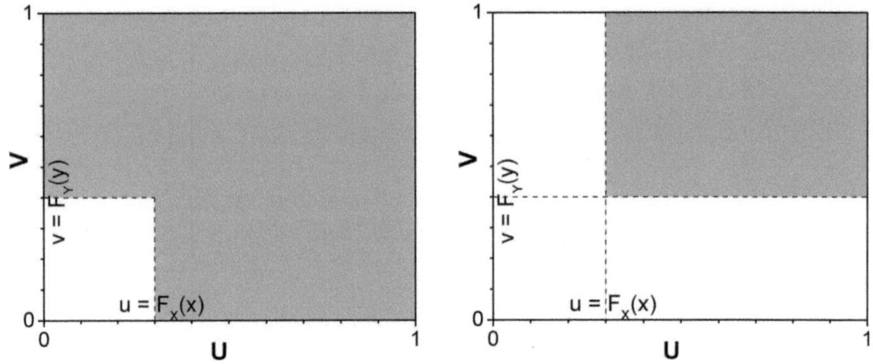

Fig. 8.2 Regions in $[0,1]^2$ (*shaded* areas) embodying the probability masses of (**a**) $P(X > x \vee Y > y)$ and (**b**) $P(X > x \wedge Y > y)$

Conditional distribution functions and return periods can also be expressed easily with copulas (see e.g. Salvadori and De Michele, 2004; Salvadori et al., 2007; Zhang and Singh, 2006). For example the conditional distribution $F_x(x|Y = y)$ of X for a given value of $Y = y$ can be calculated using Eq. (8.9) with $u = F_X(x)$ and $v = F_Y(y)$.

In Salvadori and De Michele (2004) another "secondary" return period $\rho_t^{\vee}$ is defined to emphasize the difference to the "primary" return period $T_{X,Y}^{\vee}$ $B_C(t)$ in Eq. (8.18) is the region where the return period $T_{X,Y}^{\vee}$ of all events is equal to or less than a threshold

$$\vartheta(t) = \frac{\mu_T}{(1-t)}. \tag{8.36}$$

Therefore K_C (t) is the probability that an event with the return period $T_{X,Y}^{\vee} \leq \vartheta(t) = \mu_T/(1-t)$ occurs for any realization of the random process. This implies that the survival function of K_C (t), defined as

$$\bar{K}_C(t) = 1 - K_C(t), \tag{8.37}$$

is the probability that for any given realization of the random process an event with a return period $T_{X,Y}^{\vee} > \vartheta(t) = \mu_T/(1-t)$ occurs. The secondary return period $\rho_t^{\vee}$ can be expressed as the reciprocal value of the survival function

$$\rho_t^{\vee} = \frac{\mu_T}{\bar{K}_C(t)} = \frac{\mu_T}{1 - K_C(t)}. \tag{8.38}$$

If a critical threshold $\vartheta(t)$ is defined in the design stage for a flood protection structure, the secondary return period specifies the mean interarrival time of a critical event.

The regions in $[0,1]^2$ embodying the probability masses for $K_C(t) = P(C(u,v) \leq t)$ and $\bar{K}_C(t) = P(C(u,v) > t)$ (shaded areas) are illustrated in Fig. 8.3.

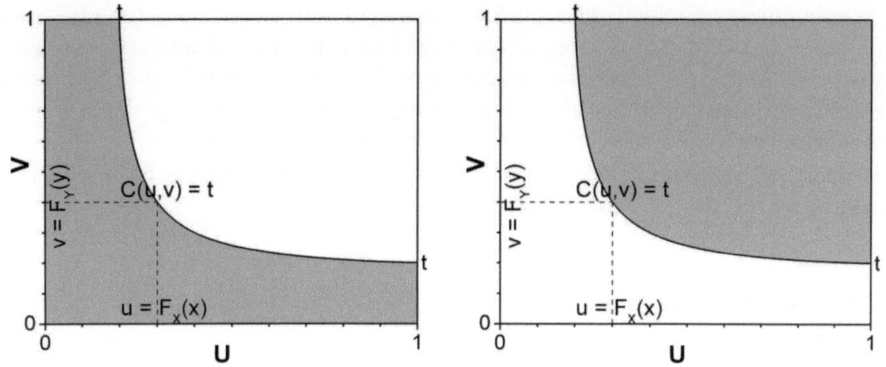

Fig. 8.3 Regions in $[0,1]^2$ (*shaded* areas) embodying the probability masses of (**a**) $K_C(t) = P(C(u,v) \leq t)$ and (**b**) $\bar{K}_C(t) = P(C(u,v) > t)$

8.3 Case Study 1: Risk Analysis for the Wupper Dam

The first case study for an application of copulas within the framework of risk analysis of dams is concerned with the Wupper dam, located in the mid-western part of Germany (see Fig. 8.4). As mentioned before it is very important for the design of flood control structures to consider the flood volume besides the flood peak.

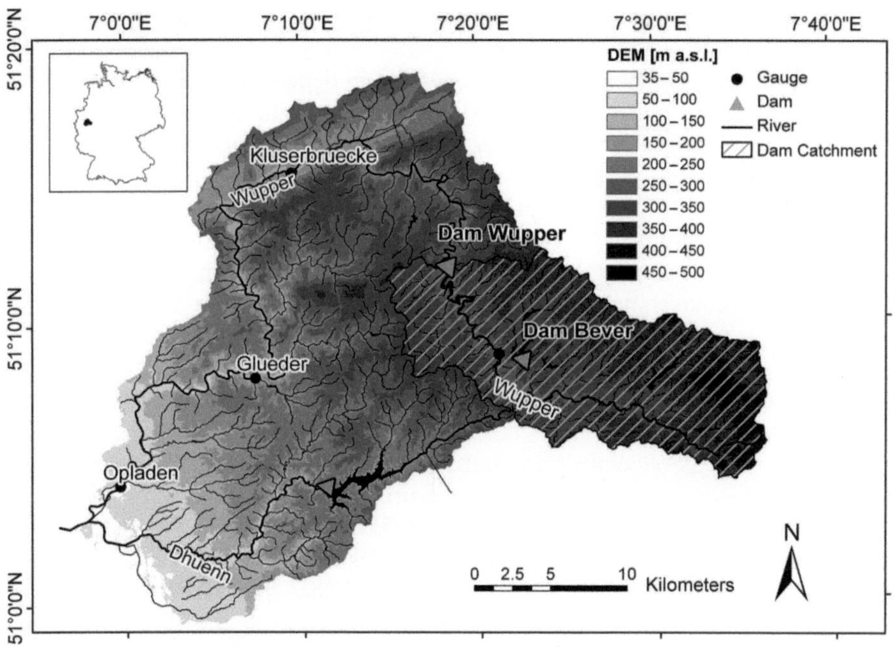

Fig. 8.4 Topographical map of the watershed of the river Wupper

Therefore these two flood variables have been used in this case study for a bivariate frequency analysis of flood events. The flood events for the risk analysis were generated by coupling a continuous stochastic rainfall generator with a deterministic rainfall-runoff model. Using this methodology a synthetic 2000 year time series of discharge with a time step of 1 h was generated. Since the observed time series was too short for multivariate frequency analysis, these 2000 years has been used as a database for bivariate frequency analysis.

8.3.1 Study Area

The Wupper dam has a watershed of 212 km^2 and a mean inflow of 4.4 m^3s^{-1}. It is located in the catchment of the river Wupper with a total area of 813 km^2 (see Fig. 8.4). Apart from the flood control function it is used mainly for low water regulation in drought periods. The available flood control storage is seasonally variable. In the summertime between May and October no flood storage is alocated. From the beginning of November until the end of January a maximum flood storage of 9.9×10^6 m^3 is provided. This flood storage is reduced continuously until the end of April. The flood spillway is a weir with a total width of 36 m and is controlled with fish belly gates. The maximum capacity of the spillway with open gates is 318 m^3s^{-1}. The two bottom outlets have a capacity of 88 m^3s^{-1} each.

In the watershed of the Wupper dam several smaller dams are located. From these dams particularly the dam Bever is important for flood control with a catchment area of 25.7 km^2 and an available flood control storage of 5×10^6 m^3 in wintertime from November until the begin of February.

During a flood the Wupper dam is operated in dependency of the discharge at the control gage Kluserbruecke located downstream in the city of Wuppertal. The discharge at this gage is an indicator for the danger of flooding of the residential areas downstream. The uninfluenced subcatchment between the dam and the gage has an area of 125 km^2. During a flood the controlled discharge at the gage should stay below 80 m^3s^{-1}. The outflow of the dam should not exceed 50 m^3s^{-1} to ensure this limit. During large floods the control level for the discharges at the gage Kluserbruecke is increased stepwise to 100, 150 and 190 m^3s^{-1}.

8.3.2 Stochastic-Deterministic Generation of Flood Events

In the first step the stochastic rainfall generator of Hundecha et al. (2009) was applied to generate continuous spatially distributed precipitation time series with a time step of 1 day and a length of 2000 years. In total 28 stations with daily precipitation with a minimum time series length of 30 years were used for model parameterization. In the second step the 2000 year time series of daily precipitation was disaggregated in hourly values using the stochastic models MuDRain (Koutsoyiannis, 2001) and HYETOS (Koutsoyiannis et al., 2003). The semi-distributed deterministic model NASIM (Hydrotec, 2007) has been used for rainfall-runoff modeling. It was calibrated with observed rainfall-runoff data for the period 1. January 2001–31 December 2006. The operation of the dams in the region has been considered in NASIM. It has been shown that the statistical properties of the rainfall and the runoff were well reproduced (Petry et al., 2008).

8.3.3 Bivariate Frequency Analysis of Annual Flood Peaks and Corresponding Volumes

The annual flood peaks and corresponding volume were selected from the generated hourly inflow time series of 2000 years for the Wupper dam. The start of the surface runoff was marked by the abrupt rise of the hydrograph and the end of the flood runoff was identified by the flattening of the recession limb of the hydrograph. Between these two points the total volume was estimated for the analysis. In Fig. 8.5 the selection of the flood volume is illustrated.

The statistical dependencies between the two random variables flood peak and flood volume have been estimated by the statistical measures listed in Table 8.2, where particularly Pearson's coefficient of linear correlation and the rank based measure of dependency Spearman's ρ as well as Kendall's τ have been considered.

Pearson's correlation coefficient provides a measure of linear dependency only and depends on the marginals, in contrast to the rank-based measure of dependency provided by Kendall's and Spearman's coefficient. The population value of

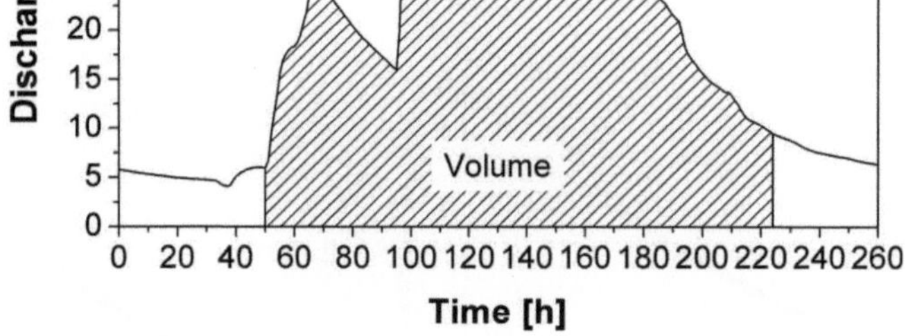

Fig. 8.5 Evaluation of the corresponding flood volume

Table 8.2 Measure of dependency for flood peaks and corresponding volumes for the Wupper dam

Sample size	Pearson's r	Kendall's τ	Spearman's ρ
2,000	0.546	0.379	0.541

the Pearson's correlation coefficient is always well defined. However, in some cases (e.g. for heavy-tailed distributions such as the Cauchy-distribution) a theoretical value of Pearson's correlation does not exist. Furthermore, Pearson's correlation coefficient is not robust. If the two variables are not linearly related the correlation is not well defined and outliers can strongly affect the correlation. On the other hand the rank based correlations such as Kendall's τ and Spearman's ρ are robust since they are not dependent on the distributional assumptions. They can describe a wider class of dependencies and are resistant to outliers. The summary in Table 8.2 shows that there is a strong positive dependency between the two random variables.

8.3.3.1 Marginal Distributions

The first step to build the bivariate probability distribution according to Eq. (8.2) consists in an estimation of the marginal distributions of the random variables. The Pearson type 3, log-Pearson type 3, Gumbel, Weibull, Gamma, Exponential, Generalized Pareto and Generalized Extreme Value (GEV) distributions were used for the univariate statistical analyses of the random variables. The goodness-of-fit of the different distributions and the different parameter estimation methods have been compared with the Kolmogorov-Smirnov and the Cramér-von-Mises $n\omega^2$ tests. Because the focus of this chapter is the bivariate analysis the results of the univariate frequency analysis are not described here in detail.

These goodness-of-fit tests revealed that the flood peak can be described in the best way by the GEV distribution. The volume is described with the Pearson III distribution optimally. For parameter estimation the product moments were used. The fitted distributions are shown in Fig. 8.6.

8.3.3.2 Copula Estimation

The bivariate probability distribution (Eq. 8.2) has been constructed using the four Archimedian copulas Ali-Mikhail-Haq, Frank, Cook-Johnson (also known as Clayton copula) and Gumbel-Hougaard (Equations see Table 8.1), since those are the commonly used families of Archimedian copulas (see e.g. Shiau et al., 2006;

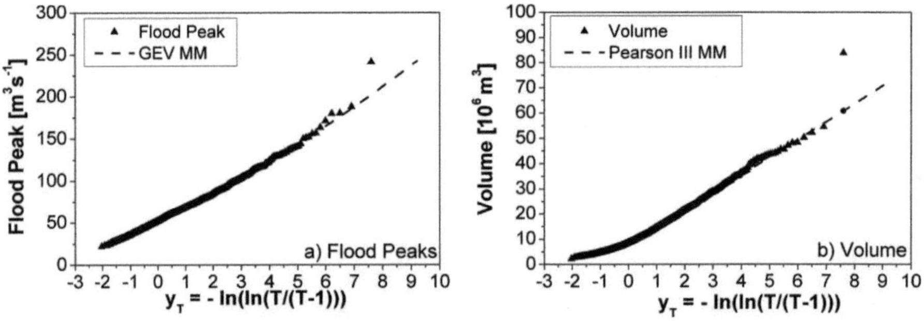

Fig. 8.6 Comparison of the (**a**) annual flood peaks and the (**b**) corresponding flood volumes of the simulated inflows to the dam Wupper using the stochastic-deterministic approach and the fitted (**a**) GEV distribution and (**b**) Pearson III distribution

Table 8.3 Parameters and goodness-of-fit measure AIC of the fitted copulas for the analysis of the flood peaks and the corresponding flood volumes at the dam Wupper

Copula	Estimate(s)	AIC
Gumbel-Hougaard	$\theta = 1.53$	–689
Frank	$\theta = 3.83$	–681
Cook-Johnson	$\theta = 0.858$	–631
Ali-Mikhail-Haq	$\theta = 0.963$	–658
BB1	$\theta_1 = 0.404$; $\theta_2 = 1.31$	–779

Zhang and Singh, 2006). The parameters have been estimated with the maximum pseudolikelihood method. The fitted parameters are summarized in Table 8.3.

In Fig. 8.7 the parametric and nonparametric values of $K_C(t)$ are plotted for the four different copulas. A copula is considered as satisfactory if the plot shows a good agreement with the straight dashed line that passes through the origin at 45°.

As a further visual test, the margins of 100,000 random pairs (U_j, V_j), chosen from the copula and transformed back into the original units using the marginal distributions $F_X(x)$ and $F_Y(y)$, are compared with the sample values. The scatter

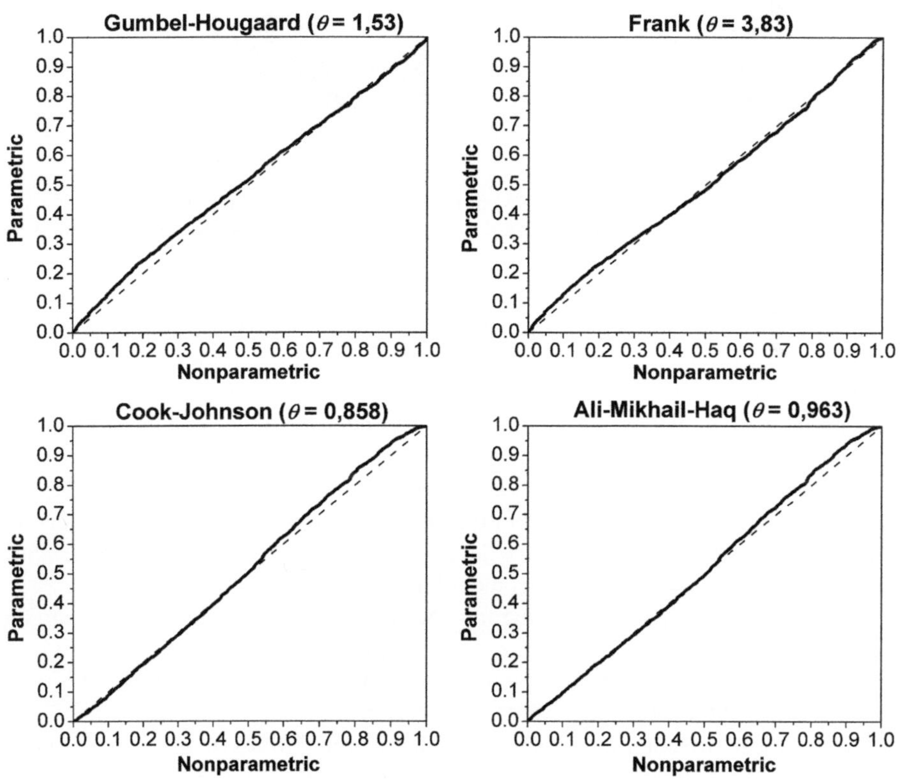

Fig. 8.7 Comparison of nonparametric and parametric estimations of $K_C(t)$ for the fitted Archimedean copulas (The *dashed* 45°-*line* represents perfect agreement)

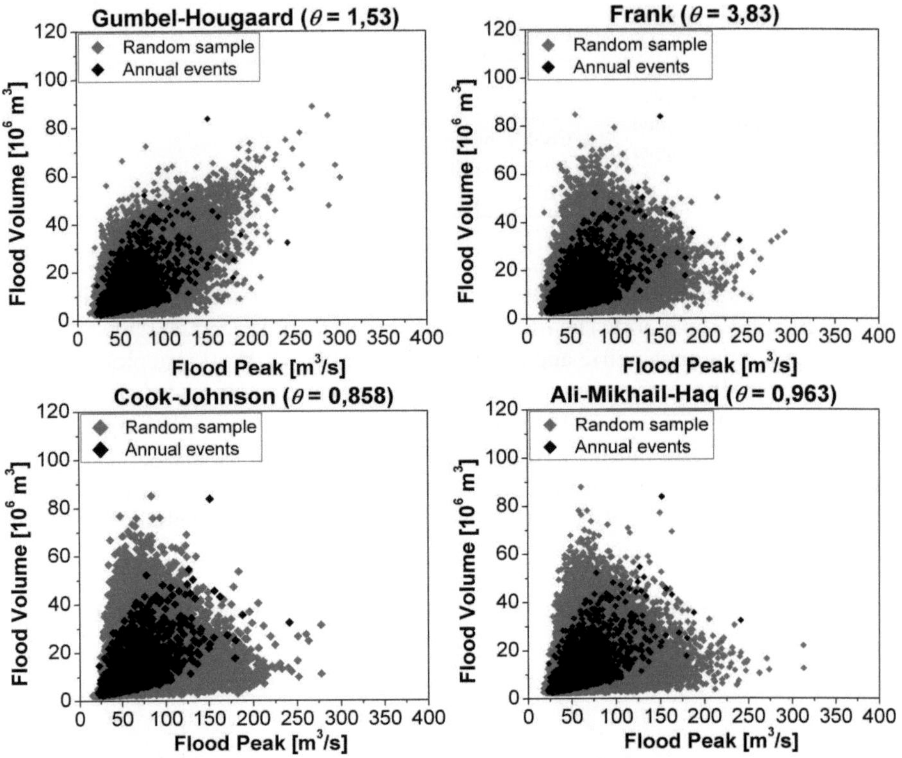

Fig. 8.8 100,000 random pairs (*grey dots*) of the annual flood peaks and the corresponding volumes at the dam Wupper chosen from the four Archimedian copulas and the events from the simulated time series (*black dots*)

plots of the four copulas are shown in Fig. 8.8. It can be seen that among these four Archimedian copulas only the Gumbel-Hougaard copula can describe the dependency structure of the sample adequately since only the random pairs generated from this particular copula (grey dots) follow the dependency structure of the sample data (black dots).

As an additional measure the Akaike's information criterion AIC (Eq. 8.25) was used to compare the goodness of fit of the different copula models and to select the appropriate model which models the dependency between the two random variables optimally. The best model provides the minimal value for this criterion.

Comparing the AIC values of the four Archimedian copulas in Table 8.3 and the parametric and nonparametric estimates it seems that the Gumbel-Hougaard copula and the Frank copula provide a similar goodness-of-fit to the database. Looking more closely at the comparison of the parametric and nonparametric estimation of $K_C(t)$ for these two copulas in the region of large probabilities (Fig. 8.9) it is obvious that the Gumbel-Hougaard provides a better fit in this region. For the Frank copula the parametric estimate is much higher than the corresponding nonparametric estimate from the sample.

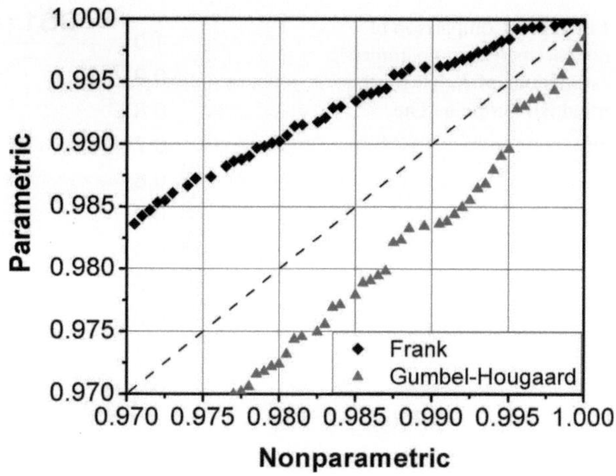

Fig. 8.9 Comparison of the nonparametric and parametric estimations of $K_C\,(t)$ for the Frank and Gumbel-Hougaard copula in the region of large probabilities

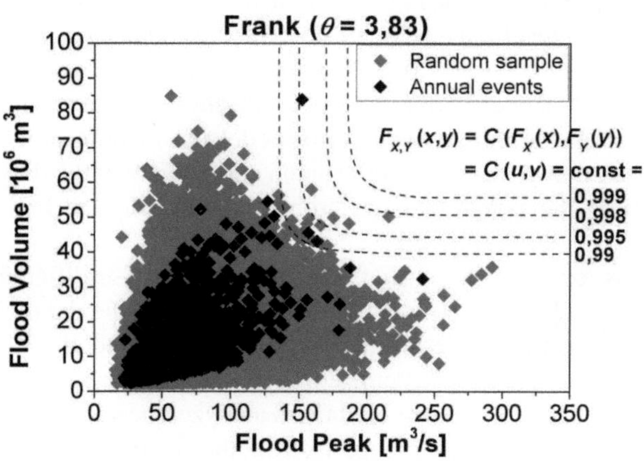

Fig. 8.10 100,000 random pairs (*grey dots*) of the annual flood peaks and the corresponding volumes at the dam Wupper chosen from the Frank copula, the events from the simulated time series (*black dots*) and level curves for different probabilities $C(u, v) = t$

E.g. for the same value t the nonparametric value provides $K_{C_n}(t) = 0.99$ and the parametric estimate of $K_{C_\theta}(t)$ for the Frank copula has a value of approx. 0.996. If the random process is modeled with the Frank copula the probability of the occurrence $P(C(u, v) \leq t) = K_{C_\theta}(t) = 0.996$ of an event with a value of the copula $C(u, v) \leq t$ is much higher than the empirical probability derived from the dataset $P(C(u, v) \leq t) = 0.99$, which was used for parameter estimation of the model. This effect could be demonstrated for the 100,000 random pairs generated from the Frank copula in Fig. 8.10.

In the random process 18 values of the random pairs (X_j, Y_j) generated from the copula are above the level curve of $C(u, v) = t = 0.99$. The parametric estimate of $K_{C_\theta}(0.99)$ for the fitted Frank copula is 0.99981. Hence generating 100,000 random

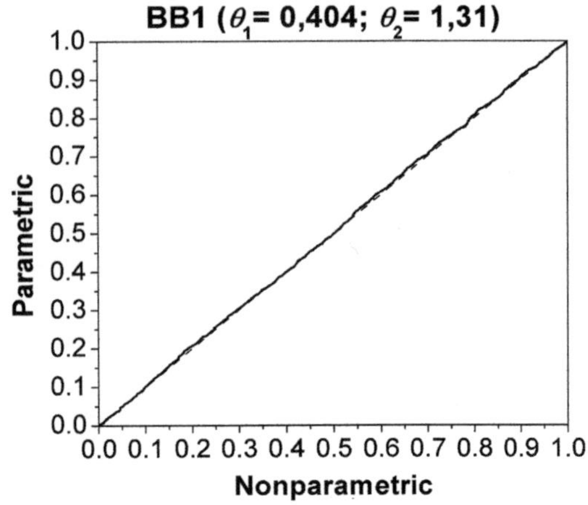

Fig. 8.11 Comparison of nonparametric and parametric estimations of $K_C(t)$ for the fitted BB1 copula (The *dashed* 45°-*line* represents perfect agreement)

pairs from the copula theoretically 99981 of the pairs should be lying on, under or to the left of the level curve $C(u, v) = t = 0.99$ und 19 random pairs above. In the sample with the size of 2,000 elements four pairs are above the level curve $C(u, v) = t = 0.99$. Therefore for 100,000 realisations of the random process 200 values should be above the level curve to model the random process adequately. This analysis has shown that the Frank copula can not model the dependency structure adequately in the region of large probabilities. Using the Frank copula only a few random pairs are generated where both variables have a small exceedance probability. This example shows that it is important to use different methods for model selection to choose the model with the best fit to the data. Using only one goodness-of-fit measure may result in selecting a wrong model.

None of the four considered copulas provides an acceptable fit to the database. Therefore the two parameter BB1 copula was analyzed besides the four one parameter Archimedian copulas. The comparison of the parametric and nonparametric values of $K_C(t)$ are shown in Fig. 8.11 and the comparison between 100,000 randomly generated pairs from the BB1 copula and the database are shown in Fig. 8.12.

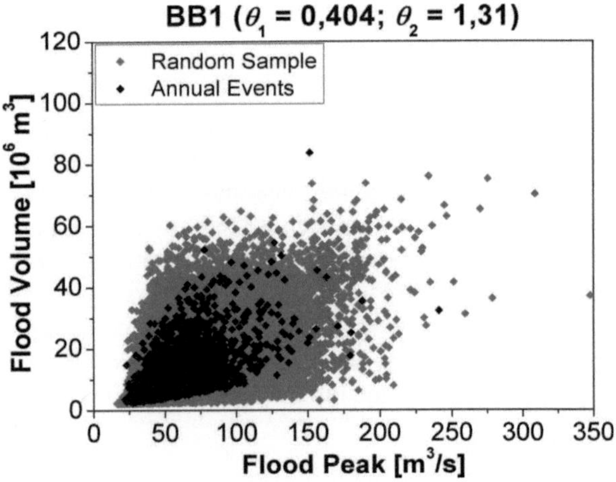

Fig. 8.12 100,000 random pairs (*grey dots*) of the annual flood peaks and the corresponding volumes at the dam Wupper chosen from the BB1 copula and the events from the simulated time series (*black dots*)

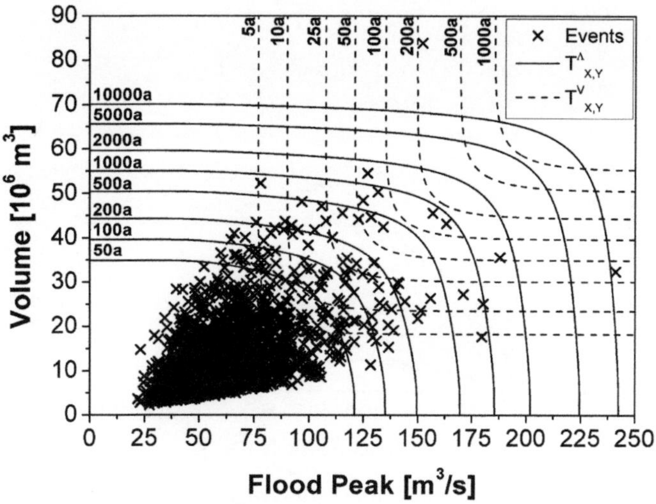

Fig. 8.13 Joint return periods $T^{\vee}_{X,Y}$ (exceeding *x* or *y*) and $T^{\wedge}_{X,Y}$ (exceeding *x* and *y*) of the corresponding flood peak and flood volume and the annual events from the simulated time series (*black crosses*)

From these figures and the AIC values in Table 8.3 it became obvious that the BB1 copula provides the best fit to the database. Therefore it was selected for further analysis.

8.3.3.3 Bivariate Frequency Analysis

For the bivariate frequency of flood events the contours of the joint return periods $T^{\wedge}_{X,Y}$ (for which *x* and *y* are exceeded) and $T^{\vee}_{X,Y}$ (for which either *x* or *y* are exceeded by the respective random variables *X* and *Y*) with respect to flood peak and corresponding volume and the annual events from the simulated time series are illustrated in Fig. 8.13.

8.3.4 Evaluation of the Effect of the Wupper Dam on Flood Control

The yearly floods from the simulated 2000 year time series of the inflow to the dam were used for the analysis of the efficiency of the Wupper dam in flood mitigation. The resulting maximum water levels and maximum releases for these events were calculated with the integrated storage module of the rainfall-runoff model NASIM considering the release rules of the dam with respect to the control gage Kluserbruecke. The discharge from the uninfluenced subcatchment area between the dam and the control gage was considered. As mentioned before it is important to consider flood volume and flood peak for the design and analysis of flood control structures. Here the joint probability of these two flood variables was used as probability measure for the single flood events.

Two different critical values for the release were considered in the analysis: (1) a first control level of 80 m^3s^{-1} and (2) a second critical control discharge

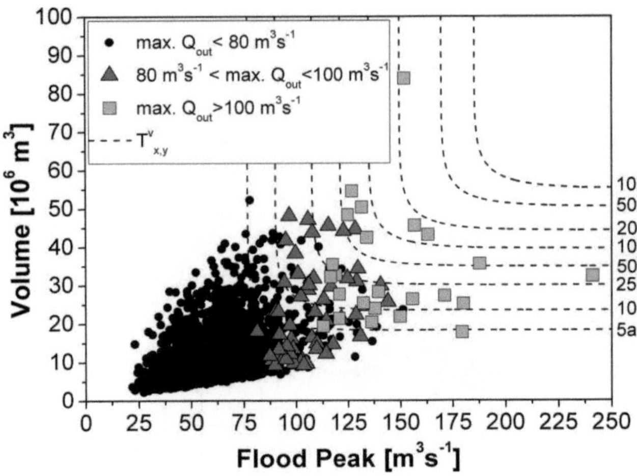

Fig. 8.14 Maximum resulting outflows at the dam Wupper from the annual events from the simulated time series and the contours of the joint return periods $T^{\vee}_{X,Y}$ (exceeding x or y)

of 100 m^3s^{-1} at the gage Kluserbruecke. If the releases of the dam exceed these values critical control levels are exceeded anyway independently from the runoff of the uninfluenced subcatchment between the dam and the gage. These two critical values were chosen to demonstrate the methodology to evaluate the effect of the Wupper dam on flood control. Though these values are defined as control levels at the gage Kluserbruecke both are not specifying critical flood conditions in this region. In Fig. 8.14 the critical events which exceed the respective thresholds are illustrated together with the contours of the joint return periods $T^{\vee}_{X,Y}$ (for which either x or y are exceeded by the respective random variables X and Y).

The contours of the joint return period $T^{\vee}_{X,Y}$ can be used to define regions of different hydrological risk where an critical event occurs. The contours which define the boundaries of the different regions are selected in relation to the secondary return period $\rho^{\vee}_{t}$ after Eq. (8.38). Thresholds ϑ (t) for the boundaries were selected for which (in the statistical mean) every $\rho^{\vee}_{t} = 25, 50, 100$ and 200 years an event with a joint return period $T^{\vee}_{X,Y} > \vartheta(t)$ occurs. E.g. the value $\vartheta(t)$ of the contour line for which in the random process every $\rho^{\vee}_{t} = 200$ years an event with a return period $T^{\vee}_{X,Y} > \vartheta(t)$ occurs (region situated to the right and above the contour line) is according to the inverse function of Eqs. (8.38) and (8.36) $\vartheta(t) = 50$ years. The tresholds ϑ (t) for the other secondary return periods $\rho^{\vee}_{t} = 25, 50$ und 100 years are summed up in Table 8.4.

If the discharge value of 80 m^3s^{-1} is defined as a critical release of the Wupper dam (triangles and squares in Fig. 8.14) 81 out of the 2,000 events are specified as critical events. The ratios of critical events in the different regions defined by their boundaries in Table 8.4 are illustrated in Table 8.5.

In the region with $T^{\vee}_{X,Y} > 8$ years 60% (45 of 76 events) of the events are critical events. The statistical return period that an event with a joint return period $T^{\vee}_{X,Y} > 8$

Table 8.4 Tresholds ϑ (t), for which every $\rho_t^{\vee}$ years an event occurs with a joint return period $T_{X,Y}^{\vee} > \vartheta(t)$

$\rho_t^{\vee}$ [a]	$\vartheta(t)$ [a]
200	50
100	26
50	14
25	8

Table 8.5 Ratio of critical events for the regions depending on the joint return period $T_{X,Y}^{\vee}$

$T_{X,Y}^{\vee}$	Number of events	Ratio of critical events	
		Critical release > 80 m^3s^{-1}	Critical Release> 100 m^3s^{-1}
> 50 a	9	100% (9 events)	89% (8 events)
26 a $< T_{X,Y}^{\vee} \leq$ 50 a	7	71% (5 events)	28% (2 events)
14 a $< T_{X,Y}^{\vee} \leq$ 26 a	25	68% (17 events)	24% (6 events)
8 a $< T_{X,Y}^{\vee} \leq$ 14 a	35	40% (14 events)	11% (4 events)
< 8 a	1,924	2% (36 events)	0.2% (4 events)

years occurs in the random process is according to Table 8.4 $\rho_t^{\vee} = 25$ years. Hence the hydrologic risk is high that an event in this region is a critical event.

If a value of 100 m^3s^{-1} is defined as critical release (squares in Fig. 8.14) 24 out of the 2,000 events are specified as critical events. Eight out of nine events with a joint return period $T_{X,Y}^{\vee} > 50$ years are critical events. The statistical return period that an event with $T_{X,Y}^{\vee} > 50$ years occurs in the random process is $\rho_t^{\vee} = 200$ years (see Table 8.4). In this region the hydrological risk is very high that an event is a critical event.

Measures to improve the efficiency of flood protection of the dam, e.g. optimized flood control or enlarged flood control storage, can be evaluated by comparing the change of the ratios of critical events in Table 8.5. For the example illustrated in Table 8.5 the flood protection of the dam is higher for a larger defined critical release.

It is obvious that the distribution of the critical events can not be described by using the univariate return period of flood peak or flood volume only. Hence it is very important to consider the joint probability of these two flood characteristics. Even with the joint probability of flood peak and flood volume the distribution of critical events can not be described completely because it depends also on other random variables such as the water level at the begin of the event, pre-events, runoff of the subcatchment, shape of the event, etc.

8.4 Case Study 2: Unstrut River Basin

The second case study for the application of copulas for the risk analysis of dams is the Unstrut River Basin. In contrast to the first case study examining only a single dam the flood protection system of the river Unstrut catchment is substantially more complex. The flood control system of the river Unstrut has been investigated, optimized and extended through an integrated and interdisciplinary flood risk assessment instrument as part of an interdisciplinary research project (Nijssen et al., 2009). For the risk-oriented approach a large variety of different hydrological scenarios was generated by coupling a stochastic rainfall generator with a deterministic rainfall-runoff model. For risk analysis a probability has to be attributed to the individual hydrological scenarios. Copulas are used to set up multivariate probability distributions which consider more than one flood characteristic in the frequency analysis of the events (Klein et al., 2010). Two different bivariate probabilities were used for the risk-analysis:

The joint probability of flood coincidences at reservoir sites at two rivers is applied to consider possible superposition of floods from the main river and one tributary. It aims to describe the spatial heterogeneity of flood events within these analyses (Section 8.4.3).

The joint probabilities of the flood peak and corresponding flood volume at the two reservoirs sites are used to characterize the flood retention in both reservoirs completely. The flood volume is very important and therefore its probability should be considered in the risk-based analysis in addition to the probability of the peak (Section 8.4.4).

8.4.1 Description of the River Basin

The flood prone catchment of the river Unstrut has a watershed of 6,343 km^2 and is situated in Mid-East Germany (see Fig. 8.15). The catchment has a variable topographic structure with elevations ranging from 104 to 982 m a.s.l. Two mountainous regions are located in the area: the Harz Mountains in the North and the Thuringian Forest Mountains in the South of the basin. Two large flood control reservoirs, the reservoir Kelbra with a catchment area of 664 km^2 and a flood control storage of 35.6×10^6 m^3 and the flood detention reservoir Straussfurt with a catchment area of 2,044 km^2 and a flood control storage of 18.6×10^6 m^3 were built to control the two main tributaries. A large polder system with a total storage volume of approximately 50×10^6 m^3 is located downstream of these two flood control reservoirs.

The reservoir Straussfurt is operated with seasonally varying flood control storage. During winter time a storage of 18.6×10^6 m^3 and in summer a storage of 12.7×10^6 m^3 is allocated for flood control. The maximum capacity of the operating outlets is approx. 300 m^3s^{-1} and the flood spillway is a crested weir with a length of 270 m.

The reservoir Kelbra is operated with seasonally varying flood control storage as well. In winter a storage of 35.6×10^6 m^3 and in summer a storage of 23.3×10^6 m^3 is available for flood control. The maximum capacity of the operating outlets is 244 m^3s^{-1}. Due to this large capacity of the operating outlets the reservoir has no flood spillway. The control gages for the combined control of the reservoirs are Bennungen, Oldisleben and Wangen. During a flood the discharges at

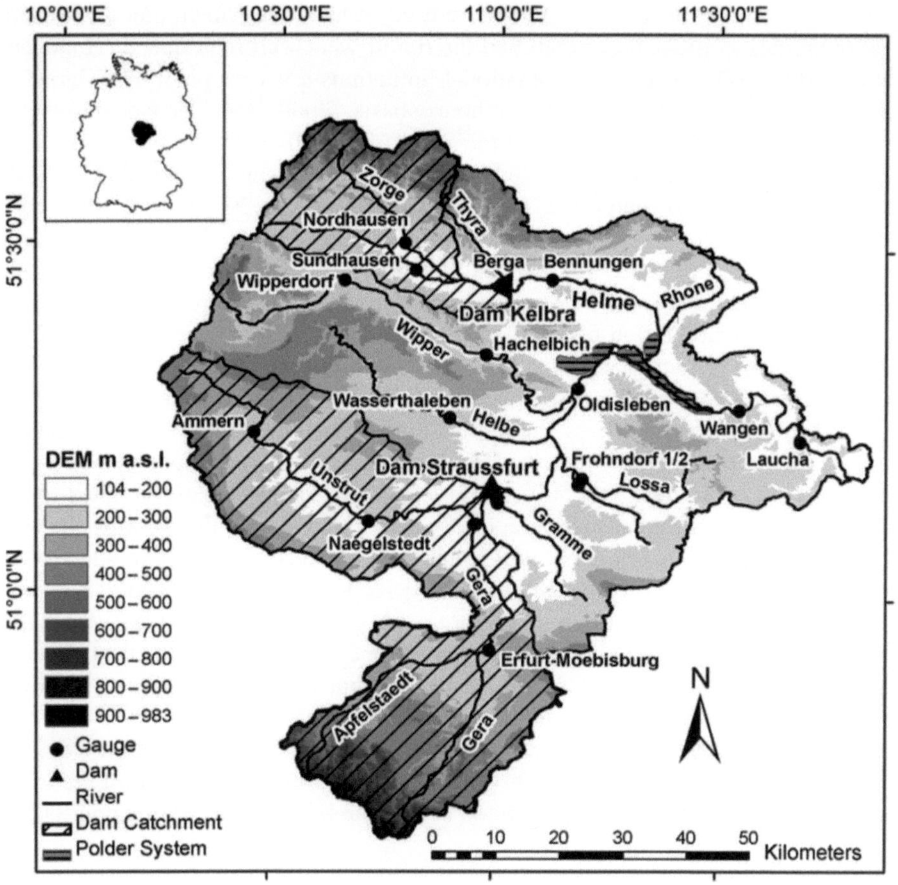

Fig. 8.15 Topographical map of the watershed of the river Unstrut, shown are important gages in the watershed, the polder system and the dams Kelbra and Straussfurt with their respective subcatchments

the gage Bennungen must not exceed 30 m^3s^{-1}, at gage Oldisleben 131 m^3s^{-1} and at the gage Wangen 152 m^3s^{-1}.

8.4.2 Stochastic-Deterministic Generation of Flood Events

A long-term daily discharge time series of 10,000 years was generated by coupling a stochastic spatially distributed daily rainfall generator (Hundecha et al., 2009) with a hydrological rainfall-runoff model following the concept of the well-known Swedish model HBV (Lindström et al., 1997). For the parameter estimation of the stochastic rainfall generator 122 stations with observed daily values from 1961 to 2003 and for the calibration of the continuous rainfall-runoff model on daily basis discharge series from 1991 to 1996 were used. It has been shown that the statistical properties of the daily rainfall and the runoff were well reproduced (Hundecha et al., 2008, 2009). Due to computational limitations it wasn't possible to build up a continuous rainfall-runoff model on hourly basis. Flood events have been selected from the generated daily discharge time series of 10,000 years. For these events the daily precipitation was disaggregated into hourly values using the models MuDRain (Koutsoyiannis, 2001) and HYETOS (Koutsoyiannis et al., 2003) and used as input for an event-based hourly hydrological rainfall-runoff model following the same concept as the continuous rainfall-runoff model on daily basis. The starting and boundary conditions for the event-based model were derived from the continuous model. For the calibration of the event-based rainfall runoff model only two flood events (April 1994 und December/January 2002/03) were available. The operation of the reservoirs has been implemented in the rainfall-runoff models.

A representative sample of the population of possible flood events is required to evaluate a flood control system. Here six different return periods were considered ($T = 25, 50, 100, 200, 500, 1,000$ years). The risk is not only related to the flood peak, since different hydrograph shapes (e.g. multi-peak floods or floods with long duration and large volume) will result in different risk levels. The selection of hydrological scenarios was therefore further supported by cluster analysis of historical events to determine typical hydrograph shapes and volumes. In total five different hydrological scenarios were selected for each return period at the reference gage Straussfurt, which encompass various spatial distributions of precipitation. Furthermore, one event with a return period of the peak of more than 1,000 years was considered. Hence in total 31 hydrological scenarios were selected from the entire 10,000 years discharge time series. A more detailed description of the event selection and the generation of the flood events can be found in Schumann (2009).

8.4.3 Bivariate Frequency Analysis of Corresponding Flood Peaks at the Reservoir Sites

Since the observed time series was too short for multivariate frequency analysis and was influenced by the control of the reservoirs and river construction works in the sixties the generated synthetic 10,000 year time series has been used for the bivariate frequency analysis shown here.

To account for the impact of different tributaries in the risk analysis, it is crucial to characterize the spatial distribution of the flood events due to the topographic structure of this particular catchment, where flooding originates mainly from mountainous regions upstream of the two reservoirs. From the analysis of the historical flood events, it was found that particularly three different rainfall distributions have to be considered in the risk-analysis: (1) heavy precipitation in the mountainous region in the North/North-West, (2) heavy precipitation in the mountainous region in the South and (3) concurrent heavy precipitation in both mountainous regions.

The inflow gages of the two reservoirs (see Fig 8.15) were henceforth used as reference locations to quantify the probability of occurrence of the 31 hydrological

Table 8.6 Measure of dependency for corresponding flood peaks at reservoirs Straussfurt and Kelbra

Sample size	Pearson's r	Kendall's τ	Spearman's ρ
12,768	0.72	0.48	0.67

scenarios that were selected from the generated 10,000 year time series for the risk analysis. These two reservoirs represent the upper boundary of the flood control system of the Unstrut catchment. The remaining flood protection structures are located downstream of these two reservoirs. The joint probability between the two flood peaks at the inflow gage of the two reservoirs was assessed by copulas to describe the overall probability of occurrence of flood events for the entire region in terms of probabilities of coincidences of a particular flood event at both tributaries. Hence the corresponding annual flood peaks had to be selected for further analysis. The annual flood peaks of the inflow time series to the reservoir Straussfurt and corresponding flood peaks of the inflow time series to the reservoir Kelbra have been determined. In cases where the identified flood peak at the reservoir Kelbra did not coincide with the annual flood peak at this reservoir, the annual flood peak at dam Kelbra and the corresponding flood peak at dam Straussfurt were added to the sample. In 2,768 years of the generated 10,000 years of discharge the annual flood peaks of the two inflow gages did not occur at the same time, and therefore the total selected sample size was 12,768.

The statistical dependencies between the two random variables have been estimated by the statistical measures listed in Table 8.6. The summary shows that there is a strong positive dependency between the two random variables.

8.4.3.1 Marginal Distributions

It was found that both variables can be described in the best way with the GEV distribution. The two parameter estimation methods L-Moments and product of moments provided nearly identical goodness of fit to the flood peaks at the two reservoirs. However, using graphical diagnostics the parameter estimation method of L-Moments (Hosking and Wallis, 1997) gave a better fit to the sample data for the distribution of the flood peaks at Straussfurt, while the method of product of moments provided a better fit for flood peaks at Kelbra, particularly in the region with small exceedance probabilities.

8.4.3.2 Copula Estimation

Only the Gumbel-Hougaard family was applicable to describe the dependency structure of the data among the four commonly applied Archimedian copulas mentioned in Section 8.3.3. This copula was chosen for the analysis, since the dependency structure between the flood peaks and the volumes at the dam Wupper is akin to the dependency structure between the two corresponding flood peaks. However, in addition to the Gumbel-Hougaard copula, the two-parameter copula BB1 was used for the construction of the bivariate distribution function. The parameters have been

Table 8.7 Parameters and goodness-of-fit measure AIC of the fitted copulas for the analysis of the corresponding flood peaks at the two reservoirs

Copula	Estimate (s)	AIC
Gumbel-Hougaard	$\theta = 1.88$	–8,254
BB1	$\theta_1 = 0.114$; $\theta_2 = 1.79$	–8,303

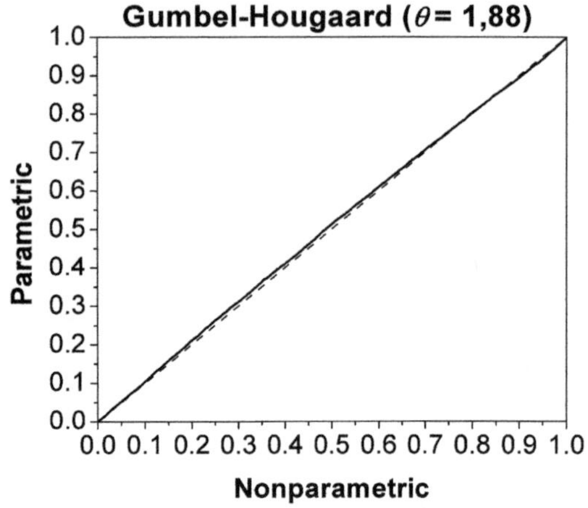

Fig. 8.16 Comparison of nonparametric and parametric estimations of $K_C\ (t)$ for the fitted Gumbel-Hougaard copula

estimated with the maximum pseudolikelihood method. The fitted parameters are summarized in Table 8.7.

The goodness-of-fit of both copulas was nearly identical (see e.g. the AIC values in Table 8.7. Hence the Gumbel-Hougaard copula, having less parameters than the BB1 copula, was selected for further analysis. For brevity only the results for the analysis of the Gumbel-Hougaard copula are shown here.

In Fig. 8.16 the parametric and nonparametric values of $K_C\ (t)$ are plotted for the Gumbel-Hougaard copula. If the plot is in good agreement with the straight dashed line that passes through the origin at 45°, then the copula is considered satisfactory. In Fig. 8.17 the margins of 1,000,000 random pairs (U_j, V_j), chosen from the copula and transformed back into the original units using the marginal distributions $F_X\ (x)$ and $F_Y\ (y)$, are compared with the sample values. Both figures show that

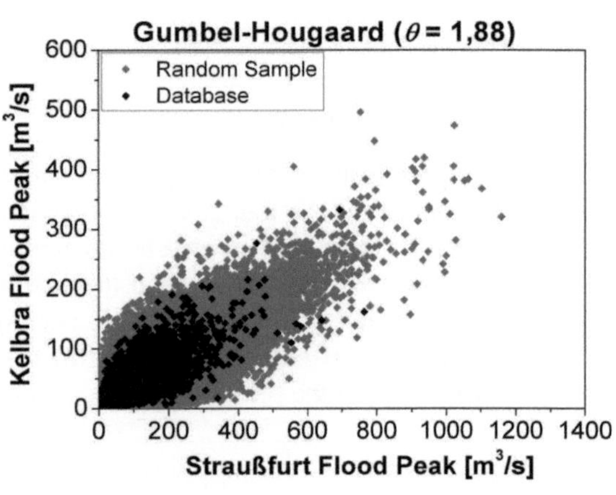

Fig. 8.17 1,000,000 random pairs (*grey dots*) of the corresponding annual flood peaks at the two reservoirs chosen from the Gumbel-Hougaard copula and the events from the simulated time series (*black dots*)

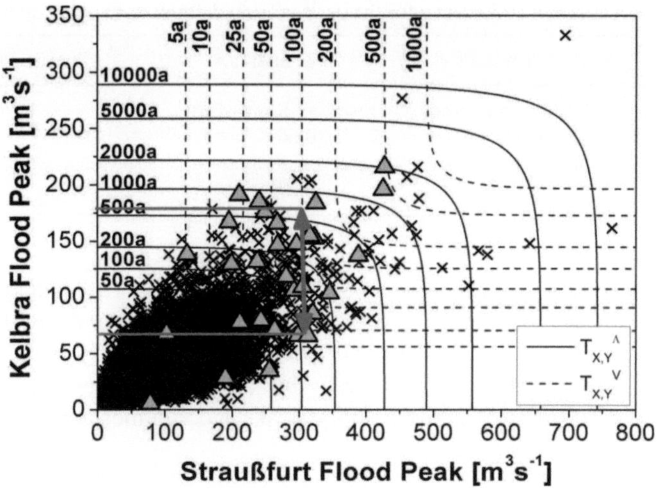

Fig. 8.18 Joint return periods $T_{X,Y}^{\vee}$ (exceeding x or y) and $T_{X,Y}^{\wedge}$ (exceeding x and y) of the annual flood peaks at the two reservoirs, the annual events from the simulated time series (*black crosses*) and the selected hydrological scenarios (*grey triangles*)

the Gumbel-Hougaard copula models the dependency structure between the two random variables adequately.

8.4.3.3 Bivariate Frequency Analysis

Figure 8.18 illustrates the contours of the joint return periods $T_{X,Y}^{\wedge}$ ("OR"-case) as well as the joint return periods $T_{X,Y}^{\vee}$ ("OR"-case) with respect to the corresponding flood peaks at the two reservoirs.

The selected 31 events represent a large variety of different hydrological scenarios. Interestingly, using for example the selected events with a flood peak of a ~100 year return period at the reservoir Straussfurt for the design of flood protection structures, the corresponding return periods of the selected flood peaks at the reservoir Kelbra range between 10 and 500 years. This additional information can be used for the planning and design of spatial distributed flood control structures in general as it is shown in Chapter 12.

8.4.4 Bivariate Frequency Analysis of the Annual Flood Peaks and the Corresponding Volumes

As mentioned before, for flood storage facilities such as reservoirs and polders it is important to consider the flood volume besides the flood peak in frequency analyses. Hence the joint return period of the corresponding flood peak and volume is used to assign a return period to the flood events at the reservoirs Straussfurt and Kelbra. Corresponding values of annual flood peaks and flood volumes were selected from 10,000 years of generated data. The corresponding flood volume to the annual flood peaks is selected according to Fig. 8.5. The estimated parameters

Table 8.8 Measure of dependency for the peak-volume datasets at the Unstrut watershed

Sample size	Pearson's r	Kendall's τ	Spearman's ρ
Corresponding flood peaks and volumes at reservoir Straussfurt			
10,000	0.81	0.66	0.85
Corresponding flood peaks and volumes at reservoir Kelbra			
10,000	0.74	0.57	0.76

which were applied to describe the dependency between flood peak and volume are given in Table 8.8. As before, there is a strong positive dependency between the corresponding flood peaks and volumes at the two reservoirs. For brevity, only the results of the analysis at the reservoir Straussfurt are presented here, since the dependency structures of the corresponding flood peaks and volumes at the two reservoirs were found to be similar.

8.4.4.1 Marginal Distributions

As for the analysis of the corresponding flood peaks at the two reservoirs, the Generalized Extreme Value (GEV) distribution was chosen as marginal distribution for the annual flood peaks at Straussfurt and Kelbra. For the corresponding flood volumes the GEV distribution provided the best fit using the method of product of moments as parameter estimation method.

8.4.4.2 Copula Estimation

As before the Gumbel-Hougaard and the BB1 copula were chosen for the analysis. The fitted parameters and the goodness-of-fit measure are listed in Table 8.9.

For both reservoirs the BB1 copula provided the better fit for the structure of dependency of the two random variables. Therefore it was chosen for further analysis.

In Fig. 8.19 the parametric and nonparametric values of $K_C(t)$ are plotted for the BB1 copula and in Fig. 8.20 the margins of 1,000,000 random pairs (U_j, V_j), chosen from the copula and transformed back into the original units using the marginal distributions $F_X(x)$ and $F_Y(y)$, are compared with the sample values. Both figures

Table 8.9 Parameters and goodness-of-fit measure AIC of the fitted copulas for the analysis of the flood peaks and the corresponding flood volumes at the dam Straussfurt and Kelbra

Copula	Estimate (s)	AIC
Corresponding flood peaks and volumes at reservoir Straussfurt		
Gumbel-Hougaard	$\theta = 2.6$	–11,787
BB1	$\theta_1 = 1.13$; $\theta_2 = 1.79$	–13,725
Corresponding flood peaks and volumes at reservoir Kelbra		
Gumbel-Hougaard	$\theta = 2.12$	–8,256
BB1	$\theta_1 = 0.986$; $\theta_2 = 1.52$	–9,818

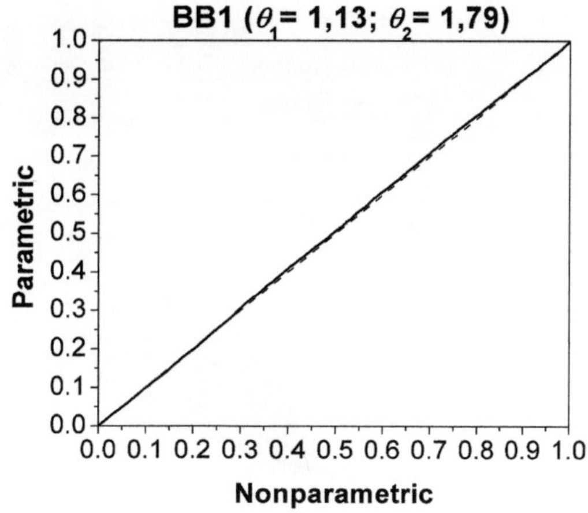

Fig. 8.19 Comparison of nonparametric and parametric estimations of $K_C(t)$ for the fitted BB1 copula (The *dashed* 45°-*line* represents perfect agreement)

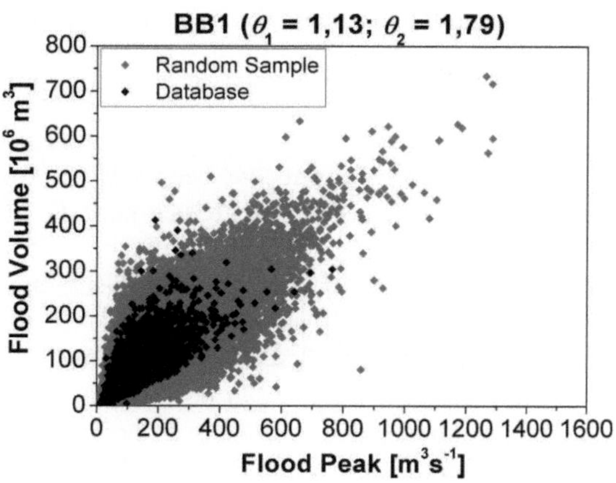

Fig. 8.20 1,000,000 random pairs (*grey dots*) of the annual flood peak and the corresponding flood volume chosen from the BB1 copula at the reservoir Straussfurt and the events from the simulated time series (*black dots*)

show that the BB1 copula models the dependency structure between the two random variables adequately.

8.4.4.3 Bivariate Frequency Analysis

Figure 8.21 illustrates the contours of the joint return periods $T_{X,Y}^{\wedge}$ ("AND"-case) and $T_{X,Y}^{\vee}$ ("OR"-case) with respect to the annual flood peak and the corresponding volume, the annual events from the simulated time series and the selected events (triangles). Again, using the events with a return period of the flood peak around 100 years as an example, the corresponding return periods of the flood volume range between 25 and 2,000 years. It is therefore recommended to use the joint probabilities for a detailed description of flood events instead of using the univariate probability of the flood peak only.

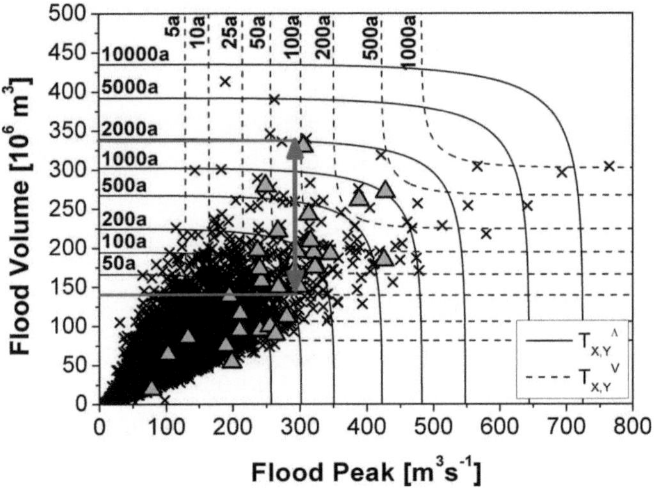

Fig. 8.21 Joint return periods $T^{\vee}_{X,Y}$ (exceeding *x* or *y*) and $T^{\wedge}_{X,Y}$ (exceeding *x* and *y*) of the flood peaks and corresponding flood volumes at the reservoir Straussfurt, the annual events from the simulated time series (*black crosses*) and the selected hydrological scenarios (*grey triangles*)

A broad range of events has been selected from the time series and categorized based on the analysis of joint return periods of the flood volume at the reservoir Straussfurt, as well as the joint return periods of the flood peak at the reservoirs Straussfurt and Kelbra. The selected events are displayed in Figs. 8.18 and 8.21 (shown as triangles). By using this approach, the spatial distribution between the two main tributaries is considered and the crucial aspect of using different combinations of flood peak and volume can be taken into consideration in flood risk analyses. Those data are in turn used for assessment of the flood control system, which hereby integrates the spatial component of probability.

8.4.5 Evaluation of the Effect of the Reservoir Straussfurt on Flood Control

The efficiency of the reservoir Straussfurt in flood mitigation is evaluated using the 31 selected hydrological scenarios. In Fig. 8.22 the maximal resulting water levels at the reservoir Straussfurt are shown to demonstrate the effects of the different hydrological loads on the flood protection structures depending on the combination of flood peak and flood volume.

Here a water level exceeding 150.3 m a.s.l. is defined as a critical hydrologic event, since it is known that the corresponding outflow of more than 200 m^3s^{-1} would cause severe damage downstream. With a joint return period $T^{\vee}_{X,Y}$ (exceeding *x* or *y*) greater than 50 years (gray area in Fig. 8.22), all considered events are critical events according to this definition. Hence the hydrological risk is very high for these events. The mean interarrival time of an event lying in the region of very

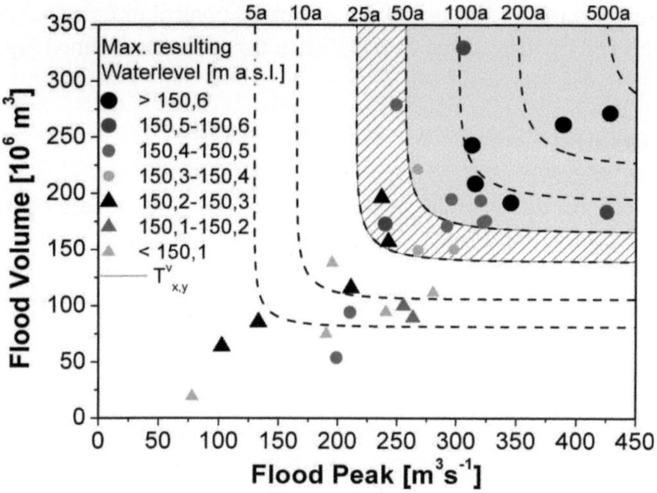

Fig. 8.22 Maximum resulting water levels at the reservoir Straussfurt from the different hydrological scenarios and the contours of the joint return periods $T^{\vee}_{X,Y}$ (exceeding x or y)

high hydrological risk, calculated with the secondary return period (Eq. 8.38) is up to 110 years. For joint return periods between 25 and 50 years (black shaded area in Fig. 8.22) five of the seven events which were considered are critical events. In this region the risk is high that an event with a return period $T^{\vee}_{X,Y}$ (exceeding x or y) between 25 and 50 years is a critical event. The return period of an event lying in this region 25 a $< T^{\vee}_{X,Y} <$ 50 a can be calculated to 105 years using following formula

$$\lambda^{\vee}_{t_1 t_2} = \frac{\mu_T}{P(t_1 < C(u,v) \le tl2)} = \frac{\mu_T}{K_C(t_2) - K_C(t_1)} \tag{8.39}$$

with

$$t = 1 - \frac{\mu_T}{T^{\vee}_{X,Y}}. \tag{8.40}$$

In the region with a joint return period $T^{\vee}_{X,Y}$ smaller than 25 years only two out of 12 events will result in substantial damages. The risk that events with these joint probabilities may cause damage is relatively low. Not all critical events can be identified by the joint return period $T^{\vee}_{X,Y}$ such, as e.g. the two critical events lying in the region with $T^{\vee}_{X,Y} < 25$ a, because the distribution of the critical events depends additionally on other flood characteristics (shape of hydrograph, antecedent conditions, flood storage at the onset of the event etc.). But not all random variables can be considered in the multivariate frequency analyses and therefore this analysis is reduced to the two most important random variables flood peak and corresponding volume.

Flood reduction measures such as an optimized control or enlarged flood storage could be analyzed by the change of the critical events in the defined regions.

8.5 Conclusions

A large variety of different hydrological loads can be generated by coupling a spatially distributed stochastic rainfall generator with a rainfall-runoff model. Especially in large catchment areas with several interacting flood protection structures it is important to consider the spatial distribution of floods. For the risk analysis the different hydrological loads have to be categorized probabilistically. In this chapter a methodology to categorize flood events for risk analysis and risk-orientated design of dams and flood control systems using copulas is presented. The advantage of using copulas is the possibility to apply a multivariate distribution function with different univariate marginal distributions. Joint return periods and also conditional return periods can be derived easily for hydrologic scenarios from copulas.

Results from two different case studies were presented: a single dam located at the river Wupper was analyzed in the first case study and a flood protection system with two dams and a flood polder system – the river Unstrut – were analyzed in the second case study. Different flood characteristics were used to derive probabilities for the risk analysis. For single flood control structures the flood volume was considered besides the flood peak to estimate the joint probability. For large river basins it is also important to consider the spatial distribution of the flood events in the risk analysis. Therefore, to reassign an overall probability of a flood event, which has different probabilities in the different tributaries, the joint return period of the corresponding flood peaks in the main tributaries can be obtained using copula analysis. For the example of the case study for the Unstrut river the flood peaks of the inflows to the two reservoirs located at two main tributaries were used for the bivariate frequency analysis to characterize the flood risk for the confluence.

Another advantage using copulas is the possibility to identify critical events for flood protection structures such as reservoirs and to evaluate the effect of flood control of these reservoirs. Using the joint return period $T^{\vee}_{X,Y}$, for which the random variables flood peak (X) or corresponding flood volume (Y) exceed the respective thresholds x or y, thresholds can be identified to evaluate the hydrological risk of critical events.

In general observed time series are too short for multivariate frequency analysis. By generating a synthetic discharge time series through coupling a stochastic rainfall generator with a deterministic rainfall-runoff model the database can be extended for the multivariate analysis. It must be noted that a simulated database can not replace observed records. On the contrary observations are required for reliable parameter estimation of the simulation model and for the verification of the results.

Copula analysis can provide valuable information to decision makers. Such informations is more adequate for flood control problems than traditional return period analysis. Consideration of multiple flood characteristics is essential for risk-based planning.

References

Akaike H (1974) A new look at statistical model identification. IEEE Trans Automatic Control AC19(6):716–723

Bergmann H, Sackl B (1989) Determination of design flood hydrographs based on regional hydrological data. In: Kavvas ML (ed) New directions for surface water modelling. IAHS Publ. no. 181, Wallingford

Blazkova S, Beven K (2004) Flood frequency estimation by continuous simulation of subcatchment rainfalls and discharges with the aim of improving dam safety assessment in a large basin in the Czech Republic. J Hydrol 292(1–4):153–172

Cherubini U, Luciano E, Vecchiato W (2004) Copula methods in finance. Wiley, Chichester

Deheuvels P (1979) Empirical dependence function and properties – nonparametric test of independence. Bull De La Classe Des Sci Acad R De Belg 65(6):274–292

De Michele C, Salvadori G (2003) A generalized pareto intensity-duration model of storm rainfall exploiting 2-Copulas. J Geophys Res Atmospheres 108(D2):ACL 15–1

De Michele C, Salvadori G, Canossi M, Petaccia A, Rosso R (2005) Bivariate statistical approach to check adequacy of dam spillway. J Hydrol Eng 10(1):50–57

Embrechts P, Lindskog F, McNeil AJ (2003) Modelling dependence with copulas and applications to risk management. In: Rachev ST (ed) Handbook of heavy tailed distributions in finance. Elsevier, North-Holland, Amsterdam

Evin G, Favre AC (2008) A new rainfall model based on the Neyman-Scott process using cubic copulas. Water Resour Res 44(3):1–18

Favre AC, El Adlouni S, Perreault L, Thiemonge N, Bobee B (2004) Multivariate hydrological frequency analysis using copulas. Water Resour Res 40(1):1–12

Genest C, Favre AC (2007) Everything you always wanted to know about copula modeling but were afraid to ask. J Hydrologic Eng 12(4):347–368

Genest C, Remillard B (2008) Validity of the parametric bootstrap for goodness-of-fit testing in semiparametric models. Annales de l'Institut Henri Poincaré – Probabilités et Statistique 44(6):1096–1127

Genest C, Rivest LP (1993) Statistical-Inference procedures for Bivariate Archimedean Copulas. J Am Stat Assoc 88(423):1034–1043

Genest C, Quessy JF, Remillard B (2006) Goodness-of-fit procedures for copula models based on the probability integral transformation. Scand J Stat 33(2):337–366

Genest C, Remillard B, Beaudoin D (2009) Goodness-of-fit tests for copulas: a review and a power study. Insur Math Econ 44(2):199–213

Goel NK, Seth SM, Chandra S (1998) Multivariate modeling of flood flows. J Hydraulic Eng 124(2):146–155

Grimaldi S, Serinaldi F (2006a) Design hyetograph analysis with 3-copula function. Hydrol Sci J-J Des Sci Hydrologiques 51(2):223–238

Grimaldi S, Serinaldi F (2006b) Asymmetric copula in multivariate flood frequency analysis. Adv Water Resour 29(8):1155–1167

Haberlandt U, Ebner von Eschenbach AD, Buchwald I (2008) A space-time hybrid hourly rainfall model for derived flood frequency analysis. Hydrol Earth Syst Sci 12(6):1353–1367

Hiemstra LAV, Zucchini WS, Pegram GGS (1976) Method of finding family of runhydrographs for given return periods. J Hydrol 30(1–2):95–103

Hosking JRM, Wallis JR (1997) Regional frequency analysis: an approach based on L-moments. Cambridge University Press, Cambridge

Hundecha Y, Pahlow M, Klein B, Schumann A (2008) Development and test of a stochastic rainfall generator for the generation of hydrologic load scenarios (in German). Forum für Hydrologie und Wasserwirtschaft, Heft 23.06, Beiträge zum Tag der Hydrologie 2008, Hannover, pp 34–41

Hundecha Y, Pahlow M, Schumann A (2009) Modeling of daily precipitation at multiple locations using a mixture of distributions to characterize the extremes. Water Resour Res 45(W12412): 1–15

Hydrotec (2007) Niederschlags-Abfluss-Modell – NASIM – Version 3.6.0. Programmdokumentation. Hydrotec Ingenieurgesellschaft für Wasser und Umwelt mbH, Aachen

Joe H (1997) Multivariate models and dependence concepts. Chapman and Hall, New York

Kao SC, Govindaraju RS (2008) Trivariate statistical analysis of extreme rainfall events via the Plackett family of copulas. Water Resour Res 44(2):1–19

Karmakar S, Simonovic SP (2008) Bivariate flood frequency analysis: Part 1. Determination of marginals by parametric and nonparametric techniques. J Flood Risk Manag 1: 190–200

Karmakar S, Simonovic SP (2009) Bivariate flood frequency analysis. Part 2: a copula-based approach with mixed marginal distributions. J Flood Risk Manag 2:32–44

Kim G, Silvapulle MJ, Silvapulle P (2007) Comparison of semiparametric and parametric methods for estimating copulas. Comput Stat Data Anal 51(6):2836–2850

Klein B, Schumann A (2006) Generation of multi-peak hydrographs for the flood design of dams (in German). Forum für Hydrologie und Wasserwirtschaft, Heft 15.06, Beiträge zum Tag der Hydrologie 2006, Band 2, München, pp 255–266

Klein B, Pahlow M, Hundecha Y, Schumann A (2010) Probability analysis of hydrological loads for the design of flood control systems using copulas. J Hydrol Eng 15(5). doi: 10.1061/(ASCE)HE.1943-5584.0000204

Koutsoyiannis D (2001) Coupling stochastic models of different time scales. Water Resour Res 37(2):379–392

Koutsoyiannis D, Onof C, Wheater HS (2003) Multivariate rainfall disaggregation at a fine timescale. Water Resour Res 39(7):1–18

Lindström G, Johansson B, Persson M, Gardelin M, Bergström S (1997) Development and test of the distributed HBV-96 hydrological model. J Hydrol 201(1–4):272–288

Nelsen RB (1999) An introduction to copulas. Springer, New York, NY

Nijssen D, Schumann A, Pahlow M, Klein B (2009) Planning of technical flood retention measures in large river basins under consideration of imprecise probabilities of multivariate hydrological loads. Nat Hazards Earth Syst Sci 9(4):1349–1363

Petry U, Hundecha Y, Pahlow M, Schumann A (2008) Generation of severe flood scenarios by stochastic rainfall in combination with a rainfall runoff model. Proceedings of the 4th International Symposium on Flood Defense, 6–8 May, Toronto, ON

Renard B, Lang M (2006) Use of a Gaussian copula for multivariate extreme value analysis: Some case studies in hydrology. Adv Water Resour 30(4):897–912

Salvadori G, De Michele C (2004) Frequency analysis via copulas: theoretical aspects and applications to hydrological events. Water Resour Res 40(12):1–17

Salvadori G, De Michele C (2006) Statistical characterization of temporal structure of storms. Adv Water Resour 29(6):827–842

Salvadori G, De Michele C (2007) On the use of copulas in hydrology: theory and practice. J Hydrol Eng 12(4):369–380

Salvadori G, De Michele C, Kottegoda NT, Rosso R (2007) Extremes in nature: an approach using copulas. Series: Water Science and Technology Library, vol 56. Springer, Dordrecht

Sandoval CE, Raynal-Villasenor J (2008) Trivariate generalized extreme value distribution in flood frequency analysis. Hydrol Sci J-J Des Sci Hydrologiques 53(3):550–567

Saunders R, Laud P (1980) The multidimensional kolmogorov goodness-of-fit test. Biometrika 67(1):237–237

Schumann A (2009) Abschlussbericht des BMBF-Verbundforschungsvorhabens "Integrative Nutzung des technischen Hochwasserrückhalts in Poldern und Talsperren am Beispiel des Flussgebiets der Unstrut". Schriftenreihe Hydrologie/Wasserwirtschaft Nr. 24, Lehrstuhl für Hydrologie, Wasserwirtschaft und Umwelttechnik, Ruhr-Universität Bochum, Bochum

Schwarz G (1978) Estimating the dimension of a model. Ann Stat 6(2):461–464

Serinaldi F, Grimaldi S (2007) Fully nested 3-copula: procedure and application on hydrological data. J Hydrol Eng 12(4):420–430

Shiau JT (2003) Return period of bivariate distributed extreme hydrological events. Stoch Environ Res Risk Assess 17(1–2):42–57

Shiau JT, Feng S, Nadaraiah S (2007) Assessment of hydrological droughts for the Yellow River, China, using copulas. Hydrol Processes 21(16):2157–2163

Shiau JT, Wang HY, Tsai CT (2006) Bivariate frequency analysis of floods using copulas. J Am Water Resour Assoc 42(6):1549–1564

Singh K, Singh VP (1991) Derivation of bivariate probability density-functions with exponential marginals. Stoch Hydrol Hydraul 5(1):55–68

Sklar A (1959) Fonctions de répartion à n dimensions et leurs marges. Publ Inst Stat Univ Paris 8:220–231

Stedinger JR, Vogel RM, Foufoula-Georgiou E (1993) Frequency analysis of extreme events. In: Maidment DR (ed) Handbook of hydrology. McGraw-Hill, New York, NY

Whelan N (2004) Sampling from Archimedean copulas. Quant Finance 4(3):339–352

Yue S (1999) Applying bivariate normal distribution to flood frequency analysis. Water Int 24(3):248–254

Yue S (2001) A bivariate gamma distribution for use in multivariate flood frequency analysis. Hydrol Processes 15(6):1033–1045

Yue S, Ouarda T, Bobee B (2001) A review of bivariate gamma distributions for hydrological application. J Hydrol 246(1–4):1–18

Yue S, Ouarda T, Bobee B, Legendre P, Bruneau P (1999) The Gumbel mixed model for flood frequency analysis. J Hydrol 226(1–2):88–100

Zhang L, Singh VP (2006) Bivariate flood frequency analysis using the copula method. J Hydrol Eng 11(2):150–164

Zhang L, Singh VP (2007a) Bivariate rainfall frequency distributions using Archimedean copulas. J Hydrol 332(1–2):93–109

Zhang L, Singh VP (2007b) Gumbel-Hougaard copula for trivariate rainfall frequency analysis. J Hydrol Eng 12(4):409–419

Zhang L, Singh VP (2007c) Trivariate flood frequency analysis using the Gumbel-Hougaard copula. J Hydrol Eng 12(4):431–439

Chapter 9
Hydraulic Modelling

Mark Musall, Peter Oberle, and Franz Nestmann

Abstract Modern flood management should attempt to analyse the full scope of a flood event, using integrative concepts and taking into account multifaceted expertise from diverse fields. Such analyses should be conducted, both at the forefront, to compile meta-concepts, and during the event, to enable direct response to unanticipated occurrences, for example, failure of protective devices. Within such a management plan, beginning with the large-scale, meteorological analysis of weather conditions, to local examination of a short dike section, the whole functional chain of a flood event should be taken into account, from its origin, to individual damage potential analysis. Hereby, a special role is given to the prediction of the expected flow situation in the affected region. This establishes an interface between the large-scale observations of the disciplines meteorology and hydrology and local catastrophe management, by providing temporal and spatial high resolution information about water levels and flow paths, and thus, the existing danger situation. Thus, based on hydraulic simulations, the identification of vulnerable areas can take place, for example, within the framework of a flood contingency plan with high regional discretisation. In addition it should be possible, even during a flood event, to produce short-term verifiable information to optimise the concrete plan of operation, taking into account ongoing hydrological and topographical conditions. In this chapter, the scientific and functional fundamentals of such hydraulic simulations will be examined. Subsequently, practice-oriented examples of tangible application will be used to define the model requirements and possible use.

Contents

M. Musall (✉)
Karlsruhe Institute of Technology, Karlsruhe, Germany
e-mail: Musall@kit.edu

A.H. Schumann (ed.), *Flood Risk Assessment and Management*,
DOI 10.1007/978-90-481-9917-4_9,

9.1 Fundamentals

9.1.1 Preface

Similar to the hydrological models, hydraulic modelling balances the input and output variables of precipitation, evaporation and discharge. In contradiction to hydrological models hydrodynamic-numeric (HN)-Models additionally consider spatial high resolution information about the terrain, such as relief and roughness. This data is needed to provide detailed predictions of hydraulic processes within the area of interest.

The mathematical description of the discharge process in flowing water is based on differential equations, which are deduced from the balance of mass and force at a control volume. Thus, flow processes in their variability in space and time within almost any area can be accurately described. Since these differential equations cannot generally be analytically solved in a direct way, approximate solutions are produced using numeric methods. Hereby a spatial and temporal discretisation in the form of discrete computational nodes is carried out. Additionally, the underlying differential equations are often significantly simplified using assumptions (e.g. perpetual density or negligence of the vertical components of flow). This approach is rational because the computational requirements to solve the complete equations for natural runoff conditions are still missing. Practical engineering experience has shown that simplifications, e.g. the utilisation of average values of flow parameters are mostly adequate.

In the emerging hydrodynamic-numerical (HN) models, the natural flow systems are abstracted, but all essential parameters and interactions are represented (DVWK, 1999).

The required input into these models comprises topographical data (terrain in various levels of abstraction), hydrological data (discharge, water levels), information about expected flow resistance characteristics (due to structures, vegetation), as well as diverse parameters to control the numeric solution and the data processing (see

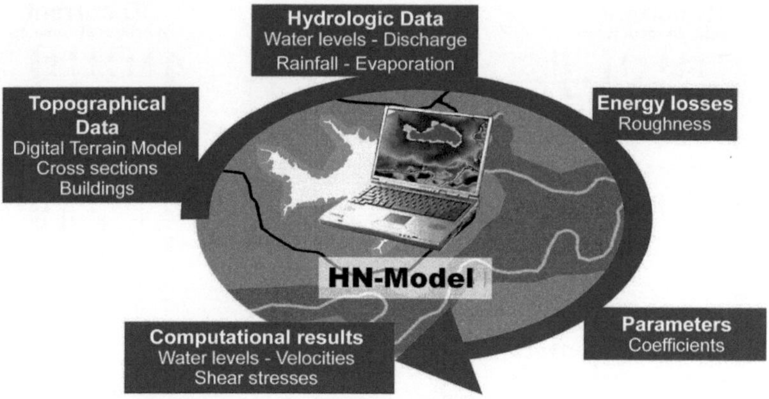

Fig. 9.1 HN-Model components

Fig. 9.1). The hydrological input data of discharge and water level at the boundaries of the modelled region constitute input constraints of numeric computation which are absolutely necessary and have to be defined preferably exact, as they determine the complete process of computation.

Observed water levels, preferably distributed across the whole model area, and corresponding discharge measurements are essential for the necessary calibration of the model.

The numeric model calculates hydraulic characteristics as water levels or flow velocities in the entire simulated area. Depending on the type of the problem and the complexity of the numeric model applied, these can also be morphodynamic or thermodynamic magnitudes.

The modeled research area can encompass river districts, individual stream stretches, e.g. river bends or inlet areas, or just the surrounding area of a hydraulic structure. Differences also exist in regard to the temporal independence of the discharge process. Prior to a numeric simulation, detailed analysis of the expected flow behaviour must be conducted to ensure correct functionality of the model.

9.1.2 Flow Characteristics

A current in naturally flowing waters, upon close observation, is always three-dimensional and transient. However on large-scale observation it becomes apparent, that the usually existing main current often features a clearly defined direction. Consequently, it can be described using one- or two-dimensional hydrodynamic-numerical approaches.

Figure 9.2 schematically illustrates, using the example of a rectangular channel, the fundamental differences between one- and multidimensional flow situations. In the case of a continuous channel, without any built-in structures, a one-dimensional flow will emanate. This is distinguished across the whole discharge area by flow

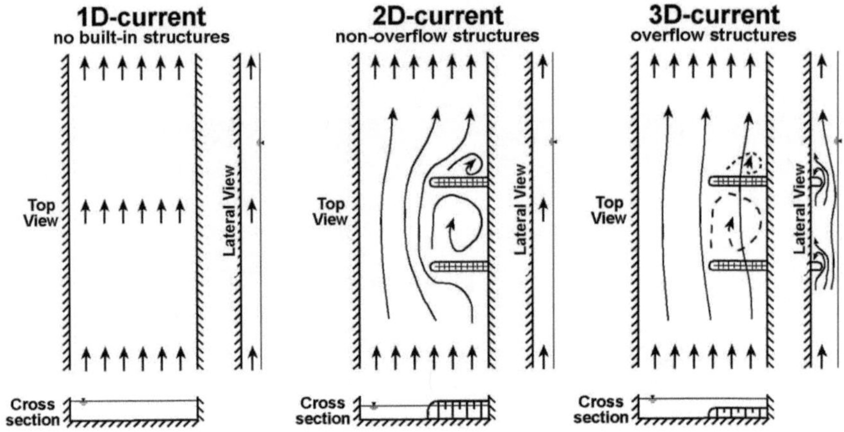

Fig. 9.2 One-, two-, and three-dimensional flow characteristics

lines parallel to the stream centreline. The current primarily flows in one direction; the flow velocities across this main flow direction are significantly smaller and can be neglected.

If there are non-overflow components, like groins, in the observed channel, the flow exhibits clear transversal current effects, and a two-dimensional examination is necessary. The flow lines then show a horizontal curvature. In this case, only the vertical flow components can be disregarded.

If an additional overflow of the built-in components occurs, completely irregular (horizontally and vertically) flow line curvatures appear and a three-dimensional examination is required.

In this type of analysis of natural flow behaviour, it is important to keep in mind that the previously described multi-dimensional flow phenomena often show local aspects. Depending on the required target size (e.g. only water level) and existing observation criteria, these local aspects needn't be analysed. Their influence on the main current, however, in the form of loss of energy, must be taken into account.

9.1.3 Model Types

Allocation of diverse numeric model types takes place similar to flow classification and depends on local current behaviour. Figure 9.3 shows a diagram of the various distributions of individual nodes of calculation, possible element shapes, and the calculated flow direction of the most important types.

Selective point distribution (calculation nodes) is applied in the one-dimensional modelling of flowing water. Thereby, every point is marked with characteristic cross-section information (overflowed cross-section area depending on velocity and discharge). Based mostly on the 1D-Saint-Venant equations, average cross-sectional water levels, and flow rates at the computation nodes are calculated (Cunge

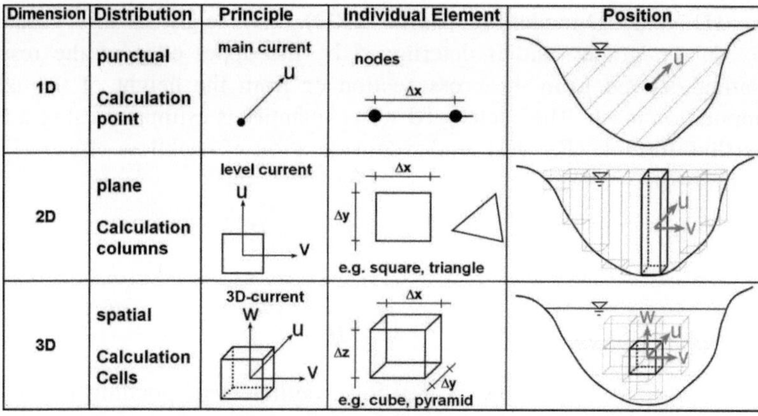

Fig. 9.3 Distribution of the calculation nodes of various models

et al., 1980). A level, plane distribution of the calculation points is the basis of the two-dimensional, depth-averaged modeling of a river section, using the 2D-shallow water equations. The area of calculation is divided into triangular or rectangular columns covering the total depth of the water. Detailed statements about the local distribution of depth-averaged velocities, water level and riverbed shear stress, etc. are calculated for the area of interest. Computational procedures applied should guarantee high stability and full compliance of the conservation equations of mass and momentum to be applicable even for simulation of ultra critical discharge conditions (Beffa, 1994). In the three-dimensional case, a completely three-dimensional distribution of the calculation points takes place in the modeled area, for example, using cubes or triangular pyramids. The water level (or pressure) and the three velocity components of all spatial dimensions are calculated. 3D-procedures in flood management are now only applied in a few special cases, for example, when planning hydraulic structures. This is due to the high discretisation necessary and the extensive computations, but also the low gain of flood relevant information.

In addition to the main types of hydraulic models described above, there are several intermediate types where certain physical characteristics of the related higher-dimensional versions are neglected without a reduction in dimensionality (e.g. 3D-shallow water approach neglecting the vertical momentum, but considering vertical velocity). Furthermore, linked models of various types are known. An innovative model, especially applied to simulate floodings of larger areas, combines the following two approaches:

- 1D-modelling of areas with distinct main stream direction across the whole range of discharge (e.g. river bed, flood channel)
- 2D-modelling of areas with strong multi-dimensional flow characteristics (e.g. flood plains, junctions, polder areas)

A 2D-HN-module and an additional 1D-HN-module constitute the basis of this interlinked model. A weir overfall is assumed to be the hydraulic connection between 1D- and 2D-modelling (Beffa, 2002). The height of this virtual weir depends on topography and is determined by the upper edge of the respective bank, either derived from the cross section or from the height of the adjacent 2D-computation mesh. The exchanged water quantity is estimated using a Poleni-like overflow formula (Eq. 9.1), under consideration of backflow effects (Eq. 9.2) (Hager, 1986):

$$Q = {}^{2}\!/_{3}\, \mu\, \varphi\, B\, \sqrt{2g}\, h^{3/2} \tag{9.1}$$

$$\varphi = \left[1 - \left({}^{H_u}\!/_{H_o}\right)^{a}\right]^{1/a} \tag{9.2}$$

For this process the parameters overflow coefficient μ, ponding factor φ, and squaring factor *a* (*a* = 4.0 for broad-crested weirs) have to be specified. The specifications concerning overflow height h and the two energy heads H_u and H_o are shown in Fig. 9.4. When water flows from 2D- into the 1D-simulation area, the energy head in the flood plain is used instead of the water surface height, whereby the direct incoming flow of the junction is considered.

The location of junctions connecting 1D- and 2D-model areas are automatically determined by the spatial proximity of individual cross sections to adjacent 2D-cells on the relative banks. If these distances exceed a defined maximum, no exchange will take place. Ideally, each cross-section is allocated an edge cell, but the connection of an edge cell with two or more cross-sections is also possible.

Beside to the described procedure, others exist that differ, concerning the basic equations or the model linking, but which use the same principle to accelerate the computation. Other methods to accelerate the computations are based on approaches with oversimplified basic equations. These, however, exhibit deficits in regard to the stipulated short time of calculation and the often necessary specification of time-dependent constraints. Promising approaches are also appearing from the computer science branch, for example, the application of substantially parallel graphic processors for computation. These, however, are still in the research stage.

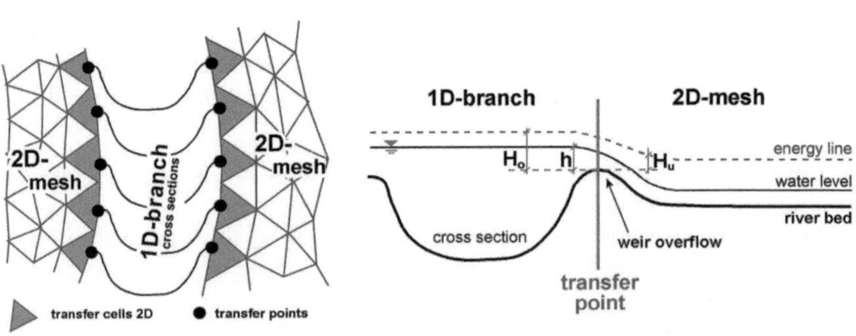

Fig. 9.4 Principle sketch for 1D/2D-model link

9.1.4 Base Data

Geodata from different sources are necessary to construct and operate an HN-model. In the following, an overview of the essential geodata for flood simulations is given, and the requirements for their applicability are discussed.

9.1.4.1 Terrain Topography (River Channel/Flood Plain)

The discrete description of the terrain topography of an investigated area can be provided in vector form, via selected ground points, contour lines, and breaking edges or as a grid matrix. The following approaches can be considered when collecting original topographical base data:

- Riverbed:
  - vessel-based (multi-beam) sounding (in navigable waters)
  - tachymetric surveys (smaller waters)
- Flood plain:
  - tachymetric surveys (terrestrial)
  - GPS-RealTime Kinematic (terrestrial)
  - aerial photogrammetry (from airplanes)
  - airborne laser-scanning-technique
  - active microwave (radar) (from airplanes)

Digitalising altitudes from geo-referenced topographical maps presents a secondary method to record flood plain geometry. This method, though, is more and more taking a back seat, due to inevitable error propagation and the growing availability and processability of digital data.

The approaches for collecting the regional river flood plain data differ in efficiency and accuracy. Thus the different data sources are applied depending on local specifications and requirements. In Fig. 9.5 the uncertainties for different altitudinal data of flood plains are summarized.

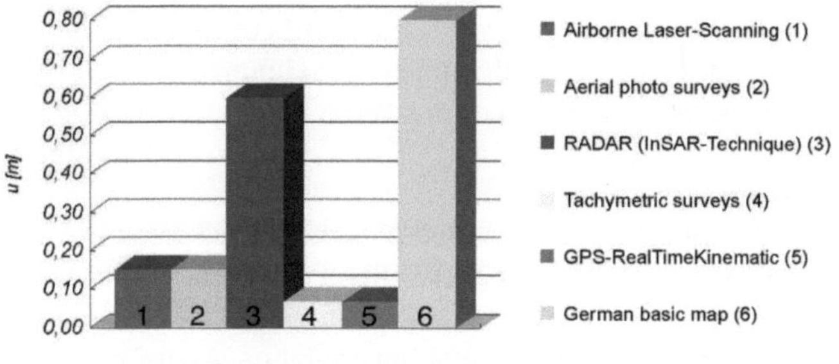

Fig. 9.5 Approach-oriented altitudinal uncertainties of the terrain surface (Brockmann, 1999)

It should be pointed out that terrain structures can also be very precisely replicated by appropriate choice of measuring points and polygons, even using only a relatively small amount of data. An appraisal and selection of the topographic information also occurs during the photogrammetric analysis of stereo-aerial pictures. In comparison to terrain investigations, reduced accuracy is in opposition to the efficiency of large-scale data collection. The laser-scanner-technique provides altitudinal information of the earth's surface in grid form with user-defined resolution. The accuracy of the individual points is primarily dependent on aircraft altitude, vegetation density and terrain gradients. Information about terrain surface, the plant coverage situation, and structural objects (e.g. buildings) can be extracted from the raw data (Vögtle and Steinle, 2001).

The plausibility and, where necessary, post-processing of the topographical data base (error detection, interpolation of data gaps, definition of breaking edge, optimisation of amount of available data, etc.) constitute an essential and often time-consuming component of model building. Often multiple data sources have to be combined to capture entire (flood-) relevant geometries.

9.1.4.2 Water Level Information and Flood Boundaries

Measured flow parameters are necessary to calibrate a HN-Model. Calibration of models for flowing waters is done by comparing calculated and measured water surface levels and corresponding model adaptation (generally by modification of surface roughness). Measured water levels, including those spreading laterally from the river axis, are needed to calibrate multidimensional HN-models, especially in areas with complex hydraulic conditions. Normally only scattered water surface measurements are available. These are, however, additionally afflicted with great insecurities, especially in regard to flood data. These measurements obtain spatial bearings through the connection to the corresponding river kilometre, respectively, the water positioning, and possibly, the respective banks. The allocation of the water level perpendicular to the flow axis line, particularly with wide flood plains and branched river systems, rarely allows a clear determination of its exact location. A data file of 3D-space coordinates, with specification of the precise time and a reference to the amount of discharge, should be aimed for in the future.

Extensive recording of the peak water levels, especially in large areas of study, poses an extraordinary challenge. Particularly difficult is the temporal and spatial coordination of an investigation at the time of maximum discharge. When formulating a realistic accuracy requirement of measured flood levels to calibrate the model, aspects such as swell, wind surge, and the uncertainty of discharge classification should be considered.

Due to the difficulties to record boundaries of flooded areas in a comprehensive way, during the passage of a flood wave, flood boundaries in maps are generally based on recordings of floating debris and statements of residents. These sources of information are often afflicted with great uncertainties. Additionally, selected information in maps is often connected to make a line, without the original data being marked accordingly. The uncertainties of flood measurements and their effects on the interpretation of simulation results should always be considered.

9.1.5 Investigation

In all practical applications of numeric models, the key activity is not primarily computation, but the preceding current analysis, selection of the model and data processing, as well as the subsequent analysis, presentation, and interpretation of the results. This procedure can be divided into three main components: pre-processing, processing, and post processing (see Fig. 9.6).

Pre-processing comprises the exact definition of the problem, analysis and plausibility of the available data and existent flow characteristics. Depending on the anticipated flow conditions, an appropriate procedure has to be chosen. Following this, specification of the model area, preparation of the terrain topography, compilation of the computational mesh and determination of the spatial distribution of roughness characteristics needed for quantification of energy losses along the flow course take place. Finally, boundary and initial conditions for the computations must be defined. In this process, discharges, and/or water levels at the boundaries of the model area, as well as initial water levels must be specified at the beginning of simulations. Subsequently the processing takes place, for the most part, without any direct user-influence. In these computations the discrete basic equations are solved. The computed results ("a large amount of numbers") are subsequently visualised, analysed, and reviewed. This procedure is called post-processing.

Pre-processing, processing, and post-processing are generally done repeatedly in the course of numeric modelling. This is the case, for example, in sensitivity analyses, the calibration of the model, and, depending on the task, in further computations of desired hydrological or geometric variations.

Geo-information systems (GIS) are often applied in these pre- and post-processing parts. GIS is an ideal tool for these tasks, due to its capabilities of geocoded superimposition and intersection of all forms of digital geo-data.

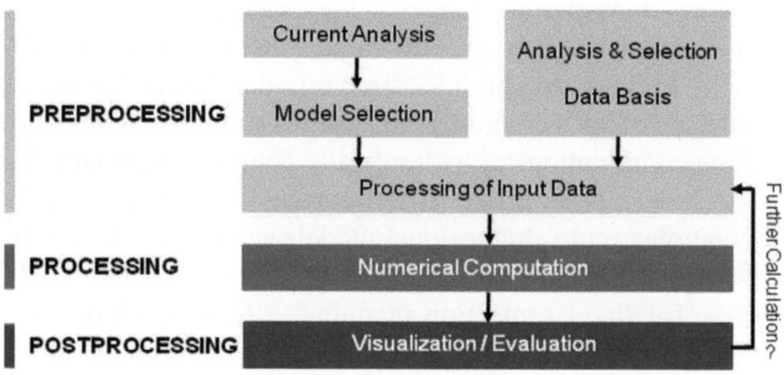

Fig. 9.6 Flowchart HN-modelling

9.2 Flood Management Models

The requirements of flood risk management on hydraulic modelling are manifold. In the planning process resilient predictions of anticipated flow conditions in the form of high resolution flood images and/or distributions flow velocity are required, for example to identify flood endangered areas or to develop an emergency management plan. On the other hand fast and robust approaches are required for operational risk management. Here real-time application during a flood event, for example to identify potential failure sites of flood protection facilities within adequate time, or to forecast flooding of important elements of infrastructure, is needed. A variety of requirements also occurs due to different characteristics of the modelled area. Differences can be made between:

- investigated areas with distinct flow characteristics which can be described hydraulically relatively easy and
- areas with complex hydraulic characteristics.

A hydraulically well defined river-flood plain system usually has a clear direction of the main flow, even during a flood, which allows the application of fast and quite efficient 1D-HN-models. Water levels calculated with this model are generally transferred orthogonally to the river axis into the flood plains. Use of correction factors may be necessary in greater curvatures or flood protected areas. These 1D-models have low calculation times that allow easy application during the planning stage as well as for operational purposes during a flood event.

The selection of model becomes more difficult if branching, islands, or structures, such as bridges, weirs, or culverts, exist within the examined river reach. Before a model can be set up an experienced hydraulics specialist should characterise the anticipated current in detail. This analysis can be used to decide on the one hand if multi-dimensional flow effects can be neglected or at least described by abstractions, and on the other hand if the anticipated main flow paths will proceed equally across the entire spectrum of relevant discharges. Doing so complex 1D-HN-models with integrated branching, island-bypassing, and energy losses at local structures can be perfectly acceptable and may often lead to desired results, even in quite complex areas. However if changes in the course of the main stream path (e.g. short-cutting a meander-loop) occur with rising water levels, such a 1D-model is no longer acceptable. In this case multiple, discharge-dependent 1D-models have to be applied. Considering today's computer capabilities and innovative numeric methods to accelerate flow computations of hydraulically complex areas such 1-dimensional modelling is no longer appropriate. If the current characteristics of a study area are obviously complex multi-dimensional models should be applied. In such cases, the modelling requirements must, again, be predetermined. If detailed modeling is necessary, e.g. for flood protection planning, a high resolution, complete 2D-HN-model should be chosen which is able to solve most of the related hydraulic problems adequately with sufficient accuracy. If the necessity arises to apply the model in operational mode during or shortly before a flood event, alternative methods, e.g. a 1D-2D-linked model approach should be chosen, especially in large study areas.

In addition to these purely technical criteria of model choice, there are often individual requirements defined by users that have to be considered. In total an increasing demand for utilisation of supposedly more accurate multi-dimensional models can be expected.

In the following, the different, site-specific requirements on hydraulic modelling in flood management are demonstrated at the examples of two contrasting model areas.

9.2.1 Case Study of a Region with Well Defined Flow Characteristics

The Neckar River has a catchment area of approximately 14,000 km^2. It is located in southern Germany and can be characterised mainly by well-defined current characteristics especially, when viewed on a large scale. In 1968, the expansion of the Neckar passage from Plochingen to Mannheim to a cascade of 27 impoundments was completed. This section of 203 river kilometres is used as a federal waterway and power plant chain. It bridges a gap in elevation of approximately 160 m, with an average inclination of 0.8 per mill before reaching the river mouth. In comparison to other German rivers with similar catchment areas, the passage of flood waves shows a markedly steep and one-dimensional characteristic. This is caused by low permabilities of soils, steep topography of large parts of the catchment and particularly highly populated flood plains which can not be used for flood retention in the Neckar valley. The one-dimensional characterisation of the current is acceptable here and enables highly efficient modelling with regard to data handling, the set-up of the model and calibration. It gives plausible calculation results and can be used easily for sensitivity analyses or variation studies (Oberle, 2004). Figure 9.7 shows an overview of the river stretch and the chosen modelling approaches.

The active discharge areas of the river channel and the flood plains were described as cross section profiles to create the systems geometry of the HN-model. For each impoundment a 1D-model was created first. Each of these models is capable of running independently from models of adjacent impoundments. In a further step, the individual models were linked via structural geometry of the weirs. Calibration of the individual models was based on the comparison of longitudinal profiles of the water surface with measured maximum water levels of past flood events. In most cases, water surface measurements from multiple flood events were available so calibration and validation of a large discharge spectrum was possible.

The retention effective areas of the Neckar valley were taken into account via a storage capacity function, depending on the calculated water levels. This was determined with the aid of a range of GIS-functions from the digital terrain model (DTM) and was made plausible by comparing calculated and measured flood plain hydrographs. Variation in flood retention volume does not affect the results of steady-state computation ($Q(t) = \text{const.}$). However, one can identify effects on the development

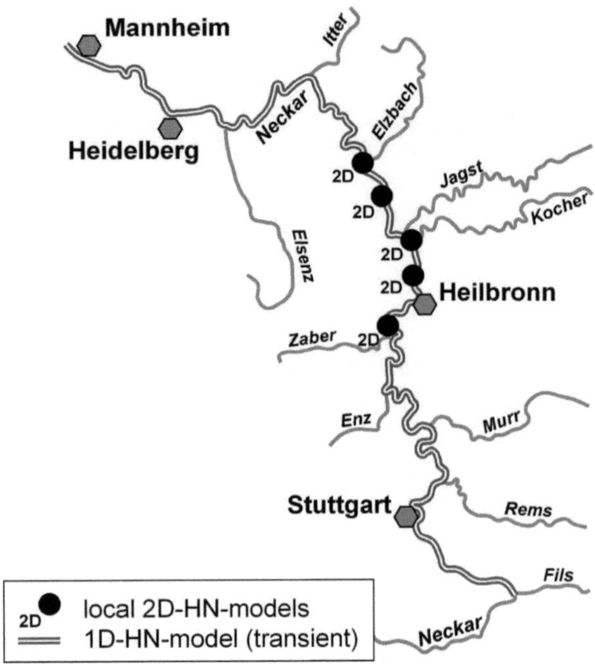

Fig. 9.7 Overview of the comprehensive model Neckar

of the discharge hydrograph based on a transient calculation. A more peaked flood hydrograph occurs if the retention volume is reduced. In this case the peak value increases, and the wave flows downstream faster.

To evaluate the informational value of the one-dimensional approach with regard to the computed water levels and flood plains, especially in areas with distinct river bends comprehensive sensitivity analyses and comparisons with the results of a two-dimensional flow model were implemented. This confirmed the model selection. Regarding operation of the hydrodynamic-numeric model in distinct flow areas, the following basic principles have to be noted:

- Steady-state modelling can be applied to calculate water levels dependent on the discharge (e.g. to determine design water levels, investigation of flood plains). Steady-state model application also allows quantification of backwater effects caused by construction in the discharge area of a flood by variation of profile geometries.
- A transient model should be applied primarily to quantify the effects of constructions on flood retention and to analyse varieties which result from different inflow conditions. It provides the deceleration or acceleration and the attenuation or peak of the flood wave (discharge- and water level differences) as results.

In a few areas with very complex current situations, which were insufficiently recorded by the one-dimensional model, the additional construction of local two-dimensional HN-models was required. The hydraulic situation around the mouth of the Kocher (103 km), near Kochendorf, and in the area of the Neckar bend near Heilbronn (108 km), changes with varying discharge amounts or discharge constellations so fundamentally that multiple 1D-HN-models must be operated parallel to adequately simulate the local current phenomena across the complete flood spectrum. Here, the results of two-dimensional calculation supplied important findings about relevant flow processes. This, in turn, made better interpretation of the 1D-results possible.

Speed and reliability of the applied approaches in connection with predicted flood levels make it possible to use the model in operational mode to support decisions during flood events. However, operational, real-time modifications of the models (variation studies of floods in dike areas, weir control, etc.) are not necessary due to the limited retention volumes in the Neckar valley and the low options for interventions during a flood event. For use in a flood event, it is sufficient to calculate various discharge constellations in advance and to provide hazard maps with reference to water levels of different scenarios to the riparian communities or to disaster control.

9.2.2 Case Study of a Region with Complex Flow Characteristics

9.2.2.1 Characterisation of the Investigated Area

An approximately 55 km long river section of the middle Elbe River with a total watershed of 140 km^2 is used here as an exemplary study area. For this reach, special requirements on applied hydraulic modelling approaches exist due to the low decline of approximately 0.2 per mill, the expansive flood plains, the meandering water passage and the mouth of the tributary Mulde River, split into several branches during a flood event. Figure 9.8 shows an overview of the investigated area.

With rising channel flow, local overflow initially appears. At high discharges, almost the entire length of the embankment is overflowed, and a large proportion (up to 50%) of the total runoff shifts into the flood plains so that it is no longer possible to distinguish between flows within the river channel and on the flood plains. In

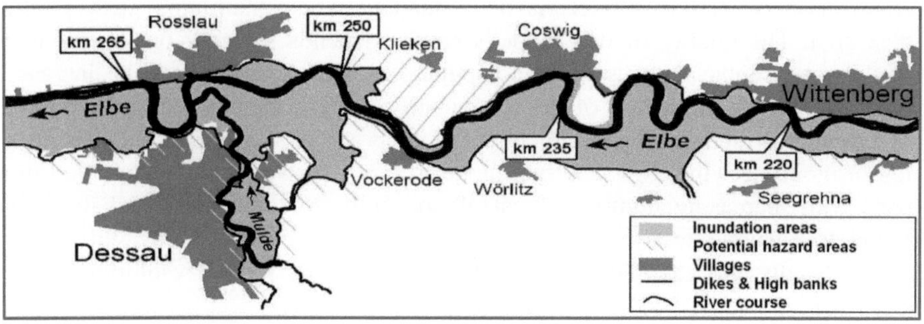

Fig. 9.8 Model area with complex current characteristics

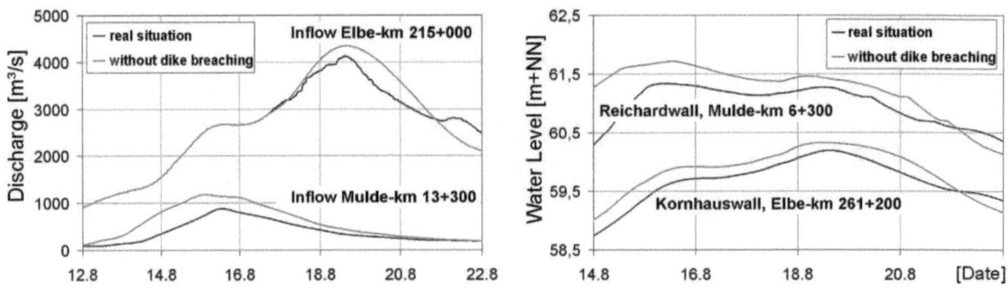

Fig. 9.9 Hydrographs of the calculation of a scenario, with and without dike breaching

extreme events there is an additional danger of dike overflows or breaches, whereby large retention volumes would be activated.

Due to numerous possible scenarios, particularly in extreme flood events, prior analyses cannot lead to the desired answers to many questions. Thus models are required which can be applied for resilient and rapid flow computations to provide operational risk analyses. These computations should take place during an event and be based on recent hydrological and topographic constraints.

This can be made clear by means of an analysis of the flood event of the year 2002. Multiple dike breaches (more than 100) occurred upstream of the study area. These were not predictable in this accumulation. Figure 9.9 shows results of an investigation of the effects of dike breaching on water levels occurring during the flood of 2002.

Here the theoretical current situations without dike breaching in the upper reach of the Elbe and Mulde rivers were calculated. The larger maximums in the inflow hydrograph and the calculated flood hydrograph can clearly be seen, exemplarily displayed for two dike points in the area of the city of Dessau. The differences of the calculated peak water levels at the dikes amount to a maximum of approximately 0.5 m. This information would have led the local catastrophe managers to significantly alter their prioritisation of actions.

9.2.2.2 Modelling Techniques

Application of 2D-HN-models is now standard when examining such a complex area. The quality of this model's results depends strongly on the accuracy of the topographic data used for computations. This accuracy can be increased through finer discretisation of the computational mesh by increasing the number of calculation cells in hydraulic relevant areas up to the point of exact imaging of the underlying digital terrain model (DTM). However, the calculation time rises with increasing number of calculation cells. So compromises have to be found to ensure effective processing. Figure 9.10 shows the sensitivity of the model results and calculation times dependent on the model resolution for the study area.

It becomes obvious that model failure is significantly increased through insufficient topography representation due to coarse resolution of calculation meshes.

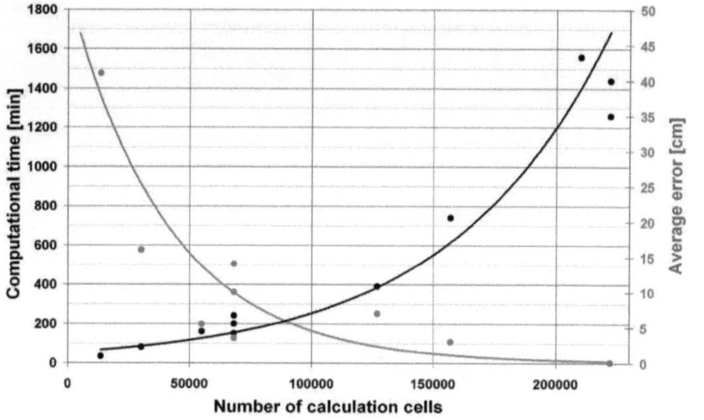

Fig. 9.10 Dependencies of calculation length, number of support points, and model errors

Even in operational applications with reduced accuracy requirements this is not acceptable at calculation lengths shorter than 30–60 min.

Although 1D-computations are very reliable and efficient for hydraulic well-defined areas these approaches do not lead to the desired results at rivers with expansive flood plains like the Elbe. Here, depending on the discharge, strongly varying flow conditions, especially on the flood plains, occur. These cannot be reliably recorded across the whole discharge spectrum, even using an elaborate meshed 1D-HN-model (Oberle et al., 2008).

Therefore, the use of today's widespread HN-models only partly fulfils the contrary requirements of operational application in the form of the reliability of computed current characteristics and short computational time for complex hydraulic conditions. Due to this fact models or model-links are increasingly developed to accelerate 2D-computation using approved simplifications. In particular the linking of 1D- and 2D-models is an effective approach (q.v. Section 1.3). Figure 9.11 compares the key data of such a coupled 1D-2D-approach and various detailed 2D-HN-models for the region of the middle Elbe. It becomes evident that, while through optimisation of the 2D-HN-model impressive increases in efficiency have been achieved, computational time reductions in appropriate magnitudes necessary for operational application in the described area were only possible with the 1D-2D-linked model approach.

Lifelike simulation of river channel (1D-area) overflowing into the flood plains (2D-area) is a very important quality factor of such models. Figure 9.12 shows examples of a model comparison, using velocity vectors, of an overflowing of a meander loop in the region of the City of Coswig during the flood of 08/2002. There is good compliance regarding the occurring flow paths. Velocities in the flood plains are minimally underestimated by the linked model (right) but this does not have a significant effect on predictability in operational application.

The fact that momentum exchange and internal friction forces at the shear zones of the interfaces of both models are not modelled is a disadvantage, compared to

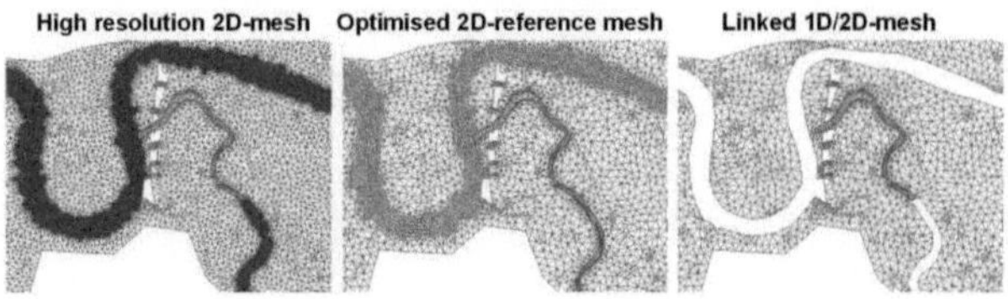

2D-cells	222.000	68.000	27.7000
1D-profiles	-	-	715
Δt flood1995	0,55 sec	0,9 sec	1,5 sec
Δt flood 8/2002	0,45 sec	0,75 sec	1,1 sec
Calculation time flood 1995	16 h	3,5 h	0,5 h
Calculation time flood 8/2002	25 h	5,5 h	1,5 h

Fig. 9.11 Key data of the investigated meshes, based on a simulation duration of 80 h (calculation on PC with Intel® Core^{TM} 2 CPU 2.4 GHz and 2 GB RAM)

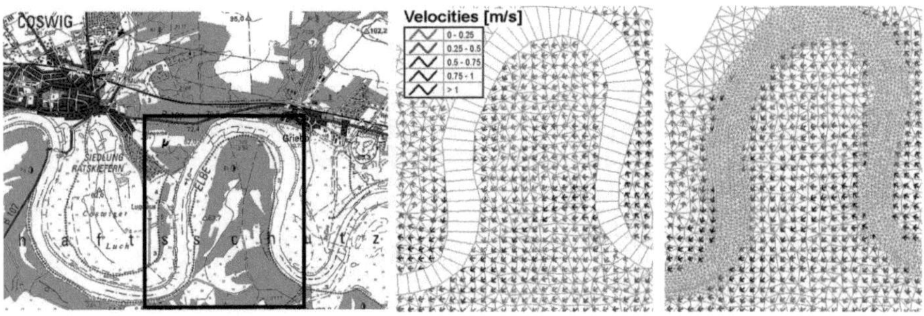

Fig. 9.12 Comparison of flow vectors in the area of a meander loop

a complete 2D-HN-model. Instead, the linked model has two additional adjustable characteristics which can possibly decisively influence calculation results:

- Location of the virtual overflow crest and
- Distance between the cross section of the 1D-area to the edge length of the bordering 2D-mesh cells.

If the modeller has adequate experience, the resulting uncertainties are small with regard to the target sizes in the flood plains or along dikes. This is especially the case in the context of a holistic evaluation of the existent model insecurities. An extensive pre-analysis of the investigated area, with a clear specification of objectives of the simulation is a precondition for application of a linked model (Musall et al., 2007).

However, to model complex flow areas without the need for operational model use or with demand for exact results, for example to compute parcel-exact inundation areas for flood risk mapping, complete 2D-HN-models should be applied to keep modelling uncertainties as low as possible.

9.3 GIS-Based User Interface

The implementation of a user-interface is a requirement for operational application of a developed and validated hydraulic simulation model. This is especially important in a flood situation in which short calculation times, intuitive operability of calculation controls (e.g. hydrological constraints, possible dike breaches or blockages) and visualisation of the results are essential. An implementation of the validated hydraulic model in a GIS-based tool, specifically designed for disaster control, therefore is very useful. Future users and decision makers from the disaster control agencies should be incorporated in development and design. Using such a personalised and practice-oriented application module, even users without consolidated hydraulic background could do HN-calculations and immediately visualise their results on PC in the GIS. These results can be superimposed with other geo-referenced data (e.g. topographical maps or aerial images) afterwards. Figure 9.13 shows a schematic diagram of operational application of such a tool.

Such a system can contain, among other things, the following functionalities (Musall et al., 2009):

- Execution of hydraulic models
- Visualisation of calculation results
- Automated freeboard analysis along dikes
- Automated inundation analysis of relevant structural facilities (e.g. dike gates)
- Superimposition with other flood relevant information
- Risk analysis of protected areas

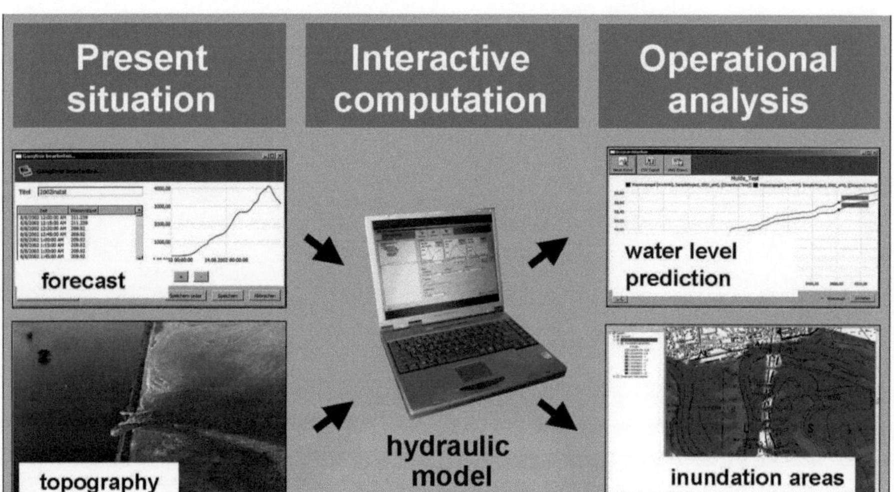

Fig. 9.13 Schematic diagram of application of an operational HN-tool

A wide variety of systems of the underlying GIS-system are on offer, beginning with OpenSource, FreeGIS-Projects, to comprehensive commercial GIS-solutions. However the functional spectrum of individual systems can differ widely. A selection, considering the requirements and possibilities of the user-oriented system should be carried out. Especially interlinks with another GIS-system should be provided using appropriate modular programming tools if necessary.

In the following, individual functions of such tools are explained by examples.

9.3.1 Hydraulic Computation

The option to perform individual HN-computations, based on current hydrological and possibly also topographical constraints, is the core of every operationally applicable tool for hydraulic simulation. For this, transient hydrographs at all model boundaries should be indicated to the user in diagram form. These should be simple to edit and easy to replace. It is also advisable to integrate the chosen calculation period in the preview of these constraints. In consideration of realistic initial conditions for the computation, access to previously calculated flow conditions should be given. By this means, the necessary simulation period and with it the requirements in computational time of transient computations may noticeably be reduced.

Model-types (for complex areas, e.g. high resolution 2D or 1D/2D-linked) and the modelled surface roughness allocation (e.g. different conditions for summer and winter) as further options for selection should be provided. For flood management options to integrate dike breaches or blockages are useful. If such impacts are newly generated, this should be done menu-driven and GIS-supported. Figure 9.14 shows an example of such a graphical user interface.

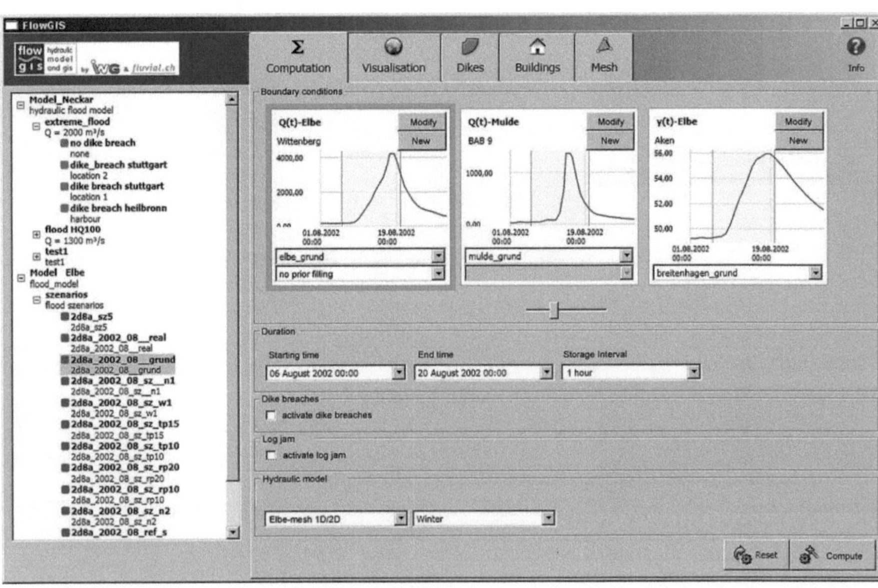

Fig. 9.14 User interface of an operational tool for hydraulic simulation

After choosing these modelling control options, the actual HN-computation should be carried out in the background, whereupon the user should be informed about the development via status display. It is not necessary for the user to become active until the desired results are to be visualised.

9.3.2 Visualisation of Results

Comprehensive visualisation of calculated flow characteristics requires both, GIS-supported two-dimensional and diagram imageries. The user should have the following results available: all essential output quantities of the HN-calculation, for example, water depths, flow velocities, or specific discharges, supplemented with other relevant data, such as the underlying original-DTM, the structure of the computation mesh, or an overview about specific features (see Fig. 9.15). The user should be able to access all data for every stored time step or view maximum values of the calculation.

GIS-output can be provided either in the individually chosen scanning resolution or mesh dependent, that means the exact calculated values are presented at the individual mesh points. For the diagram display, essential cross-sectional profiles (e.g. longitudinal cut) and gauging sites should be stored within the system; any further values should be producible via control buttons in a graphical user interface.

The automatic compilation of grided values of water depths by intersecting the calculated water levels with elevation data from the original Digital Terrain Model (DTM) is useful, as it applies all available terrain data for the most realistic representation of results. An additional alternative ASCII-export of data (∗.asc- or ∗.xyz-format for grid data; ∗.csv- or ∗.xml-format for diagram data) enables a convenient re-utilisation of the calculation results, in other software systems.

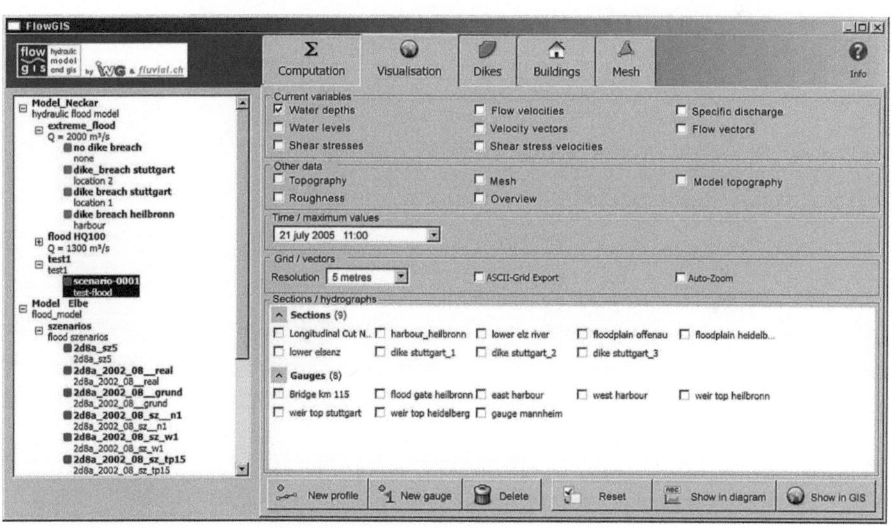

Fig. 9.15 User interface of a GIS-application module for visualisation of hydraulic calculation

9.3.3 *Specific Flood Analysis Tools*

Specifically for the requirements of flood risk management and operational application special adaptive functions are necessary. During this process, complex functions should be controlled by simple user input and run automatically in the background. Possible areas of application are shown in the following:

9.3.3.1 Freeboard Analyses Along Dikes

The dike heights of the entire analysed area should be deposited in the system. It should be as simple as possible to update these data in case of dike reconstructions. On demand an operational risk report showing the differences between computed water levels along the dikes and the elevation of dikes for all dike sections stored within the GIS should be generated automatically. This enables users to get a quick overview on the whole scenario. Alternatively also the maximum values of calculation should be provided. In the table of GIS-vector files, all further information, for example, dike heights, water levels, times, and maximum distances between water levels and dikes can be stored and provided to users during their analyses.

9.3.3.2 Hazard Analysis of Buildings

Attributes of structural elements that have been stored in the system can be compared with calculated water levels. The results can be provided in form of hydrographs in adapted diagrams or as two-dimensional hazard maps in GIS. All essential information concerning time and intensity of potential floods can be stored here in the GIS-dataset (see Fig. 9.16). This constitutes important information for the scheduling of risk control operations (e.g. in connection with the activation of mobile task forces).

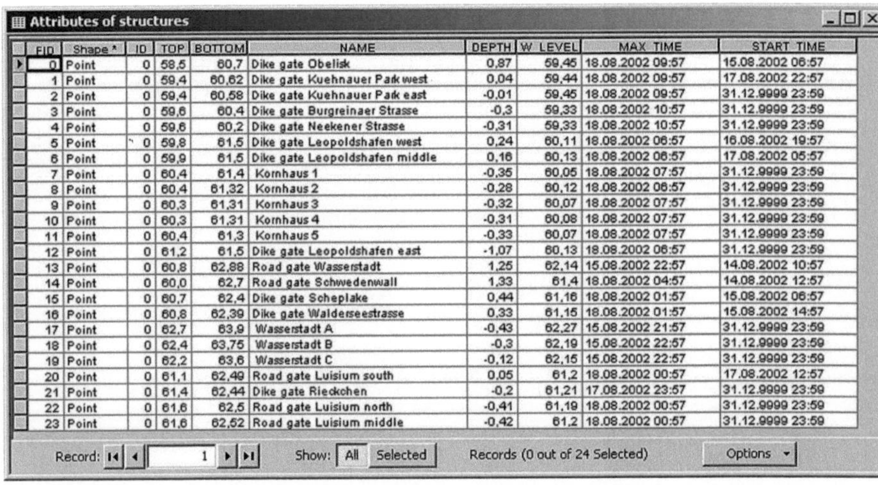
Attributes of structures

FID	Shape *	ID	TOP	BOTTOM	NAME	DEPTH	W LEVEL	MAX TIME	START TIME
0	Point	0	58,5	60,7	Dike gate Obelisk	0,87	59,45	18.08.2002 09:57	15.08.2002 06:57
1	Point	0	59,4	60,62	Dike gate Kuehnauer Park west	0,04	59,44	18.08.2002 09:57	17.08.2002 22:57
2	Point	0	59,4	60,58	Dike gate Kuehnauer Park east	-0,01	59,45	18.08.2002 09:57	31.12.9999 23:59
3	Point	0	59,6	60,4	Dike gate Burgreinaer Strasse	-0,3	59,33	18.08.2002 10:57	31.12.9999 23:59
4	Point	0	59,6	60,2	Dike gate Neekener Strasse	-0,31	59,33	18.08.2002 10:57	31.12.9999 23:59
5	Point	0	59,8	61,5	Dike gate Leopoldshafen west	0,24	60,11	18.08.2002 06:57	16.08.2002 19:57
6	Point	0	59,9	61,5	Dike gate Leopoldshafen middle	0,16	60,13	18.08.2002 06:57	17.08.2002 05:57
7	Point	0	60,4	61,4	Kornhaus 1	-0,35	60,05	18.08.2002 07:57	31.12.9999 23:59
8	Point	0	60,4	61,32	Kornhaus 2	-0,28	60,12	18.08.2002 06:57	31.12.9999 23:59
9	Point	0	60,3	61,31	Kornhaus 3	-0,32	60,07	18.08.2002 07:57	31.12.9999 23:59
10	Point	0	60,3	61,31	Kornhaus 4	-0,31	60,08	18.08.2002 07:57	31.12.9999 23:59
11	Point	0	60,4	61,3	Kornhaus 5	-0,33	60,07	18.08.2002 07:57	31.12.9999 23:59
12	Point	0	61,2	61,5	Dike gate Leopoldshafen east	-1,07	60,13	18.08.2002 06:57	31.12.9999 23:59
13	Point	0	60,8	62,88	Road gate Wasserstadt	1,25	62,14	15.08.2002 22:57	14.08.2002 10:57
14	Point	0	60,0	62,7	Road gate Schwedenwall	1,33	61,4	18.08.2002 04:57	14.08.2002 12:57
15	Point	0	60,7	62,4	Dike gate Scheplake	0,44	61,16	18.08.2002 01:57	15.08.2002 06:57
16	Point	0	60,8	62,39	Dike gate Walderseestrasse	0,33	61,15	18.08.2002 01:57	15.08.2002 14:57
17	Point	0	62,7	63,9	Wasserstadt A	-0,43	62,27	15.08.2002 21:57	31.12.9999 23:59
18	Point	0	62,4	63,75	Wasserstadt B	-0,3	62,19	15.08.2002 22:57	31.12.9999 23:59
19	Point	0	62,2	63,6	Wasserstadt C	-0,12	62,15	15.08.2002 22:57	31.12.9999 23:59
20	Point	0	61,1	62,49	Road gate Luisium south	0,05	61,2	18.08.2002 00:57	17.08.2002 12:57
21	Point	0	61,4	62,44	Dike gate Rieckchen	-0,2	61,21	17.08.2002 23:57	31.12.9999 23:59
22	Point	0	61,6	62,5	Road gate Luisium north	-0,41	61,19	18.08.2002 00:57	31.12.9999 23:59
23	Point	0	61,6	62,52	Road gate Luisium middle	-0,42	61,2	18.08.2002 00:57	31.12.9999 23:59

Record: 1 Show: All Selected Records (0 out of 24 Selected) Options

Fig. 9.16 Flood information in the GIS-Shapetable

9.3.3.3 Intervention in Model Topography

For the realistic assessment of the current hazard situation and for the preparation of emergency planning, it can be necessary to modify model topography, both anticipatory as well as adapted to a specific situation. Modern tools for such analyses offer options to interactively generate dike breaches or blockages in the model area. The user has options to define menu-driven hazard areas, while automatically receiving relevant information (e.g. dike grade, dike toe heights) from the data base, and to visually review possible actions within the GIS. The subsequent analysis of scenario calculations shows possible effects on the flow situation.

9.3.3.4 Analysis of Protected Areas

By projection of calculated water levels in adjacent areas, non-flooded during computation, terrain can be identified that is indeed lower than the water surface but has no open connection to the river. Experience shows that in many cases these areas are flooded by seepage or hidden culverts. Additionally, it is often perfectly appropriate to record potentially endangered areas, too (risk mapping) (Oberle, 2004).

9.3.3.5 Superposition of Other Flood-Related Data

The GIS-internal options of imagery and superposition of calculation results with existing data can be usefully supplemented with a user-adapted interface. A menu-driven systematic access to existing geocoded flood data like for example flood debris lines or photographs of previous flood events could be useful to quickly gain an overview over potentially affected areas. Figure 9.17 shows an example of superposition using a GIS-workstation of the water management administration Baden-Wuerttemberg for the Neckar valley.

9.4 Summary

Application of hydraulic model tools to simulate flow characteristics in the investigated area has meanwhile become an indispensable component of modern flood management. For cross-disciplinary evaluation, optimisation, and implementation of sustainable risk management concepts, there is additional need for tools that allow decision-makers to predict flood development and effects, depending on the specific condition of the river valley of interest. Essential requirements are:

- Assessment of the river and the flood plains as a complete system
- Possibility to simulate any discharge scenario
- Realistic visualisation of flood risk
- Identification of endangered areas
- Simple operation and robust performance

Modularly structured GIS-application modules are of high practical importance for water management. These consolidate the model-, data-, and dialogue

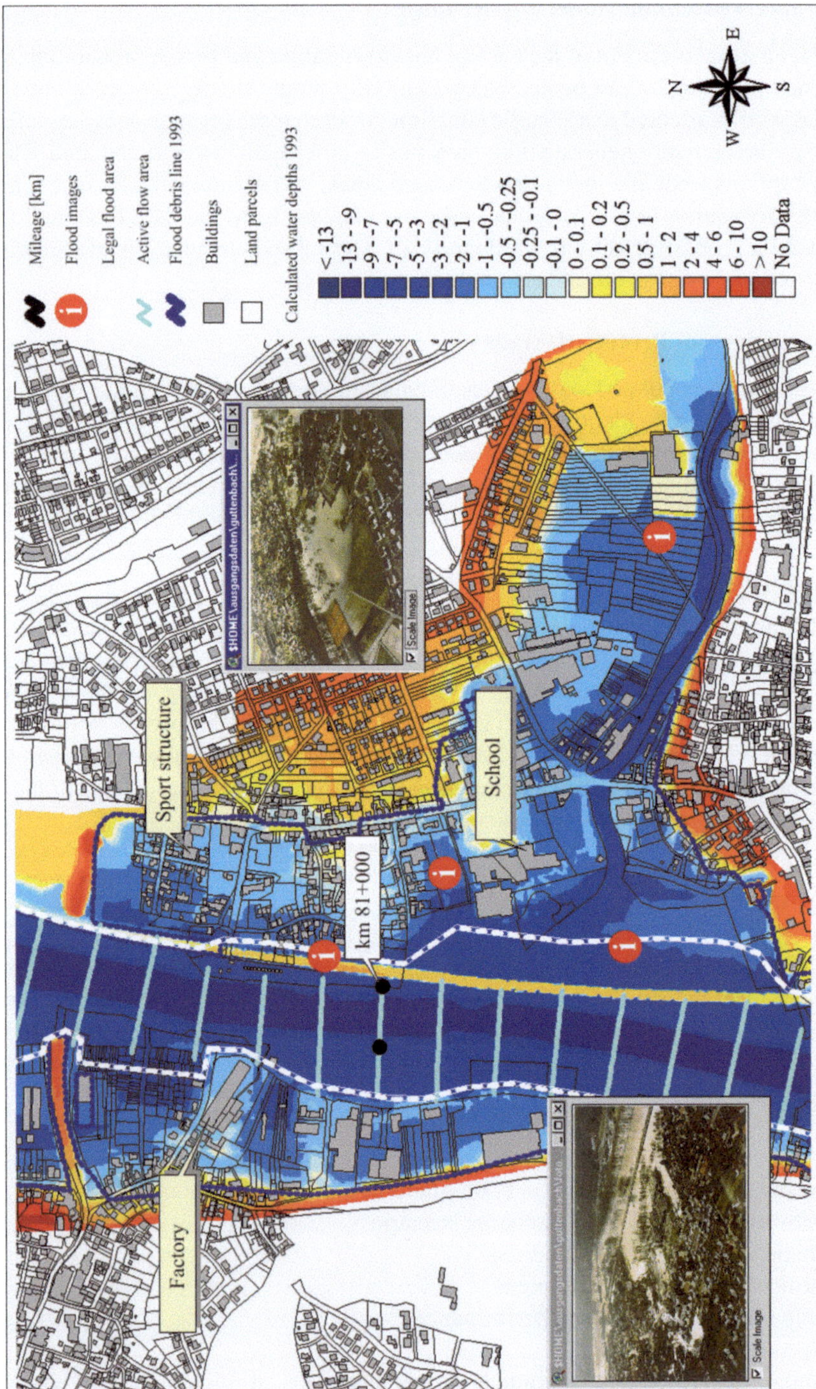

Fig. 9.17 Example of superposition of flood relevant data in the GIS, of the river Neckar

components of the flood simulation system and make it possible for the user to conduct complex operational sequences, using automated, logically structured user-interfaces.

The presented case studies for the Neckar and the Elbe rivers show examples of such application modules. They are distinguished by their approach with regard to the respective current characteristics and can be transferred to hydraulically comparable regions.

References

Beffa C (1994) Praktische Lösung der tiefengemittelten Flachwassergleichungen. Mitteilung 133 der Versuchsanstalt für Wasserbau, Hydrologie und Glaziologie. ETH Zürich

Beffa C (2002) Integration ein- und zweidimensionaler Modelle zur hydrodynamischen Simulation von Gewässersystemen. In: "Moderne Methoden und Konzepte im Wasserbau". Mitteilung 174 der Versuchsanstalt für Wasserbau, Hydrologie und Glaziologie. ETH Zürich

Brockmann H (1999) Einsatz flugzeuggestützter Fernerkundungstechniken zur Bearbeitung hydrologischer Fragestellungen, in: Beiträge und Unterlagen zum Workshop Ermittlung von Überflutungsflächen an Fließgewässern, Karlsruhe

Cunge JA, Holly FM, Verwey A (1980) Practical aspects of computational river hydraulics, Pitman advanced publishing program, Londra

DVWK (1999) Numerische Modelle von Flüssen, Seen und Küstengewässern. Schriften Nr. 127

Hager WH (1986) Discharge measurement structures. Communication 1, Chaire de Construction Hydrauliques. EPFL Lausanne

Musall M, Kron A, Oberle P (2007) Floodplain modelling for real-time use based on an interlinked numerical 1D/2D model, http://www.cedim.de/download/62_Musall_etal.pdf, 8. DKKV-Forum Katastrophenvorsorge

Musall M, Kron A, Oberle P, Nestmann F (2009) GIS-gestütztes HN-Simulationswerkzeug für das operationelle Hochwassermanagement. In: Unterlagen zur DWA-Tagung GIS in der Wasserwirtschaft, Kassel

Oberle P (2004) Integrales Hochwasserinformationssystem Neckar – Verfahren, Werkzeuge, Anwendungen und Übertragung, Mitteilung 226 des Instituts für Wasser und Gewässerentwicklung, Karlsruhe

Oberle P, Musall M, Kron A, Nestmann F (2008) 1D/2D-gekoppelte Simultion von Hochwasserabflüssen an der mittleren Elbe, Wasserwirtschaft 7-8/2008, S. 46–50

Vögtle T, Steinle E (2001) Erfahrungen mit Laser-Scanner-Daten zur automatisierten 3D-Modellierung von Gebäuden. In: Deutscher Verein für Vermessungswesen (DVW), Landesverein Baden-Württemberg, Mitteilung H. 2, 48 Jg

Chapter 10
Groundwater – The Subterranean Part of Flood Risk

Thomas Sommer

Abstract Floods cause significant damages not only on, but also beneath the earth's surface. Infiltration of surface water into deeper soil, flooding of the urban sewer system and, in consequence, rising groundwater levels are main causes of subsurface damages. There are various reasons for high groundwater levels during and after flood events. Two different processes can be specified. The direct infiltration of surface water into the aquifer during a flood is the most important process. However the inflow of surface runoff into sewer systems and subsequently from the sewer system into the groundwater aquifer is another, more indirect cause for groundwater rise. Here a coupled modelling of flood scenarios with recently developed software was applied, which combines individual modules under consideration of different model geometries, time synchronization and data exchanges. The coupled model was applied for the City of Dresden (Germany). It allows a comprehensive description of the impact of floods on groundwater. As a result of this study it became obvious that surface flooding is the dominating process for flood damages in the study region. Nevertheless are risk assessments for rising groundwater levels essential for subterranean infrastructure and buildings. Maps of subsurface flood hazard are helpful for urban planning in flood endangered regions. Measures to mitigate damages can be applied gradually accordingly to the state of risk.

Contents

T. Sommer (✉)
Dresden Groundwater Research Centre, Dresden, Germany
e-mail: tsommer@dgfz.de

A.H. Schumann (ed.), *Flood Risk Assessment and Management*,
DOI 10.1007/978-90-481-9917-4_10,

10.1 Introduction

Flood events cause significant damages in urban areas due to great accumulated assets in these regions. Flood risk assessment and flood risk management became important issues of research in the last decade (Schanze, 2007; Merz, 2006; Kreibich et al., 2006). However, the main focus of flood risk research is directed towards damages and losses caused by flash floods, coastal floods and river floods.

Subsurface water can result in another category of damages. Losses caused by high groundwater levels are seldom considered separately in loss assessment studies (Kreibich et al., 2009). During the Elbe flood in August 2002 (which was the highest within the last 100 years), 16% of all damages on premises within the Federal State of Saxony were caused by groundwater (Table 10.1). Groundwater and water which was transferred through the sewerage system caused 45% of all damages. Therefore, rising groundwater is one of the factors that must be considered to account for all losses.

The reasons for high groundwater levels during and after flood events can be various, yet three main causes can be specified. The direct infiltration of surface water into the aquifer during a flood is the most important process. The inflow of surface water into sewer systems and an outflow from the sewer system into the groundwater can be another indirect cause for rising groundwater. Additionally, increased percolation and inflow from the groundwater recharge areas after concurrent rainfall events is a third way for rising groundwater levels.

Fast rising groundwater level results often in instability of buildings by buoyancy effects (Beyer, 2003), increased infiltration of groundwater into the sewerage (Karpf and Krebs, 2004), a recontamination of soils and sediments by polluted groundwater or a remobilization of pollutants and endangering of drinking water (Marre et al., 2005).

Table 10.1 Causes of damages by the flood event August 2002 on premises of the Free State of Saxony (Germany) (Huber et al., 2003)

Cause	Percentage of total loss (%)
Water through sewerage	29
Overland flow	26
Precipitation and soil water	21
Groundwater	16
Rooftop water (precipitation)	8

In this chapter an overview about processes and impacts of rising groundwater in succession of a river flood is given. Furthermore, the article characterises possibilities for forecasts applicable for flood risk management in cases of rising groundwater. It is based on experiences of the large flood event in August 2002 in the city of Dresden (Sommer and Ullrich, 2005, BMBF, 2008)

10.2 Flood and Groundwater – Characteristics, Impacts and Parameters

10.2.1 Characteristics

The interactions between flood and groundwater are a research field for a long time (Ubell, 1987a, b). During the flood in Dresden in August 2002 it became evident that the groundwater levels of observation wells lying further away from the flooded river rise later than in observation wells near the river. In transects across the Elbe river valley three different types of groundwater dynamics were differentiated during the flood event (Fig. 10.1). Near by the river and adjacent inundation areas the dynamic of the groundwater was dominated by the dynamics of river discharge. Here the groundwater responded directly to the development of the flood (Fig. 10.1a). In the surrounding area the groundwater level was dominated by groundwater recharge from the catchment area. Here the maximum groundwater levels were delayed up to half a year in comparison with the time of the flood peak. Moreover, the groundwater table remained on a high level for more than 1 year (Fig. 10.1b). Here the effects of groundwater recharge from the catchment were dominating. Finally also a combined type of hydrograph was observed (Fig. 10.1c). This type results from the impacts of the flooded stream in combination with the effect of the groundwater recharge.

10.2.2 Impacts

Rising groundwater levels can have manifold impacts on buildings and their subsurface structures. A first group of damages is caused directly by water and moisture.

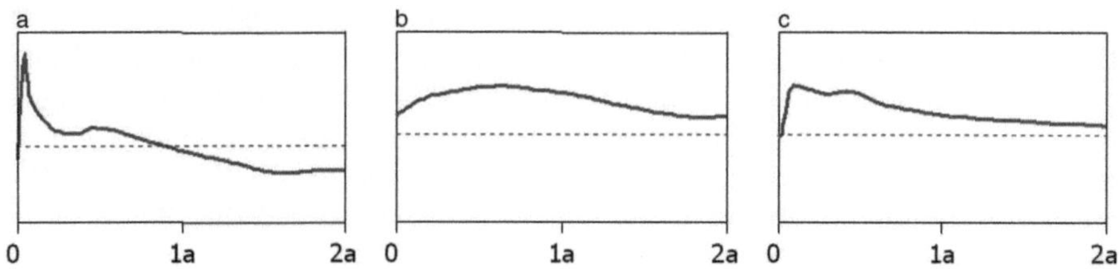

Fig. 10.1 Types of groundwater hydrographs during flood events. (**a**) dominated by receiving stream, (**b**) dominated by recharge in the catchment area, (**c**) combined type. The *dotted line* is the mean annual groundwater level

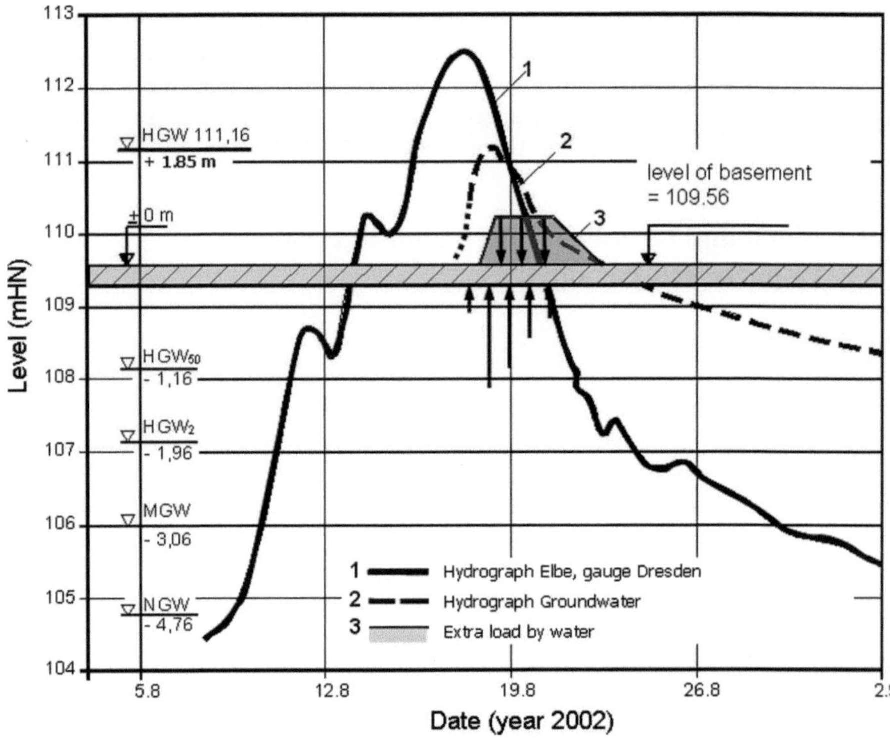

Fig. 10.2 Impact of rising groundwater on the basement of a building with low load (Beyer, 2003). The building can be secured by extra load with water filled "Quick dams"

One reason for these damages is direct groundwater intrusion through walls or floors. Another reason is flooding through leaky house service connections. The second group of damages to buildings results from changing load-bearing capacities of the subsurface caused by rising groundwater. The increase of the water content in the underground can lead to increasing buoyancy. E.g. the floor of a sports hall in Dresden was destroyed due to buoying upwards caused by rising groundwater in August 2002 (Fig. 10.2).

Furthermore, dissolutions of pollutants from contaminated soils and chemical reactions at polluted sites can have negative impacts on groundwater quality. In Dresden two main problems were identified through analyses of the groundwater quality during and after the flood in August 2002. First the groundwater chemistry was affected by infiltrated floodwater for around three months. Secondly the trend of the groundwater quality downstream of pollutant sites was not consistent. This effect was strongly dependent on the kind of the contaminated site. (Marre et al., 2005).

Also the interactions between groundwater and sewerage systems may result in impacts on subsurface building structures. If the groundwater table lays below the sewerage system the water from the sewer can exfiltrate into the groundwater. At the other side groundwater can infiltrate into the sewerage if the groundwater table rises above the top of the sewerage system. Therefore, the impact of discharge in the sewerage system on the groundwater was studied in Dresden after the flood of August 2002 in greater detail. These investigations focussed on the impact of surface flooding and the flooded sewer network on groundwater dynamics and groundwater quality (Karpf and Krebs, 2004).

10.2.3 Parameters

Flood risk depends on the flood hazard and the vulnerability of flood exposed subjects. Flood hazard characterises potential damages to property resulting from flooding. Vulnerability describes the susceptibility to damages in endangered areas, e.g. the values of buildings and inventory. Based on these definitions, flood-damage-functions are in use for risk assessment of floods (Merz, 2006). Flood hazards result from the product of intensity of the flood multiplied with the probability of the event. These definitions were developed for surface-floodings. For damages by fast increasing and long lasting groundwater levels new characteristics are needed for flood risk assessment.

A "Groundwater flood" is connected with (fast) rising groundwater and a groundwater status on a high level over a long period which is caused by a river flood. After the flood in August 2002 a meaningful methodology has been developed to characterise the potential of subsurface hazards resulting from high groundwater levels. The *minimum depth to groundwater table* is here the most important parameter for assessments of groundwater floodings. Therefore, it is necessary to observe the groundwater levels over a sufficient long time period after the flood. For this reason the observation period after the August flood 2002 in Dresden was expanded over 1 year. However, the depth to groundwater table is only a static characteristic and does not include the dynamic aspects of groundwater flow. Therefore, the potential hazard of subsurface water flow has to include the *duration* of groundwater levels above a predefined limit as well. This limit can be the mean high water level of groundwater or another fixed groundwater level e.g. 3 m below surface (Fig. 10.3).

With these two parameters a matrix of the flood hazard potential caused by groundwater was developed. This matrix was based on the matrix of damage potential which was developed in Switzerland (BWG 2001), the so- called "Swiss Model" (IKSR, 2002).

In Fig. 10.4 the parameter "duration of groundwater flood" is shown on the x- axis, the parameter "minimal depth to groundwater table" is plotted on the y-axis. Within this matrix three zones of groundwater based hazards are specified. A zone of "low hazard" assigns the area with high depths of groundwater table and short durations of high groundwater levels (yellow zone). The zone of middle hazard (coloured in blue) specifies the area with medium depths of groundwater table and medium duration of high groundwater levels. The red area shows the zone with a high hazard. This area is mainly characterised by low depths to groundwater levels.

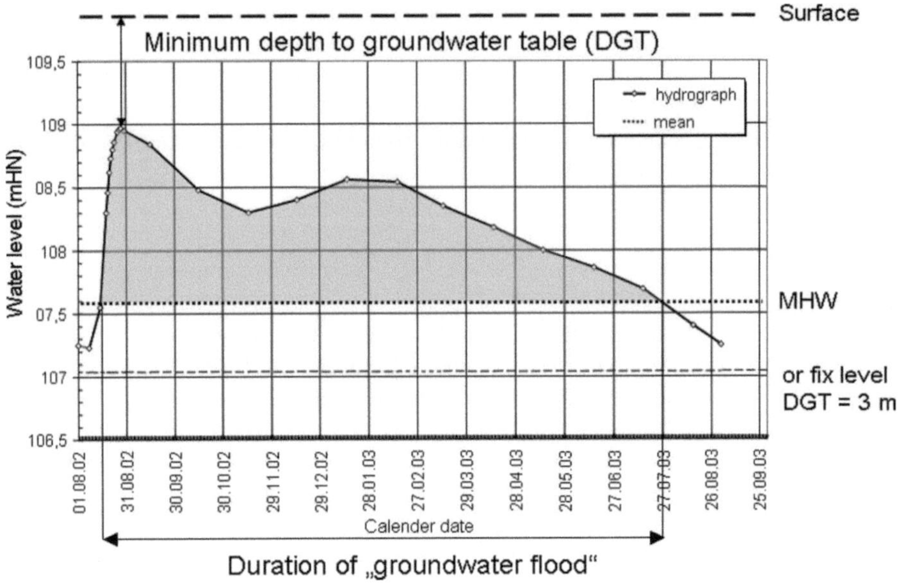

Fig. 10.3 Parameters "groundwater flood" shown at the example of a groundwater hydrograph during the flood in August 2002 in Dresden

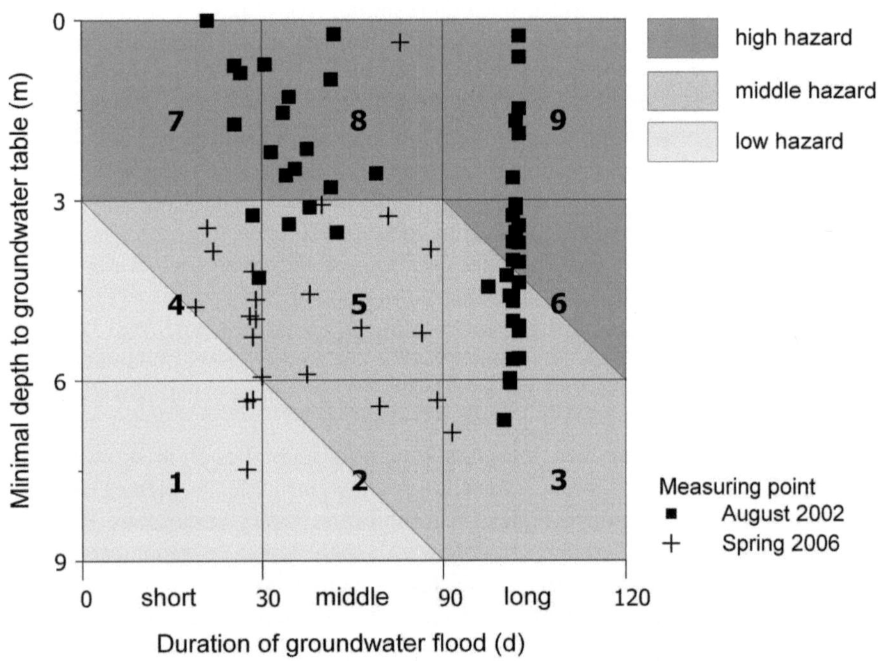

Fig. 10.4 Matrix of hazard potential (Sommer and Ullrich, 2005, updated with data of spring flood 2006)

10.3 Model Coupling

10.3.1 Coupling Concept

Forecasting of groundwater flood events with regard to the maximum groundwater level and the temporal variability of the groundwater table is not corresponding with forecasts of river floods. Multiple factors determine the groundwater dynamics during and after flood events. Therefore, any predictions of groundwater have to be based on scenarios which include the various boundary conditions. Integrated flood risk management in urban areas requires simulation of all relevant flow processes including runoff, drainage through sewerage, and groundwater flow. While individual solutions for one or at most two coupled processes exist, there is still a lack of modelling of more complex systems which integrate all three flow regimes. For this reason a coupled modelling system with three components was developed. This system includes the most important subsurface water fluxes occurring during and after flood events (Sommer et al., 2009). For the coupling previously established simulation programs for each one of the three domains were combined. These components were coupled with specific software (MpCCI software, s. http://www.mpcci.de). It manages communication, temporal synchronization and data exchange between the individual codes (Fig. 10.5). The technique of coupling is described in Sommer et al. (2009). The coupled modelling was applied for the whole area of the quaternary aquifer in the city of Dresden (s. Fig. 10.6).

The first step in coupling models consists in the identification of coupling parameters. Therefore, it is necessary to identify all relevant water fluxes between surface flooding and subsurface water. The influence of surface water on groundwater can be assessed using data from groundwater measuring wells. Close to the receiving

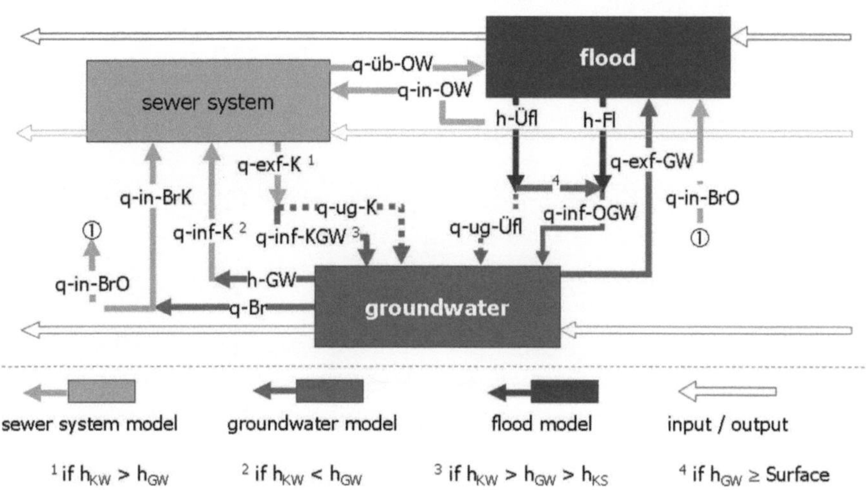

Fig. 10.5 Relevant water fluxes as coupling quantities

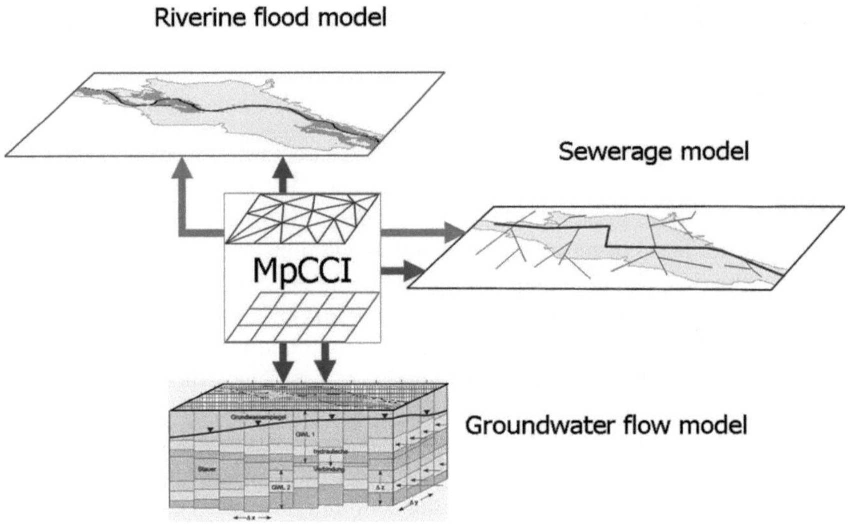

Fig. 10.6 Strategy of coupled modelling (BMBF 2008)

stream the groundwater dynamic corresponds directly with the changes of surface water levels. Here the dominating effect is the direct water flux from the surface water body into groundwater. Further away from rivers and flooding areas the groundwater rise slowly but the groundwater table remains on a high level for a considerable amount of time.

In the underground interaction between the various subsurface water fluxes has to be considered. This means either infiltration into the sewerage network caused by rising groundwater or an increase in exfiltration of sewage water into the surrounding soil due to increased water levels in the sewerage system. All in all, the coupled models have to consider a number of exchange processes between surface flood and subsurface water which are shown schematically in Fig. 10.7.

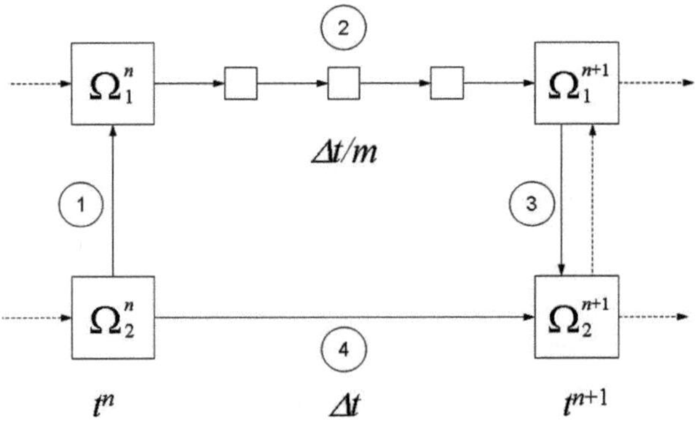

Fig. 10.7 Example of the computation flow for a 3-Code Coupling (Peetz et al., 2007)

Infiltration of groundwater and inflow of drainage and surface water into the sewerage systems form basic components of the flow in sewerage system. These inflows influence significantly the costs and the operation of drainage systems and waste water treatment plants.

10.3.2 Model Coupling

Here established simulation programs for each of the three domains (surface water, groundwater, sewer system) were coupled.

Depending on the task, two different codes were used for each of the surface flooding (RisoSurf (Ettrich, 2003), TrimR2D (Fulford, 2003)] and the sewerage systems (HAMOKA (Universtiy of Kaiserslautern), Hystem-Extran (Fuchs et al., 2004)) at different scales. Because the groundwater code (PCGEOFIM (Sames et al., 2005)) is able to calculate selected regions with a finer spatial resolution, it can be used for large scale simulations, coupling TrimR2D and Hystem-Extran. For small scale simulations, Risosurf and Hamoka were coupled.

Due to its flexible grid structure, RisoSurf is especially suited for small scale simulations and was used in local study areas. Since the sewer model HAMOKA and the RisoSurf system had already been coupled (Ettrich, 2003), HAMOKA was also chosen for modeling in the local scale.

RisoSurf and TrimR2D are based on the 2-dimensional Shallow Water Equations. Adaptive Triangles (RisoSurf) and a Cartesian grid (TrimR2D) are used for discretisations in space. Hamoka and Hystem Extran apply the 1-dimensional Shallow Water Equations. PCGEOFIM simulates the groundwater flow described by Darcyś Law; the grid consists of 3-dimensional Cartesian cubic volumes with different mesh sizes. A detailed description of the models is given in Sommer et al. (2009).

10.3.3 Spatial and Time Step Coupling

In order to ensure communication between the coupled software packages, the geometrical part of simulation models and the quantities which are exchanged have to be specified for each one of the bilateral exchanges. For the coupling between the surface water code and the groundwater code the geometrical part of the model is defined by the potentially flooded elements (triangles or rectangles) of the surface water model and the top side of the cells nearest to the surface of the groundwater model.

The quantity which is provided by the surface water code is the water level above ground, estimated in meters for the center of each cell. The quantity which is given return by the groundwater code is the water velocity (or water flux per area) measured in meters per second. The same procedure is applied to each one of the other two combinations: surface water code and sewerage simulation program as well as groundwater code and sewerage simulation (Fig. 10.8).

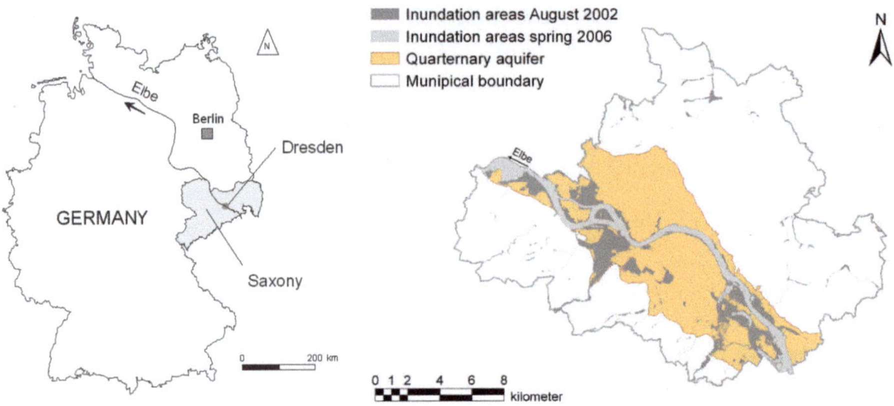

Fig. 10.8 Map of Dresden. The quaternary aquifer is the area of the groundwater model. Data source: Environmental Office, City of Dresden

Concerning the unification of time steps, different velocities of surface and subsurface flows have to be considered. While typical flow velocities at land surfaces and within the sewer are around 1 m s^{-1}, the flow velocities in groundwater are much smaller (around to 10^{-6} cm s^{-1}). Therefore, the corresponding codes work with very different time steps. The coupling process considers these different velocities, while the codes for the surface runoff and the sewerage system with their faster dynamics couple more frequently with each other before both are coupled with the groundwater code. Figure 10.8 shows the principle of the coupling (Peetz, 2008)

For more details about the implementation of the coupling see Sommer et al. (2009) and Peetz et al. (2007).

10.4 Case Study Dresden

10.4.1 Introduction of Study Area

The analyses of flood impacts on groundwater were performed for the City of Dresden, (Germany). The city is situated within the Elbe valley. The river valley is of tectonic nature ("Elbe basin") and is roughly 10 km wide. Cretaceous sediments (sandstone, limestone) are the footwall of the Quaternary aquifer in the Elbe valley. The main sediments of this aquifer are gravel and sand of glacio-fluviatile series with a thickness of less than 10–60 m from South to North. The aquifer can be seen as a uniform sediment complex since the aquicludes are not widespread over the whole area of the quaternary aquifer. Sand and fine gravel with a thickness less than about 10 m and permeability between 2×10^{-4} and 1×10^{-5} ms^{-1} are situated below. The upper sediments consist of sand and fine gravel form the low-terrace with a thickness of about 12 m. Holocene alluvial clay with a thickness of $1-4$ m represents the upper end of the Quaternary profile. These sediments are not widespread over the aquifer. The existence of alluvial clay and silt of the lower terrace has an important effect on infiltration of flood water into the aquifer.

The annual mean water level of the river Elbe at the gauge Dresden is 1.98 m. In August 2002 the maximum water level at the gauge Dresden was 9.40 m. One week before the river Elbe flooded the city, its tributaries Weißeritz and Lockwitz flooded large parts of Dresden. At the same time extreme precipitation values up to 200 mm/day were measured in Dresden. In spring of 2006 another flood of the river Elbe occurred, with a maximum water level of 7.49 m at gauge Dresden. It was a typical winter flood caused by snowmelt in the upper regions of the basin. Figure 10.6 shows the inundation areas of both floods in Dresden.

10.4.2 Flood and Groundwater in the Study Area

The interactions between flood and groundwater have been studied in Dresden during the flood in August 2002 and during the flood event in spring 2006. In August 2002, a close relationship between rising of the flood and the groundwater level was observed (Fig. 10.9a). The impacts of two separate flood events – the flash flood of Weißeritz and the Elbe flood – could be identified in hydrographs of observation wells near the rivers and close to their inundation areas. Simultaneous heavy rainfall resulted in strong infiltration in the surrounding area which caused additional groundwater rising. This effect could be observed in wells located further away from the inundation area (Fig. 10.9b). In summary, in the region of Dresden the characteristics of the groundwater dynamic was characterized by fast rising of groundwater levels (until 0.3 m/h), a high groundwater level far away from the river and a long duration of high groundwater tables. Figure 10.10 shows some hydrographs of groundwater observation wells at different distance to the river Elbe during the flood in August 2002.

In March of 2006 the groundwater levels in Dresden were untypical low. At long observed groundwater measuring points the groundwater levels were around 30 cm

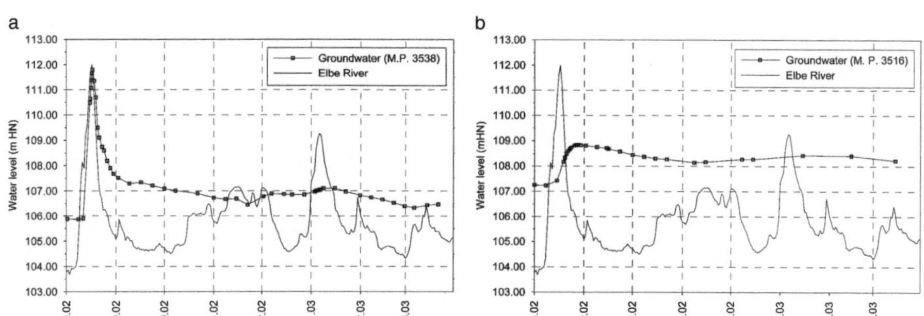

Fig. 10.9 Hydrographs of river Elbe and at observation wells, (**a**) observation well 100 m far from inundation border, (**b**) observation well approx. 1,000 m far from inundation border

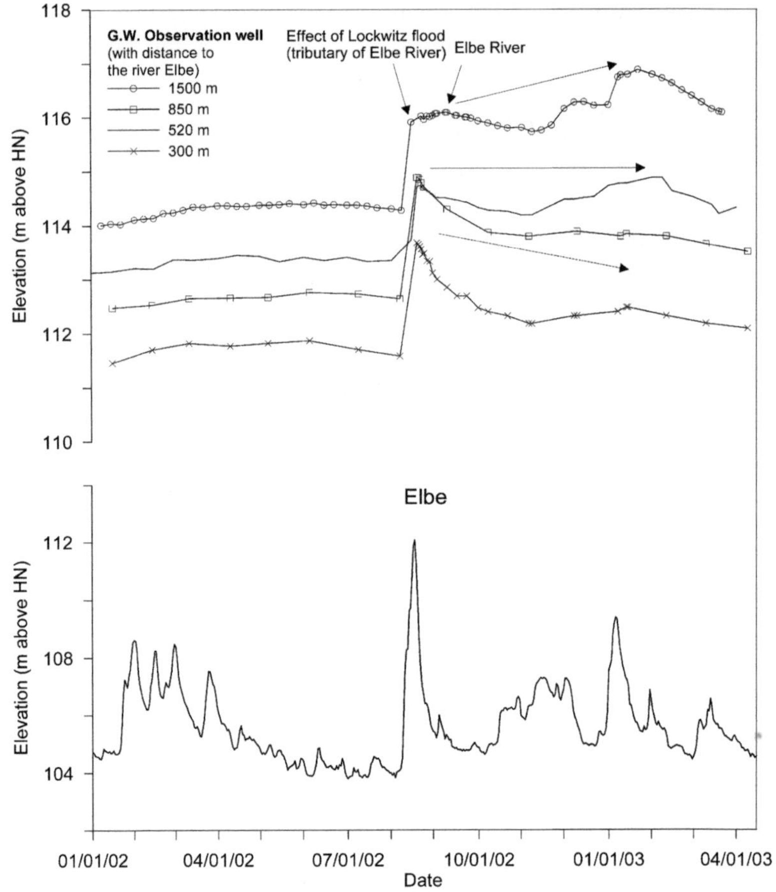

Fig. 10.10 Hydrographs of groundwater observation wells at different distances from the Elbe river during the flood in August 2002

lower than the long time averages for March. By spatial analyses of the rise of groundwater levels it could be shown that the flood had an impact on the groundwater level within a distance of 800 m from the border of the inundation areas. Within these areas the groundwater rose in average around 1 m. The maximum rising was 2.36 m. In a distance of more than 800 m from the inundation areas the groundwater was increased in average by around 10 cm only.

The depth to water table is an important parameter for subsurface flood risk assessment in urban areas. Normally in Dresden the depth to the groundwater table is between 3 and 8 m. During the August 2008 flood event the depth to water table decreased until 4–0 m. In spring 2006 the minimum depths of groundwater table achieved values from 2 to 5 m. Figure 10.11 shows the development of the depths to groundwater table from the beginning of the flood until 10 days after the flood peak by a histogram derived from a large number of groundwater observation points in Dresden.

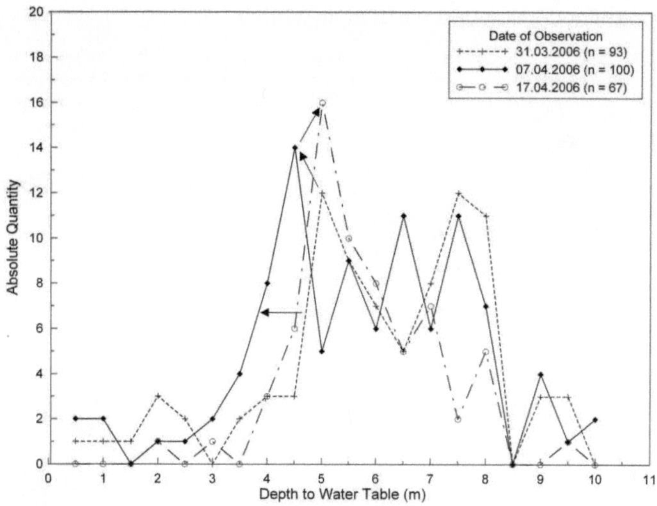

Fig. 10.11 Histogram of classes of depth to water table at n groundwater observation points at spring 2006 at several dates of observation. 31.03.2006: before the begin of flood; 07.04.2006: peak of riverine flood; 17.04.2006: after the flood

The sewer system catchment area covers 98 km^2 with approximately 470,000 inhabitants; the sewerage network consists of 900 km of combined sewers, 380 km of foul water pipes and 340 km of storm water pipes (Sommer et al., 2009).

10.4.3 Results of Modelling

The modelling of the surface flooding generated boundary condition for the sewerage model. Infiltration rates into the sewerage network are the results of the coupling modelling concerning surface flood and the discharge in the sewerage. The modelling of the surface flooding generated boundary condition also for the groundwater model. The impacts on the sewerage and on the groundwater are significant. Otherwise, the feedback from the groundwater and the sewerage had no substantial influence on the spatial and temporal distribution of the surface inundation during a flood event. The results can be demonstrated with the scenario of a 100-year flood.

Concerning the sewer system the results of the coupled modelling have shown that the area flooded by rivers is approximately congruent to the area where an overload of the sewerage system can be expected (Fig 10.12). But by contrast, a simultaneous rain event during the flood can endanger areas which are normally not affected by the river flood.

The coupled modelling allows a comprehensive description of the impact of floods on the groundwater. The focus of groundwater modelling was directed towards a description of the rising groundwater and of maximum groundwater levels. As the minimum depth to groundwater is an important parameter of groundwater flood maps, it supports risk management planning for subsurface water dynamics significantly. Maps of the duration of high groundwater levels are another

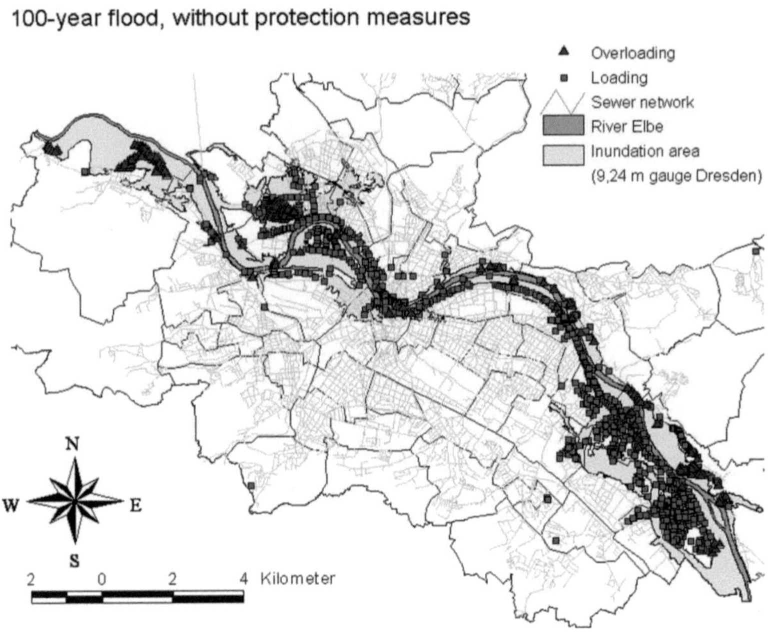

Fig. 10.12 Backwater effects of sewerage during 100-year flood in Dresden (Karpf, 2008)

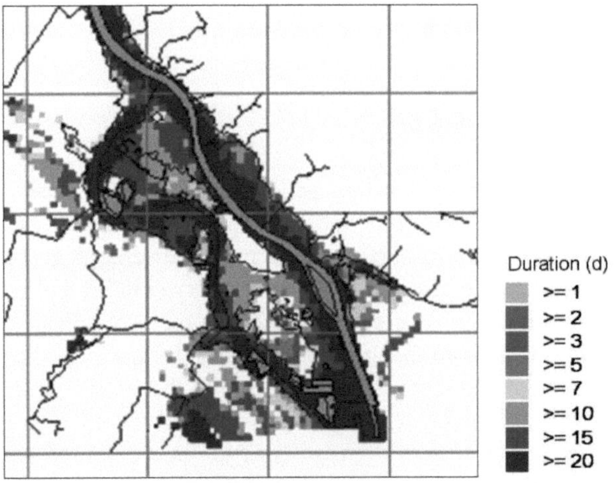

Fig. 10.13 Map of duration of groundwater levels with depths to groundwater table ≤3.0 m at a 100-year flood scenario (Eulitz, 2008)

practical result of modelling. As an example Fig. 10.13 shows a map of the duration of groundwater depth less than 3 m at a 100-year flood scenario for a local area in the south-eastern part of Dresden. Further results are described in Sommer et al. (2009).

10.5 Conclusions

The results of groundwater observation during and after flood events in conjunction with groundwater modelling on the basis of flood scenarios are essential for a groundwater risk assessment. Urban land-use planning has the aim of a long-term development of urban infrastructure, residential areas and industrial areas. Thereby, the urban land-use planning also has to ensure the livelihood and a sustainable urban development. It can be shown that land-use planning requires the consideration of the impacts of groundwater on nature as well as on built infrastructure. On the one hand, the permanent possibility of a natural subsurface drainage of groundwater is a basic demand of sustainable urban land-use planning. On the other hand, high groundwater levels, which are caused by floods, can have a damaging effect on residential houses and industrial buildings as well as on subsurface infrastructure. Therefore, these impacts have to be considered also in urban land-use planning. In this respect, urban landuse planning plays an important role in flood risk mitigation. Resulting from increasing prices of building areas the importance of subsurface constructions (e.g of subterranean car parks or shopping centres) is increasing. However, in urban regions which are situated along rivers such subsurface buildings could be in conflict with the eventuality of temporary high groundwater levels during flood events.

Therefore, maps of subsurface flood based hazard can be an helpful tool for urban land-use planning in flood endangered cities. According to the differences in potential hazards, mitigation measures can be applied gradually (Fig. 10.14).

All mitigation measures can be differentiated between temporary and permanent measures (Sommer and Ullrich, 2005). Examples for temporary subterranean flood

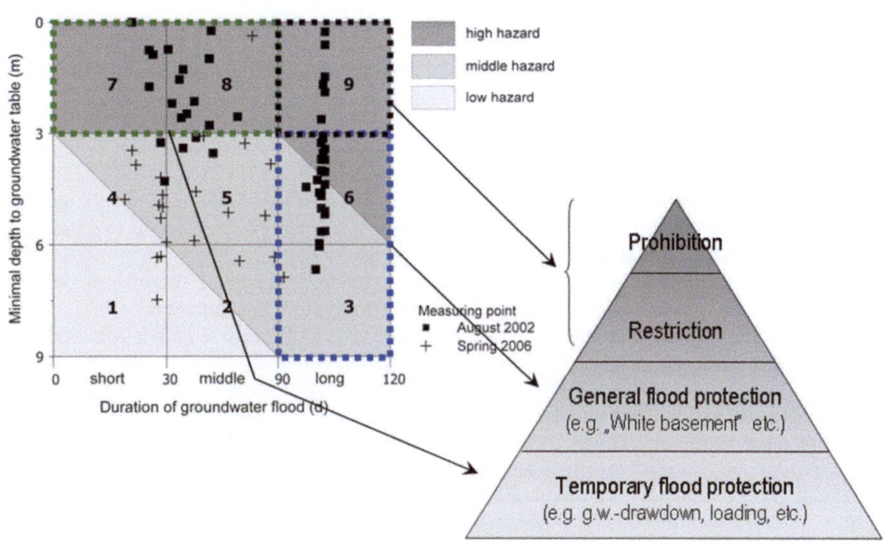

Fig. 10.14 Graduation of protection measures against groundwater flood

protection measures are the lowering of groundwater by flood protection wells or the flooding of basements in case of rising groundwater to create counter pressure. These measures are appropriate for areas with a short duration of high groundwater levels. Damage mitigation via such temporary measures relies on functioning of flood warning systems and a high degree of preparedness (Thieken et al., 2007; Kreibich and Thieken, 2008; Kreibich et al., 2009). In areas where high groundwater levels are expected over a long time permanent measures are required. Examples are waterproofed constructions of cellars by tanking (e.g. waterproof concrete tanking or constructions with a waterproof skin i.e. bitumen sealing) or flood adapted constructions of basements (e.g. basement garages which can be flooded and exhausted easily). The last steps of risk mitigation are recommendations, design specifications and restrictions for subterranean constructions. These measures are appropriate in areas with low depths to groundwater table over a long duration.

Acknowledgements The results are based on research projects "Impact of the August-2002-Flood on the Groundwater Body in the Area of Dresden City – Solutions and Activity Recommendations" (Proj. No. 0330493), "Development of a 3-Zone-Model for the Groundwater Management and the Infrastructure Management after Extreme Flood Events in Urban Areas (3ZM-GRIMEX)" (Proj. No. 02WH0557) as well as "MULTISURE" (Proj. No. 0330755) – sponsored by Federal Ministry of Education and Research of Germany.

References

Beyer K-D (2003) Erhalt der Gebäudesicherheit – Sofortmaßnahmen und Dauerlösung Beispiel St.-Benno Gymnasium Dresden. In: LH DD and DGFZ (2003): Hochwassernachsorge Grundwasser Dresden. Tagungsband zum Statusseminar am 08.10.2003, pp 63–68

BMBF (2008) Development of a 3-Zone-model for the groundwater management and the infrastructure management after extreme flood events in urban areas ("3ZM-GRIMEX") (Final report – in German). Proj. Number: 02WH0557 (URL http://edok01.tib.uni-hannover.de/edoks/e01fb09/594029457.pdf Last access: 2009-07-13)

BWG (2001) Hochwasserschutz an Fließgewässern. Wegleitung des Bundesamtes für Wasser und Geologie, Bern

Ettrich N (2003) Surface-sewer coupling and detailed elevation models for accurate urban drainage modelling. Proceedings Cost Session AquaTerra Conference: special aspects of urban flood management Institut für Wasserbau, TU Hamburg-Harburg, pp 183–195

Eulitz K (2008) Grundwasser. In: BMBF (2008) Development of a 3-Zone-model for the groundwater management and the infrastructure management after extreme flood events in urban areas ("3ZM-GRIMEX") (Final report – in German), pp 156–159

Fuchs L, Scheffer C, Verworn H-R (2004) Model description of Hystem-Extran 6. (in German) Institut für technisch wissenschaftliche Hydrologie GmbH (ITWH), Hannover

Fulford J M (2003) Computational technique and performance of transient inundation model for rivers – 2 Dimensional, Report, USGS, Open-File Report No. 03-371

Huber G, Hiller G, Braune, A. (2003) Concepts of flood control measures for the buildings of free state saxony in the historical town of Dresden (in German). In: Flood-aftertreatment groundwater. Proceedings of the Status Workshop 8 October 2003, Dresden, pp 57–61

IKSR (2002) Hochwasservorsorge – Maßnahmen und ihre Wirksamkeit. Internationale Kommission zum Schutz des Rheins (IKSR) 2002 (ISBN 3-935324-44-8)

Karpf C (2008) Ergebnisse des Kanalnetzmodells. In: BMBF (2008) Development of a 3-Zone-model for the groundwater management and the infrastructure management after extreme flood events in urban areas ("3ZM-GRIMEX")(Final report – in German), pp 130–142

Karpf C, Krebs P (2004) Sewers as drainage systems – quantification of ground-water infiltration. Proceedings 5th international conference NOVATECH, Lyon, pp 969–976

Kreibich H, Thieken AH (2008) Coping with floods in the city of Dresden, Germany. Nat Hazards. doi 10.1007/s11069-007-9200-8, Nat Hazards 51:423–436

Kreibich H, Thieken AH, Grunenberg H, Ullrich K, Sommer T (2009) Extent, perception and mitigation of damage due to high groundwater levels in the city of Dresden, Germany. Nat Hazards Earth Syst 9:1247–1258

Kreibich H, Thieken AH, Petrow T, Merz B (2006) Flood loss reduction of private households due to building precautionary measures – lessons learned from the Elbe flood in August 2002. Nat Hazards Earth Syst 5:117–126

Marre D, Walther W, Ullrich, K (2005) Influence of the flood 2002 on groundwater chemistry in Dresden, Germany (in German). Grundwasser 10:146–156

Merz B (2006) Hochwasserrisiken. Grenzen und Möglichkeiten der Risikoabschätzung. Schweizerbart'sche Verlagsbuchhandlung, Stuttgart

Peetz JV (2008) Kopplungssoftware MpCCI, Kopplungsgrößen und räumliche Kopplung, In: BMBF (2008) Development of a 3-Zone-model for the groundwater management and the infrastructure management after extreme flood events in urban areas ("3ZM-GRIMEX") (Final report – in German), pp 95–99

Peetz JV, Steckel B, Sommer Th, Eulitz K, Ettrich N, Müller M (2007) 3-Code coupling in flood simulation. Proceedings MpCCI 8th User Forum, Bonn, 13–14 February, pp 108–117 (URL: http://www.mpcci.de/fileadmin/mpcci/Userforum/MpCCI_8th_UserForum.pdf Last access: 2009-07-17)

Sames D, Boy S, Brückner F (2005) PCGEOFIM, programsystem for computation of GEOFIltration and GeoMigration, unpublished, 2005 (in German)

Schanze J (2007) A conceptual framework for flood risk management research. In: Schanze J (ed) Flood risk management research. From extreme events to citizens involvement. Proceedings European Symposium on flood risk management research (EFRM 2007), 6–7th February, Dresden, pp 1–10

Sommer T, Ullrich K (2005) Influence of the flood 2002 on groundwater. Research Report (in German). Capital Dresden, office for environment, Dresden, 68 p (ISBN 3-00-016634-9) (URL: http://www.dresden.de/media/pdf/umwelt/gw_forschungsbericht.pdf Last access: 2009-07-17)

Sommer T, Karpf C, Ettrich N, Haase D, Weichel T, Peetz JV, Steckel B, Eulitz K, Ullrich K (2009) Coupled modelling of subsurface water flux for an integrated flood risk management. Nat Hazard Earth Sys 9:1277–1290

Thieken AH, Kreibich H, Müller M, Merz B. (2007) Coping with floods: a survey among private households affected by the August 2002 flood in Germany. Hydrol Sci J 52:1016–1037

Ubell K (1987a) Austauschvorgänge zwischen Fluß- und Grundwasser (Teil I) Deutsche Gewässerkundliche Mitteilungen 31:119–125

Ubell K (1987b) Austauschvorgänge zwischen Fluß- und Grundwasser (Teil II) Deutsche Gewässerkundliche Mitteilungen 31:142–148

Chapter 11
Quantification of Socio-Economic Flood Risks

Bruno Merz, Annegret Thieken, and Heidi Kreibich

Abstract This chapter gives an overview on the assessment of direct economic losses as consequence of flooding. The basic concepts of damage assessments are introduced and the factors that influence flood damage are discussed. Finally, the damage model FLEMOps is described. FLEMOps estimates flood losses for private households. It has been recently developed based on extensive surveys of flooded households in Germany.

Contents

11.1 Increasing Demand for Flood Damage Assessments

Traditionally, design standards and structural flood defence measures were the dominant flood risk management approaches. Structural defense measures, such as dikes

B. Merz (✉)
Helmholtz Centre Potsdam, German Research Centre for Geosciences, Potsdam, Germany
e-mail: bmerz@gfz-potsdam.de

A.H. Schumann (ed.), *Flood Risk Assessment and Management*,
DOI 10.1007/978-90-481-9917-4_11,

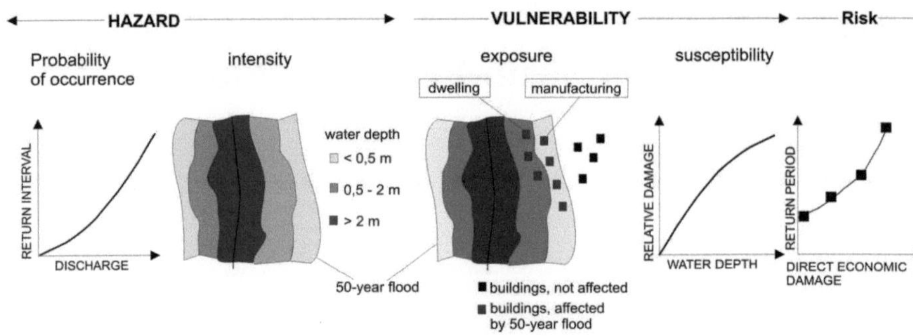

Fig. 11.1 Illustration of the terms flood hazard, vulnerability and risk (from Merz and Thieken, 2004)

and retention basins, were designed in order to control a predefined design flood, e.g. the 100-year flood. In recent years, this "flood control policy" has increasingly been challenged. New concepts are developing, usually referred to as "flood risk management". The level of protection is determined by broader considerations than some predefined design flood, and more emphasis is put on non-structural flood mitigation measures. An essential development in this context is a shift from flood hazard to flood risk. Traditionally, flood policies concentrated on the control or reduction of flood hazard, i.e. decreasing the probability of occurrence of flood discharges and inundations.

Figure 11.1 illustrates the terms hazard, vulnerability and risk. Flood hazard is described by inundation scenarios associated with certain return periods, in Fig. 11.1 exemplified by a flood frequency curve and the 50-year inundation area. The impact of the flood on the elements at risk, i.e. humans, the built environment and the natural environment, depends on their vulnerability. In its simplest form, vulnerability is composed of exposure (Who/what will be affected?) and susceptibility (How will the affected elements be damaged?). The combination of hazard and vulnerability leads to risk, defined as the probability of the occurrence of certain damage within a certain time period.

Flood damage assessments are gaining more importance within this evolving context of decision-making in flood risk management. They are needed for:

- *Assessment of flood vulnerability*. Elements at risk in flood-prone areas, e.g. households or communities, are variably vulnerable to floods. For instance, communities which experience floods on a more or less regular basis develop strategies for coping with such events. Communities or households which are not flood-experienced often neglect risk mitigation and, hence, develop higher vulnerability (Thieken et al., 2007). Knowledge about vulnerability of elements at risk is necessary for identifying risk reduction measures, e.g. development of emergency plans and the undertaking of emergency exercises.
- *Flood mapping*. Flood risk mapping is an essential element of flood risk management. In many countries, risk mapping is regulated by law. The European Union

Flood Directive [2007/60/EC], enacted in November 2007, requires member states to create both flood hazard and flood risk maps. Although flood mapping is frequently limited to mapping of flood hazard, there is a lively discussion on flood risk mapping, including adverse effects as well (de Moel et al., 2009).

- *Optimal decisions on flood mitigation measures.* Safety against floods requires resources, among others, large amounts of tax money, and it should be shown that these resources are well used. Hence, cost-benefit analyses of flood mitigation measures are needed. The current flood risk has to be estimated, the potential risk reduction options have to be determined, and benefits and costs of the different options have to be quantified and compared. For all these steps towards cost-effective risk management damage assessments are essential.
- *Comparative risk analysis.* In a wider context flood risk reduction competes against other policy fields. For example, a municipality may be prone to different types of natural hazards. Different risks within a community or a region, e.g. risks due to flooding, windstorms and earthquakes, can be compared quantitatively on the basis of consistent damage estimates (Grünthal et al., 2006).
- *Financial appraisals for insurance and reinsurance companies.* To guarantee solvency, the probable maximum loss (PML) of insurers' portfolios has to be estimated.
- *Financial appraisals during and immediately after floods.* Disaster management and governments need information on the flood damage, in order to direct and co-ordinate decisions about loss compensation.

Figure 11.2 illustrates exemplarily the results of flood loss estimation for the Seckach catchment in southwest Germany. The direct economic damage, broken down into economic sectors, is given at the municipality level for the 100-year flood scenario. The upstream municipalities are affected to a small extent compared to the downstream communities. The largest municipal damage is caused by flooding of industrial premises in Roigheim. Flood retention measures which are mainly implemented within areas of the upstream municipalities reduce significantly the damage in the downstream municipalities. Therefore, the main beneficiaries of flood defense measures in upstream areas are the downstream communities. Damage assessments quantify the spatial distribution of damage in the catchment and are therefore a necessary basis for decisions on flood mitigation.

This chapter introduces the basic concepts of direct economic flood damage assessments, discusses the factors that influence flood damage and presents the damage model FLEMOps for private households. This chapter is limited to direct economic flood damage. Floods are also associated with indirect economic damage and adverse social, psychological, political and environmental consequences. Ideally, flood risk assessments comprise all damage dimensions, in order to obtain a complete damage picture. However, risk analyses are frequently limited to direct economic losses, either because other dimensions are seen of minor importance or because the available methods do not exist to derive reliable statements. In case risk assessments do not take into account the complete spectrum of damage dimensions, the missing dimensions should at least be listed.

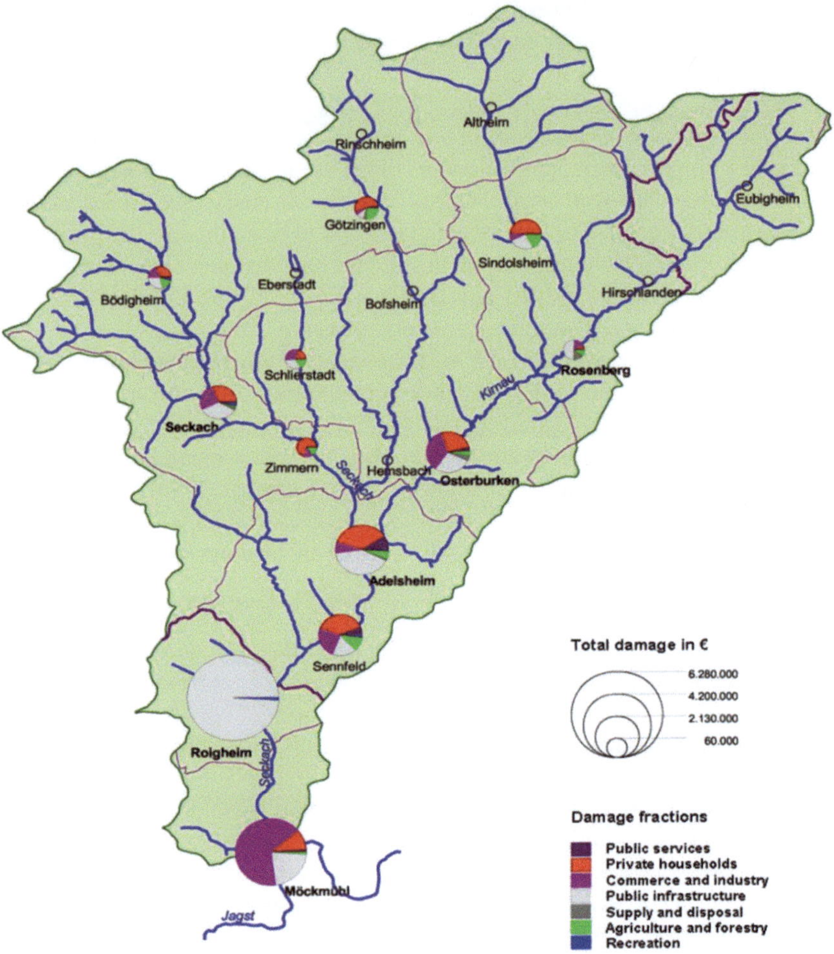

Fig. 11.2 Direct economic damage at the municipality level for the 100-year flood scenario, broken down into economic sectors

11.2 Basics of Direct Economic Damage Assessment

11.2.1 Types of Flood Damage

Flood damages can be classified into direct and indirect damage. Direct damages are those which occur due to the physical contact of the flood water with humans, property or any other objects. Indirect damages are damages which are induced by the direct impacts and occur – in space or time – outside the flood event. Both types of damages are further classified into tangible and intangible damage, depending on whether or not these losses can be assessed in monetary values. Tangible damage are damage dimensions which can be easily specified in monetary terms, whereas intangible damage is damage to goods and services which are not traded in a market and are difficult to transfer to monetary values. Some examples are:

- *Direct, tangible.* Damage to private buildings and contents; destruction of infrastructure such as roads, railroads; erosion of agricultural soil; destruction of harvest; damage to livestock; evacuation and rescue measures; business interruption inside the flooded area; clean up costs
- *Direct, intangible.* Loss of life; injuries; loss of memorabilia; psychological distress; damage to cultural heritage; negative effects on ecosystems
- *Indirect, tangible.* Disruption of public services outside the flooded area; induced production losses to companies outside the flooded area (e.g. suppliers of flooded companies); cost of traffic disruption; loss of tax revenue due to migration of companies in the aftermath of a flood
- *Indirect, intangible.* Trauma; loss of trust in authorities

Although the differentiation in direct and indirect and tangible and intangible damage is commonplace, interpretations and delineations differ. The costs of direct impacts are generally easier to quantify than indirect costs. One reason is that indirect impacts may have effects on time scales of months and years.

11.2.2 Spatial and Temporal Scales

Flood damage assessments are performed on different spatial scales:

- *Micro-scale.* The assessment is based on single elements at risk. For instance, in order to estimate the damage to a community for a certain flood scenario, damages are calculated for each affected object (building, infrastructure object, etc.).
- *Meso-scale.* The assessment is based on spatial aggregations. Typical aggregation units are land use units, e.g. residential areas, industrial areas or administrative units, e.g. zip code areas.
- *Macro-scale.* Large-scale spatial units are the basis for damage estimation. Typically, administrative units are used, e.g. municipalities, regions, nations.

The classification in micro-, meso- and macro-scale level has no clear-cut boundaries, and different groups may set the boundaries in a different way. Closely linked to the spatial scale is the context of the damage assessment (purpose, required reliability, available data, available resources etc.). Local studies, e.g. cost-benefit analysis for a single flood defence structure, usually employ the micro-scale view and derive damage estimates for each flood-prone object. Since this approach requires detailed, local input data and a large effort per unit area, meso- and macro-scale approaches are frequently chosen. Messner et al. (2007) give recommendations for the choice of the appropriate approach.

The classification in micro-, meso- and macro-scale is, on the one hand, related to the spatial extent of the damage assessment. On the other hand, there is a methodological distinction: Meso- and macro-scale approaches differ from micro-scale approaches in their need for aggregation. The steps of the damage assessment have to be performed for aggregations of damage objects. In order to compare different-scale methods and to transfer data and methods between scales, upscaling and downscaling procedures for the different steps of damage assessments are necessary. Section 4 exemplifies upscaling and downscaling approaches.

The results of a damage assessment depend on the spatial and temporal boundaries of the study. A flood might devastate a community, and at the same time, nearby communities might actually experience economic benefits, since the flood might trigger businesses and orders that cannot be performed by flood-affected companies. For example, the 1993 US Midwest floods impeded barges to navigate the river. Because of this lack of barge traffic, several trucking companies gained about US$13 million in additional revenue due to the increased demand for road transportation (Pielke, 2000). Similar considerations hold concerning the temporal scale. Flood can cause long-term consequences, such as health effects, which are not captured if a short time horizon of the damage assessment is chosen.

11.2.3 Procedure for Direct Economic Damage Estimation

The procedure for estimation of direct economic flood damage can be broken down in four steps:

- *Classification of elements at risk.* Pools elements at risk into "homogeneous" damage classes.
- *Exposure analysis.* Assesses the number and type of elements at risk which are affected by a certain flood scenario.
- *Asset analysis.* Estimates the value of the affected elements at risk.
- *Susceptibility analysis.* Relates relative damage of elements at risk to flood impact via the use of damage functions.

This four-step procedure holds for the relative damage approach. In this case the value of all elements at risk is estimated and their susceptibility is described by relative damage functions, expressing loss as a percentage of the total asset value of the elements at risk. In the absolute damage approach absolute damage functions integrate both aspects, value and susceptibility, within a single function.

There are also examples for simpler methods. The Rapid Appraisal Method (RAM), derived for fast estimates of flood damages for Australian businesses, assigns an average loss value to individual damaged buildings, regardless of damage influencing flood characteristics (Gissing and Blong, 2004).

11.2.4 Classification of Elements at Risk

Flood vulnerability varies between elements at risk. Depending on the size of the study area, the spatial extent of the flood scenarios and the necessary resolution, flood damage to a very large number of objects has to be estimated. Typically, elements at risk are classified and pooled into groups for which damage assessments are performed. For example, in the case of damage to buildings in a municipality, all buildings with certain characteristics are pooled into a group for which a unit value (e.g. related to the building area, €/m^2) and a relative damage function may be applied. The classification in groups varies between damage assessments and depends on the data availability, the necessary detail of the study and the socio-economic structure of the study area. For example, Smith (1994) argues that there are broad similarities between house types and average contents throughout much of Australia, and that this does not hold for the UK where dwelling types vary markedly, or for countries with wide variations in household income.

In the majority of cases, the classification differentiates in economic sectors, such as private households, companies, infrastructure and agriculture. This classification is based on the following arguments:

- Different economic sectors show different characteristics. In the residential sector the elements at risk are mainly buildings; this is only partly the case in other sectors like the commercial, industrial, agricultural or public sector.
- Flood impact varies between sectors. For example, flood damage to residential buildings is strongly dependent on inundation depth, whereas for damage to agricultural crops the seasonality and the duration of the flood are decisive.
- Economic data which is of utmost relevance for damage assessments is usually provided according to economic sectors.

Table 11.1 gives a classification according to economic sectors and highlights that for many sectors little data are available. Merz et al. (2004) analyse a data set of approximately 4,000 damage records, covering nine floods in Germany. They show that damage data have large variability which differs between economic sectors. Classification into sub-groups is recommended, however, sufficient data must be available for a refined classification.

The detail of the classification of a damage assessment should be in line with the relevance of the sectors or classes. There is a tendency to use more detailed approaches for sectors for which more data are available. For sectors with little data very simple approaches are frequently applied. This is problematic if data- and model-scarce sectors possess a high damage potential. A small share of flooded objects often causes a large share of damage. A single large industrial plant can incur direct flood damage that exceeds that for several hundred nearby dwellings subject to the same flood risk. For instance, the winter flood in 1993 in the Seckach catchment caused damages at several hundreds of objects in 19 communities. 40% of the direct damage emerged from a single industrial premise. A Pareto-like distribution of losses, e.g. 20% of the affected objects is responsible for 80% of the total loss, is frequently observed in damage data.

11.2.5 Exposure and Asset Analysis

Exposure analysis identifies objects that are affected by a certain flood scenario. Exposed objects are extracted by intersecting land use data with inundation data

Table 11.1 Classification of elements at risk according to economic sectors

Sector	Examples	Remarks
Private households	Residential buildings including contents, garages, summer houses etc., privately used vehicles	Majority of data sets and approaches exist for this sector. Variation of vulnerability is rather low compared to other sectors
Industry, manufacturing	Mining, metal processes, car and mechanical engineering industry, chemical industry, construction industry, installers workshop, carpentry etc.	High variability and little data available. Transfer of asset values and damage functions within sector is problematic. Booysen et al. (1999) argue that it is not possible to develop standard damage function for industries and that questionnaires have to be provided for each industrial plant
Services sector	Retail trade, wholesale trade, credit and insurance institutions, hotel and restaurant industry, lawyers, software companies etc.	Rather high variability and little data available. Transfer of asset values and damage functions within sector has to be done with care
Public sector	Education and culture (schools, universities, theaters etc.), recreation and sports (campsite, sports hall etc.), administration, health care and social welfare (hospitals, nursing home etc.), churches	High variability and little data available. Transfer of asset values and damage functions within sector is problematic
Lifelines and infrastructure	Water supply, sewerage and drainage, gas supply, power supply, telecommunication, transportation	Little data available. Within certain groups transfer of asset values and damage functions possible, e.g. unit values and damage functions for roads of certain characteristics
Agriculture	Loss of crops, damage to buildings, contents, machinery; soil erosion, loss of livestock	Methods and data availability comparatively good. Average values per element at risk might be suitable in countries with a small damage potential compared to other sectors
Others	Damage to flood defense structures; clean-up costs, evacuation and disaster management costs	Average values seem appropriate, e.g. average costs of evacuation (Penning-Rowsell and Green, 2000)

by means of GIS operations. For all flood-affected objects asset values have to be assigned. Asset values depend not only on the type of the elements at risk, but also vary in time and space. The variation in time can be attributed to economic trends, e.g. inflation, new investments and innovation. While inflation can be corrected by price indices, other changes in time can only be absorbed by a regular update of the data base. Variation in space occurs because the same object type has a different asset value in one region than in another due to regional specifications or differences in material costs, wages, etc. This variation can be covered by the use of regional or local data instead of national data.

Within one element at risk several categories of assets, e.g. buildings, machinery and equipment, or inventory, such as products or goods, can be identified. As their susceptibility varies (e.g. in case of a flood, fixed assets cannot be removed from the flooding zone, whereas moveable items such as products can be secured) and since they contribute with different proportions to the total asset value, the asset values of these categories should be estimated separately.

There are not many risk assessment studies in the literature that explicitly explain approaches for the estimation of assets. This might be due to the fact that in many risk analyses no quantitative risk indicators are used or that damage modelling is done with absolute damage functions.

Rather crude approaches were used for meso-scale assessments in Germany by MURL (2000) and Grünthal et al. (2006). Data on the gross stock of fixed assets were combined with land use data or land register data to derive unit values [€/m^2] for different economic sectors. The value of residential buildings, however, was estimated by multiplying the number of buildings with their mean insurance value, which in general represents the replacement costs of the buildings. This estimate was transformed to a unit value by relating it to the total settlement area in the study area. In a study focusing on the loss potential of the whole valley of the river Rhine (ICPR, 2001) the approach of MURL (2000) was adapted to a larger scale. Instead of land register data, CORINE land cover data (CORINE stands for Coordination of Information on the Environment) was used. It was therefore necessary to split the five economic activities and to allocate them to CORINE land cover classes. The values were calculated for each federal state in Germany. With the help of matching coefficients derived from the gross domestic product, unit values were also derived for Switzerland, France and the Netherlands. Furthermore, the approach differentiated immobile (e.g. buildings) and mobile (e.g. machinery, inventory) asset values. These crude approaches have the disadvantage that e.g. differences in building types are not considered.

Kleist et al. (2006) developed a refined methodology for the asset estimation of residential buildings in Germany. The methodology is applicable to Germany and reflects the difference of the specific construction costs per building type and the regional difference of the building stock. It links available information on standardised construction costs for residential buildings in Germany with census data about the building stock and the living area per community. The total, as well as the per-capita replacement costs for residential buildings, differentiated by type, were calculated for all communities in Germany. In parallel, Thieken et al. (2006) applied dasymetric mapping to disaggregate the community values of Kleist et al. (2006) and to provide a spatially-distributed inventory of asset values throughout Germany. The approach uses land cover data (CORINE) as ancillary variable. A strength of this approach is the relatively high resolution (minimum mapped unit of CORINE land cover data covers 25 ha) and, at the same time, the coverage of large areas such as Germany. A similar approach has been developed by Seifert et al. (2010) for commercial and industrial assets in Germany. Despite this modelling effort, the approaches are still coarse in comparison to the resolution and detailedness of the flood hazard modelling.

In the damage estimation tool HAZUS-MH, which is used in the USA, a register of commercial and industrial structures was derived from a private commercial database. Multiplying the total floor size of a building occupancy in a census block, which reflects to a certain degree the type of economic activity, with the building replacement costs per square foot in this census block provided the building asset value for the chosen building occupancy and census block (FEMA, 2003). This was done for each census block in the US and 16 different building occupancies in the commercial and industrial sector. Within a census block a uniform distribution of the buildings and, thus, of the asset values over the whole area is assumed. The smallest unit in the HAZUS-MH asset data base is therefore the census block. Further spatial disaggregation was not reported. As each census block should cover approximately the same number of inhabitants, the census blocks vary extremely in extent, i.e. their extent ranges from a few city blocks in urban areas to several square miles in rural areas. In urban areas with high building density the assumption of an uniform building distribution holds true with few exceptions (e.g. roads or parks), but in rural areas the building density is low and the assumption is questionable and may lead to a large error in the spatial distribution of asset values. For the estimation of flood losses to buildings instead of full replacement costs depreciated values are used in HAZUS-MH. They are derived from data about building costs and use the age and the condition of the structure to calculate the depreciated asset values (Scawthorn et al., 2006). Contents asset values are estimated as a fixed percentage of the building asset value (FEMA, 2003).

In another study in Japan, Dutta et al. (2003) calculates unit economic values for different elements at risk. For non-residential objects eight types of economic activity (mining; construction; production; electricity/gas/water; wholesale and retail sale; finance and insurance; real estate; services) are distinguished. To estimate the monetary values for property and inventory, the number of workers per type is multiplied by unit prices per worker and type. The values of residential buildings are estimated by the product of the unit area with the structure value per unit area and the content value per unit area, respectively. These calculations are done on ward-level. For further spatial disaggregation on a grid cell basis land cover data is used. The floor area per grid cell is determined considering land cover type, building ratios (i.e. the percentage of area covered by buildings in a given area) and floor area fractions (i.e. the total area of all storeys of a building divided by the ground surface area of the building; thus for a one-storey building the floor area fraction amounts to 1). The latter two parameters are derived from aerial photographs. This approach is feasible for small or medium sized areas, but not for a countrywide approach, since the analysis of aerial photographs for a huge area would be too time-consuming.

For an Australian risk analysis Blong (2003) uses construction costs (replacement costs) per square meter that have been published by Australian authorities. The basic idea of Blong (2003) is not to estimate the total value of all structures by using different unit values for each building type, but to relate all different building types to a medium-sized family house. Cost ratios are derived by relating the standardized construction costs of all other buildings to the construction costs of a medium-sized family house. A replacement ratio RR considers differences in the building size (floor area) and is defined as RR = [(Cost Ratio ∗ Floor area)/Floor area of a medium-sized family house]. The replacement ratios are then used in the damage model which calculates damage as house equivalents. Monetary loss can be achieved by multiplying the house equivalents by the asset value of a medium-sized family house. The advantage is that the method can be easily updated and that the damage model can be used for different hazard types. Unfortunately, it only covers the asset type of buildings and neither of machinery, equipment and inventory, which can account for substantial losses especially in industry and commerce.

This overview shows that the methods for asset estimation vary considerably in terms of detail concerning the stratification in economic classes and the spatial disaggregation of areal values. For example, HAZUS-MH does not use land use data for spatial distribution of asset values, but assumes a uniform distribution of buildings within census blocks, which is the smallest spatial unit used in HAZUS-MH. Dutta et al. (2003) use very precise information for the spatial distribution of asset values applying aerial photographs – an interesting approach which is however not feasible for a countrywide application. This broad spectrum of methods illustrates that the detail of asset estimation depends strongly on the size of the study area, the available input data and the required accuracy of the risk assessment.

11.3 Susceptibility Analysis

Most damage models have in common that the damage is obtained from the type or use of the object and the inundation depth. Other parameters, like flow velocity or previous flood experience of affected people, are rarely taken into account. Thus, simple averaging and stage-damage curve loss estimation methods ignore the large influence of other variables and result in uncertain estimations (Merz et al., 2004). Despite their uncertainty such stage-damage curves are seen as the essential building blocks upon which flood damage assessments are based. They are accepted as the standard approach to assess urban flood damage (Smith, 1994).

11.3.1 Damage Influencing Flood Characteristics

It is obvious that flood damage depends, in addition to the type of object and water depth considered by stage-damage curves, on many factors. Some of these factors are flow velocity, duration of inundation, sediment concentration, contamination of flood water, availability and information content of flood warning, and the quality of external response in a flood situation. Although a few studies give some quantitative hints about the influence of these factors (Smith, 1994; Wind et al., 1999; Penning-Rowsell and Green, 2000; Kreibich et el., 2005; Thieken et al., 2005), there is no comprehensive approach for including such factors in damage modelling.

These factors can be differentiated into impact and resistance parameters (Thieken et al., 2005). Impact parameters reflect the specific characteristics of a flood event for the object under study. Whereas impact parameters depend on the kind and magnitude of the flood, resistance parameters depend on the flood prone objects. They depict the capability or incapability of an object to resist the flood impact. Resistance parameters can be the object size or the type and structure of a building. Further, also mitigation measures, former flood experience and fast transmission of a warning influence the resistance (Kreibich et al., 2007).

Most of these damage influencing factors are neglected in damage modelling, since they are very heterogeneous in space and time, difficult to predict, and limited information on their (quantitative) effects are available. For instance, a gate being opened or closed could make the difference between high and low flow velocities and, as a consequence, scour undermining a foundation or not (Kelman and Spence, 2004). Floating and destruction of an oil-tank or not can make the difference between total damage of a building due to severe contamination or only marginal damage due to water contact only.

11.3.2 Damage Functions

In developing flood loss models two main approaches can be distinguished: empirical approaches which use loss data collected after flood events and synthetic approaches which use loss data collected via what-if-questions. An example for the first approach is the German flood damage data base HOWAS (Merz et al., 2004), from which the models of MURL (MURL, 2000) and Hydrotec (Emschergenossenschaft and Hydrotec, 2004) derived their loss functions. What-if analyses estimate the damage which is expected in case of a certain flood situation, e.g.: "Which damage would you expect if the water depth was 2 m above the building floor?" Examples for this approach are the damage functions for United Kingdom (Penning-Rowsell et al., 2005). It is possible to combine both approaches, e.g. to extend empirical data with synthetic data which was done by the US Army Corps of Engineers to develop their loss functions (USACE, Galveston District, Texas, personal communication, 2006) or to evaluate synthetic models with empirical data. Both approaches have advantages and disadvantages (Table 11.2).

Table 11.3 compares the advantages and disadvantages of relative and absolute damage functions. Which of both approaches is chosen may depend on the kind of available data, e.g. on the availability of data on the value of assets.

11.4 The FLEMOps Model

The rule-based Flood Loss Estimation Model for the private sector FLEMOps was developed to fulfil the following requirements:

- The new model should take into account more influencing factors, not only the water level.

Table 11.2 Advantages and disadvantages of empirical and synthetic flood damage models

	Advantages	Disadvantages
Empirical damage models	Real damage information possesses a greater accuracy than synthetic data (Gissing and Blong, 2004) Effects of damage mitigation measures can be quantified and taken into account in damage modelling (Kreibich et al., 2005; Thieken et al., 2008) Variability within one category and water depth is reflected by the data and uncertainty can be quantified (Merz et al., 2004)	Detailed damage surveys after floods are uncommon, so that models may be based on poor quality data (Smith, 1994) Paucity of information about floods of different magnitude and often a lack of damage records with high water depth require extrapolations (Smith, 1994; Gissing and Blong, 2004). Transferability in time and space is difficult due to differences in warning time, flood experience, building type and contents (Smith, 1994)
Synthetic damage models	In each building, damage information for various water levels can be retrieved (Penning-Rowsell and Chatterton, 1977) Approach does not rely on information from actual flood events and can therefore be applied to any area (Smith, 1994)	High effort is necessary to develop detailed data bases (inventory method) or undertake large surveys (valuation survey method) to achieve sufficient data for each category/building type (Smith, 1994) What if analyses are subjective, resulting in uncertain damage estimates (Gissing and Blong, 2004; Soetanto and Proverbs, 2004) Mitigation actions are not taken into account (Smith, 1994) Premises within one classification can exhibit large variations which are not reflected by the data (Smith, 1994)

- The model is to be based on loss ratios (instead of absolute losses) so that a combination with various asset stocks (e.g. total asset of residential buildings, insured assets/portfolios) is possible.
- Different scales of model application (such as buildings and land use units) should be enabled.

The model is based on detailed statistical analysis (e.g. Mann-Whitney-U tests, principal component analyses) of data from a survey of 1,697 private households that were affected by the flood in August 2002 (Kreibich et al., 2005; Thieken et al., 2005). Principal component analysis showed that the loss influencing parameters can be classified into the following main components (Thieken et al., 2005): (1) building characteristics (size, type and value of the affected building/contents), (2) household structure (size and age structure), (3) static flood impact (water depth, duration, and contamination) as well as (4) precaution and flood experience. Other components like emergency measures undertaken by the residents, socio-economic status of the household or dynamic flood impact (flow velocity) are only relevant

Table 11.3 Advantages and disadvantages of relative and absolute damage functions

	Advantages	Disadvantages
Relative damage functions	Simplicity, because many data sources on the value of properties are available (Messner et al., 2007) Better transferability in space and time, since they are independent of changes in market values of individual structures which may result from inflation, shifts in local economy or development status (Krzysztofowicz and Davis, 1983)	Values of the object assets are necessary
Absolute damage functions	No need for asset values	Need for regular re-calibration, e.g. damage functions of Penning-Rowsell and Chatterton (1977) were re-calibrated, reflecting larger investments in properties and contents (Penning-Rowsell and Green, 2000)

in some sub-datasets. Since mainly the components of the static flood impact and of the building characteristic were significantly correlated with the building as well as with the contents damage (Thieken et al., 2005), a two-stage model considering the following parameters was developed: In the first model stage (core model), losses are estimated according to the water level (≤ 20 cm, 21–60 cm, 61–100 cm, 101–150 cm, >150 cm), building type (one-family homes, (semi-)detached houses, multifamily houses) and building quality (low/medium quality, high quality). For all sub-categories mean loss ratios per loss type (building, contents) were derived from the empirical data (Fig. 11.3).

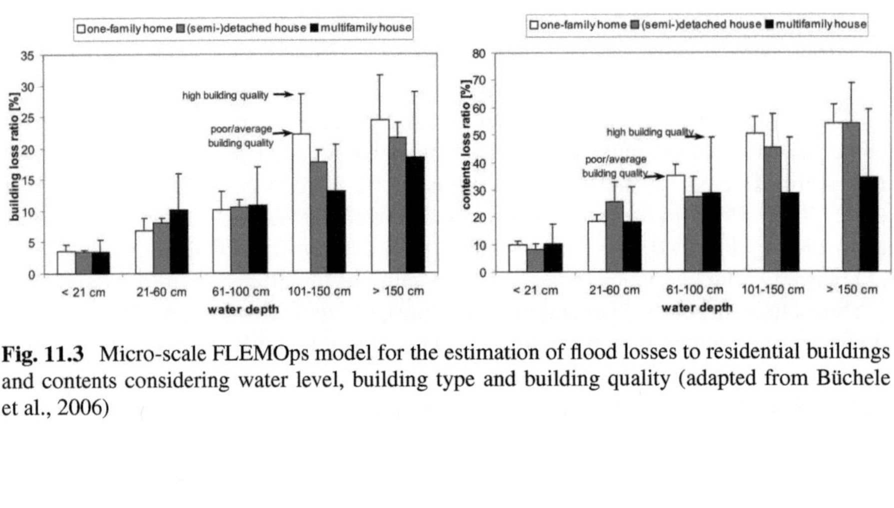

Fig. 11.3 Micro-scale FLEMOps model for the estimation of flood losses to residential buildings and contents considering water level, building type and building quality (adapted from Büchele et al., 2006)

Table 11.4 Scaling factors for different levels of contamination and precaution (adapted from Büchele et al., 2006)

	Scaling factors for damage ratios of	
	Buildings	Contents
No contamination and no precaution	0.92	0.90
No contamination and medium precaution	0.64	0.85
No contamination and very good precaution	0.41	0.64
Medium contamination and no precaution	1.20	1.11
Medium contamination and medium precaution	0.86	0.99
Medium contamination and very good precaution	0.71	0.73
High contamination and no precaution	1.58	1.44

In the second model stage (further FLEMOps+), also the influence of the contamination of the floodwater and precaution of private households can be considered by scaling factors (Table 11.4). For instance, very good precaution and no contamination reduce building loss by 41%, while heavy contamination and no precaution augment it by 58%. Since only a very limited number of households which had undertaken precautionary measures experienced high contamination ($n = 21$), it is suspected that precautionary measures are largely able to avoid contamination and that these cases can be neglected.

The model can be applied to the micro-scale, i.e. to single buildings as well as to the meso-scale, i.e. to land cover units (see Thieken et al., 2008; Apel et al., 2009). For the latter, a scaling procedure based on census data and a dasymetric mapping technique was developed (Thieken et al., 2006): By means of INFAS Geodaten (2001) and cluster analysis the mean building composition and the mean building quality per municipality was derived for whole Germany (Fig. 11.4). With the help of this classification, a mean flood loss model was set up by weighting the flood loss model for three different building types by the mean percentages of these building types in each cluster. Thieken et al. (2008) and Apel et al. (2009) demonstrate that FLEMOps is theoretically within the range of other stage-damage functions used in Germany. However, the advantage is that it takes into account the building characteristics of the area under investigation.

In addition, a dasymetric mapping approach was applied to disaggregate building asset values (Thieken et al., 2006, see also Section 2.5). Such exposure data are commonly provided at the municipal level; for loss estimations they have to be disaggregated to a finer spatial scale. To get a realistic distribution of the asset values, land cover data are used as ancillary data. By assigning a weight to each land cover class, the total municipal asset value is disaggregated within the municipality under study. In FLEMOps, the mapping technique of Mennis (2003) was adapted.

For the application of the meso-scale model FLEMOps in Germany a web-service has been set up (http://nadine.helmholtz-eos.de/FLEMO.html). Users can upload their flood scenarios (ASCII grids of water depths) and calculate the expected residential building losses. Interested parties can get a login from the section hydrology at GFZ (FLEMO@gfz-potsdam.de).

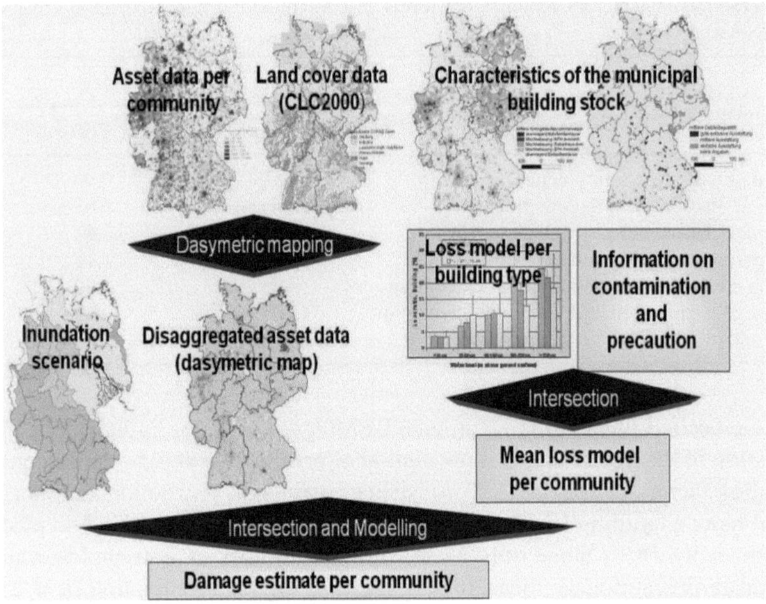

Fig. 11.4 Overview of the FLEMOps model for the estimation of flood losses to residential buildings in Germany

Table 11.5 Input data and model results of the damage assessment in the community of Eilenburg (Saxony, Germany) for the flood event in August 2002 (adapted from Apel et al., 2009)

Information of the Saxon Relief Bank (Sächsische Aufbaubank – SAB)	
Total eligible repair costs for damage to residential buildings in August 2002	€77.12 Million
Number of buildings to be repaired	765
Telephone survey after the flood event in August 2002 (Kreibich et al., 2005; Thieken et al., 2005)	
Number of surveyed households in Eilenburg	37
Share of households not affected by contaminated floodwater	24.3%
Share of households affected by heavily contaminated floodwater	64.9%
Share of households that performed NO precautionary measures	89.1%
Share of households that performed MORE THAN ONE precautionary measure	5.4%
Loss estimation for the flood in August 2002 (Scenario: 1D/2D-modelling with LISFLOOD-FP (Bates and de Roo, 2000)	
Loss function of IKSR (2001)	€34.50 Million
Loss function of MURL (2000)	€9.78 Million
FLEMOps, first stage	€48.68 Million
FLEMOps+, second stage (high contamination, no precaution)	€76.92 Million

Validations of the model have shown that FLEMOps+ outperforms other stage-damage functions that are usually applied in Germany (Thieken et al., 2008; Apel et al., 2009), which confirms the assumption that uncertainty in flood loss estimation can be reduced when more parameters, besides the water depth, are taken into consideration.

In the following it is exemplarily shown that FLEMOps is able to calculate realistic flood losses in a test community, namely Eilenburg at the Mulde River in Saxony, Germany. Input data and model results for the flood in August 2002 are summarized in Table 11.5. Besides FLEMOps, two other loss models (IKSR, 2001; MURL, 2000) were applied. The August 2002 flood caused losses to residential buildings of €77.12 Million in the community of Eilenburg (Saxon Relief Bank (SAB), personal communication as of February 2005). On basis of a hydraulic simulation with LISFLOOD-FP (Bates and de Roo, 2000) FLEMOps applied in both model stages achieves the best estimation (Table 11.5).

11.5 Conclusions

Flood damage assessment has not received much attention. Compared to the wealth of available information on flood hazard, flood damage data are scarce and damage estimation methods are crude. However, damage assessments are increasingly asked for within the policy shift from flood control to flood risk management. The models used today consider only few damage-influencing factors. Models and data are transferred in time, space and processes without sufficient justification. Additional damage data, more comprehensive models and validation studies are needed to provide more reliable flood damage assessments. In particular, the relationships between flood damage and damage-influencing factors need to be disentangled for different situations and regions.

Acknowledgements This work is part of the research project MEDIS (Methods for the evaluation of direct and indirect flood losses) which has been funded by the German Federal Ministry for Education and Research (BMBF) (No. 0330688) within the research programme RIMAX (Risk management of extreme flood events).

References

Apel H, Aronica GT, Kreibich H, Thieken AH (2009) Flood risk assessments – How detailed do we need to be? Nat Hazards 49(1):79–98

Bates PD, de Roo APJ (2000) A simple raster-based model for floodplain inundation. J Hydrol 236:54–77

Blong R (2003) A new damage index. Nat Hazards 30(1):1–23

Booysen HJ, Viljoen MF, de Villiers GduT (1999) Methodology for the calculation of industrial flood damage and its application to an industry in Vereeniging. Water SA 25(1):41–46

Büchele B, Kreibich H, Kron A, Thieken A, Ihringer J, Oberle P, Merz B, Nestmann F (2006) Flood-risk mapping: contributions towards an enhanced assessment of extreme events and associated risks. Nat Hazards Earth Syst Sci 6:485–503 de Moel H, van Alphen J, Aerts JCJH (2009) Flood maps in Europe – methods, availability and use. Nat Hazards Earth Syst Sci 9:289–301

Dutta D, Herath S, Musiake K (2003) A mathematical model for flood loss estimation. J Hydrol 277:24–49

Emschergenossenschaft and Hydrotec (2004) Hochwasser-Aktionsplan Emscher, Kapitel 1: Methodik der Schadensermittlung. Report

FEMA (Federal Emergency Management Agency) (2003) HAZUS-MH Technical Manual. Department of Homeland. Technical Report

Gissing A, Blong R (2004) Accounting for variability in commercial flood damage estimation. Aust Geogr 35(2):209–222

Grünthal G, Thieken AH, Schwarz J, Radtke K, Smolka A, Merz B (2006) Comparative risk assessments for the city of Cologne – storms, floods, earthquakes. Nat Hazards 38(1–2):21–44. doi:10.1007/s11069-005-8598-0

IKSR (Internationale Kommission zum Schutz des Rheins) (2001) Atlas der Überschwemmungsgefährdung und möglicher Schäden bei Extremhochwasser am Rhein. Koblenz

INFAS Geodaten (2001) Das Data Wherehouse. Bonn, INFAS GEOdaten GmbH, Status: Dezember 2001

ICPR (International Commission for the Protection of the River Rhine) (2001) Übersichtskarten der Überschwemmungsgefährdung und der möglichen Vermögensschäden am Rhein. Report. ICPR, Koblenz

Kelman I, Spence R (2004) An overview of flood actions on buildings. Eng Geol 73: 297–309

Kleist L, Thieken AH, Köhler P, Müller M, Seifert I, Borst D, Werner U (2006) Estimation of the regional stock of residential buildings as a basis for comparative risk assessment for Germany. NHESS 6(4):541–552

Kreibich H, Müller M, Thieken AH, Merz B (2007) Flood precaution of companies and their ability to cope with the flood in August 2002 in Saxony, Germany. Water Resour Res 43:W03408. doi:10.1029/2005WR004691

Kreibich H, Thieken AH, Petrow T, Müller M, Merz B (2005) Flood loss reduction of private households due to building precautionary measures – Lessons Learned from the Elbe flood in August 2002. NHESS. Nat Hazards Earth Syst Sci 5:117–126

Krzysztofowicz R, Davis DR (1983) Category-unit loss functions for flood 21 forecast-response system evaluation. Water Resour Res 19(6):1476–1480

Mennis J (2003) Generating surface models of population using dasymetric mapping. Prof Geogr 55(1):31–42

Merz B, Kreibich H, Thieken AH, Schmidtke R (2004) Estimation uncertainty of direct monetary flood damage to buildings. Nat Hazard Earth Syst Sci 4:153–163

Merz B, Thieken AH (2004) Flood risk analysis: concepts and challenges. Österreichische Wasser- und Abfallwirtschaft 56(3–4):27–34

Messner F, Penning-Rowsell E, Green C, Meyer V, Tunstall S, van der Veen A, Tapsell S, Wilson T, Krywkow J, Logtmeijer C, Fernández-Bilbao A, Geurts P, Haase D, Parker D (2007) Evaluating flood damages: guidance and recommendations on principles and methods, Floodsite project, www.floodsite.net

MURL (2000) Potentielle Hochwasserschäden am Rhein in NRW. Ministerium für Umwelt Raumordnung und Landwirtschaft des Landes Nordrhein-Westfalen, Düsseldorf

Penning-Rowsell EC, Chatterton JB (1977) The benefits of flood alleviation: a manual of assessment techniques. Gower Technical Press, Aldershot

Penning-Rowsell EC, Green C (2000) New Insights into the appraisal of flood-alleviation benefits: (1) Flood damage and flood loss information. J Chart Inst Water E 14:347–353

Penning-Rowsell E, Johnson C, Tunstall S, Tapsell S, Morris J, Chatterton J, Green C (2005) The benefits of flood and costal risk management: a manual of assessment techniques. Middlesex University Press, London

Pielke RA (2000) Flood impacts on society. In: Parker DJ (ed) Floods Routledge hazards and disasters, Series. Routledge Press, London, pp 133–155

Scawthorn C, Flores P, Blais N, Seligson H, Tate E, Chang S, Mifflin E, Thomas W, Murphy J, Jones C, Lawrence M (2006) HAZUS-MH flood loss estimation methodology. II. Damage and loss assessment. Nat Hazards Rev 7(2):72–81

Seifert I, Thieken A, Merz M, Borst D, Werner U (2010) Estimation of industrial and commercial assets values for (natural) hazard risk assessment. Nat Hazards 52(2):453–479

Smith DI (1994) Flood damage estimation – A review of urban stage-damage curves and loss functions. Water SA 20(3):231–238

Soetanto R, Proverbs D G (2004) Impact of flood characteristics on damage caused to UK domestic properties: the perceptions of building surveyors. Struct Surv 22(2):95–104

Thieken AH, Müller M, Kreibich H, Merz B (2005) Flood damage and influencing factors: new insights from the August 2002 flood in Germany. Water Resour Res 41(12):W12430. doi:10.1029/2005WR004177

Thieken AH, Müller M, Kleist L, Seifert I, Borst D, Werner U (2006) Regionalisation of asset values for risk analyses. NHESS 6(2):167–178

Thieken AH, Kreibich H, Müller M, Merz B (2007) Coping with floods: preparedness, response and recovery of flood-affected residents in Germany in 2002. Hydrol Sci J 52(5):1016–1037

Thieken AH, Olschewski A, Kreibich H, Kobsch S, Merz B (2008) Development and evaluation of FLEMOps – a new Flood Loss Estimation model for the private sector. In: Proverbs D, Brebbia CA, Penning-Rowsell E (eds) Flood recovery, innovation and response. WIT Press, Southampton, pp 315–324

Wind HG, Nierop TM, de Blois CJ, de Kok JL (1999) Analysis of flood damages from the 1993 and 1995 Meuse floods. Water Resour Res 35(11):3459–3465

Chapter 12
Application of Scenarios and Multi-Criteria Decision Making Tools in Flood Polder Planning

Andreas H. Schumann and David Nijssen

Abstract Effectiveness of technical flood control measures depends strongly on multiple characteristics of floods. Copulas can be applied for multivariate statistical descriptions of flood scenarios. However, the parameterisation of these multivariate statistical models involves many uncertainties. With regard to these known unknowns the multivariate statistical characteristics of flood scenarios can be handled as imprecise probabilities. Such imprecise probabilities can be specified by Fuzzy Numbers and integrated in a Multi Criteria Decision Making framework. Their application in a Multi Criteria Decision Making framework, which was developed for flood retention planning in a river basin, is demonstrated here with a case study.

Contents

A.H. Schumann (✉)
Ruhr-University Bochum, Chair of Hydrology and Water Management, Bochum, Germany
e-mail: andreas.schumann@rub.de

A.H. Schumann (ed.), *Flood Risk Assessment and Management*,
DOI 10.1007/978-90-481-9917-4_12,

12.1 Introduction

Flood risk management implies three aspects: risk analysis, risk evaluation and risk reduction. These aspects have a hierarchic order. Risk analysis is a precondition for risk evaluation and risk reduction depends on risk evaluation. If a human influenced hydrological system is considered, the impacts of risk reduction activities of the past have to be considered if new measures of flood protection are planned. The remaining risks depend strongly on the performance of existing flood protection facilities, which is affected by hydrological loads. Decision makers are forced to accept the remaining risks as an indispensable criterion of making decisions, since an absolute protection against flooding cannot be reached by technical measures. The risk of failure of a technical flood protection structure, resulting e.g. from the hydrological risk of an extreme flood which exceeds the design flood, has to be considered in such planning approaches (Plate and Meon, 1988). Here assessments of the effectiveness of flood protection measures, but also of possible consequences of failures are needed. In safety oriented planning it is sufficient to demonstrate the reliability of a technical structure under well defined hydrological conditions but in risk-based planning the consequences of possible failures have to be considered. The goal of risk reduction depends on the efficiency of flood control measures and their reliability under manifold hydrological loads. An ensemble of hydrological loads can be used to demonstrate under which conditions the performance of the planned flood control system would be insufficient and to show impacts of possible failures. The effectiveness of flood retention facilities depends on multiple flood characteristics and how those are combined within a flood event. Flood protection may be ensured under favourable conditions, but in other cases the system may fail, even if a certain flood characteristic, e.g. the flood peak, remains below the value which was assumed for the design flood. Often hydrological time series are too short to get a representative sample of the large range of hydrological loads which is needed for such analyses. Long hydrological time series can be generated with stochastic means to overcome this problem as it was discussed in Chapter 7. With regard to the impact assessments of flood retention facilities, modelling of hydrological loads demands a coupling of stochastic and deterministic components. Most of those attempts are based on complex simulations, starting with stochastic generation of precipitation as it was discussed in Chapter 7 and transforming the results with a deterministic hydrological model into a runoff time series (e.g. Blazkova and Beven, 2002; McMillan and Brasington, 2008). This methodological approach implies many uncertainties (Lamb and Kay, 2004; Cameron et al., 1999). A probabilistic characterisation of the results is difficult, as several stochastic interdependencies are incorporated. For example the meteorological load as well as the initial state of the deterministic hydrological model have a certain probability, the model parameters are uncertain, the behaviour of the model for extreme events, which are often higher than any observed flood, is uncertain, the impact of technical flood retention measures depends on unknown operation schemes and so on. These problems aggravate if such analyses are done for a large river basin with spatially distributed hydrological loads, where many different combinations of influencing factors are possible.

In practice floods are characterised by the return period of the peak only. However, such a probabilistic characterisation is often insufficient for risk estimations of flood protection systems. The efficiency of a reservoir depends on the shape of the hydrograph and the volume of the flood as it is discussed in Chapter 9, the stability of a dyke depends on flood peak and duration of the flood etc. Multivariate statistics can be applied to specify the relevant hydrological loads. However, the effectiveness of flood control measures has to be characterised with probabilities which have to be compared with the risk perception of decision makers. Hydrological scenarios can be easily derived from stochastic-deterministic simulations, but a scenario-based evaluation of flood control structures requires probabilistic measures. With regard to multiple criteria of floods, which have to be considered simultaneously (e.g. peak, volume and spatial distribution of runoff) multivariate statistics (Chapter 8) can be applied to specify at least important combinations. It should be considered that the utilisation of simulated data for this purpose limits the statistical representativeness of such characterisations. To handle uncertainties which result from insufficient statistical information or missing data we can mitigate the difficulties of precise probabilities and use imprecise ones (Klir, 1999). There are several options to express the imprecision of probabilities, e.g. by Random Sets or by Fuzzy Sets. In the following Fuzzy Sets are applied to characterise the uncertainties of hydrological loads.

For risk analysis the harmful consequences of floods have to be specified. These consequences depend on the vulnerability of landscapes and the specific characteristics of flood events, e.g. water depth, flow velocity and duration of flooding. The impacts of these characteristics differ with land-use, social structure, prosperity, season etc. Here hydraulic modelling of flood events (Chapter 9) has to be combined with socio-economic analyses (Chapter 11) to estimate damages and to specify flood risk. This risk depends on the probabilities of the hydrological loads which were used to estimate the damages. For flood protection systems this assumption is a simplification as the relationship between hydrological loads and consequences depends also on their ability to resist a flood. In risk oriented planning the risk awareness of decision makers plays an important role. As discussed in Chapter 9, the risk of failures of flood control systems depends on multiple characteristics (e.g. flood peaks, volumes and shapes of hydrographs). Multivariate statistics can be applied to characterise this complexity but decision makers can often not handle them in an adequate way. They will be confronted also with multiple failure mechanisms. There is a smooth transition from events where flood control may be ensured completely to events where all attempts to manage a flood may fail completely.

Another problem of decision making in flood control consists in unclear and multiple preferences of decision makers. Decision makers may differentiate for example between damages according to the spatial distribution, the frequency and options to get compensations. Simple cost- benefit relationships are insufficient if complex damage probabilities have to be considered. The benefits of flood control are uncertain as they depend stochastically on the occurrence of parlous combinations of flood characteristics. On the contrary, the costs of flood control can be specified in detail. Uneven distributed burdens and benefits between upstream/downstream riparian communities pose yet another problem in the decision process. Here multi-objective analyses can be helpful for structuring flood management planning. These methods involve the quantification of objectives, the generation of alternatives and the selection of a preferred one. Multi-objective techniques can be classified into three groups of methods: methods for generating a non-dominated set of solutions, methods with prior articulation of preferences and methods with progressive articulation of preferences (Goicoechea et al., 1982). For flood management a prior, but flexible articulation of preferences seems to be most appropriated with regard to the multiple participants of the planning process. If protagonists get the opportunity to explore the decision space, to balance their weighting system under consideration of the possible outcome and to bring in their personal risk perception, then the result of such analyses are more likely to be accepted.

12.2 Estimation of Flood Scenarios and Their Plausibility

As mentioned above a large variety of flood events has to be considered to analyse the performance of flood control systems for risk based planning. Scenarios have to be specifically defined in order to cover the wide range of possible circumstances. These scenarios are not "real" in the sense of being expected, but they are examples of what could occur (Kachroo, 1992). The selection of these scenarios is difficult as the members of this ensemble should be realistic and representative for all circumstances of the flood control system. Here it is not sufficient to specify only the range of flood events. Simulated failures of a flood protection system have to be evaluated with probabilistic means. A return period is the main characteristic of any design flood. However, in the conventional way it characterises the probability of the flood peak only. This characteristic is not representative for failures of flood retention systems. The assumption that flood events with similar peaks but different volumes or different spatial distributions of precipitation within the river basin would have the same probabilities is not correct. Volume and shape of the hydrograph, which are related to flood peaks in a statistical way, are other and, often more crucial characteristics for flood retention. On the one hand a flood retention facility may fail even if the flood peak of the inflow remains below the peak with a return period of the design flood. On the other hand it may perform well if a flood wave with a higher peak occurs under favourable conditions. Such complex interdependencies between flood characteristics can be specified easily by scenarios (e.g. for interactions of tributaries or for different shapes of a hydrograph with a specified volume) but it is difficult to estimate their probabilities. Normally only one characteristic of a flood scenario, the flood peak, will be defined by its probability. For combinations of other characteristics, multivariate statistical measures can be derived. However, if simulated data are used to derive these probabilities the result will depend on the modelling approach which has to be applied to estimate a long time series by stochastic-deterministic simulations. Probabilities which are specified in this way are imprecise as they depend on the way how they were estimated. Consideration of imprecision in expressing probabilities, which was strongly stimulated by Walley (1991), adds a new dimension to the formalization of uncertainty and uncertainty-based information. Walley demonstrated that reasoning and decision making based on imprecise probabilities satisfy the principles of coherence and avoidance of sure loss.

In the following an approach to consider imprecise probabilities in flood protection planning is presented. The starting point are events which belong to one set of floods according to the return period of their flood peaks, but which differ in their other characteristics (volume, shape of the hydrograph and spatial distribution of runoff). The interdependencies between these characteristics are considered with Copula statistics (Chapter 9). The resulting statistical measures are used as additional information to specify the events in a possibilistic way. Multivariate probabilities are used here to differentiate between more and less plausible scenarios. Large differences between probabilities of different flood characteristics are indicators of non-plausible scenarios. If e.g. the flood peak has a return period of 10 years, but the flood volume has a return period of 100 years, in the traditional way the return period of the flood event would be defined by its peak with 10 years. This is not plausible. The plausibility of the event seems to be high if the deviations between the different return periods are small. There are typical events where the return period of the peak and the return periods of other characteristics are similar and less typical events where these probabilities differ significantly. However, such a specification of plausibility is uncertain as the probabilities, which are used to derive it are uncertain. These uncertainties can be specified e.g. by fuzzy sets. Fuzzy sets are a way to consider uncertainties of set memberships (Klir and Smith, 2001). According to fuzzy theory the degree of membership of a single flood event to a set of scenarios with a certain return period of the flood peak can be specified by a triangular function. This degree of membership for each flood scenario is defined by the multivariate probabilities of its characteristics.

This approach is consistent with the generalised theory of uncertainty in application of fuzzy sets for possibilistic modalities of generalised constraints (Zadeh, 2005).

A generalised constraint is a constraint of the form:

$$X \text{ is } rR \tag{12.1}$$

where X is the constrained variable, R is a constraining relation and r is an indexing variable, which is blank for possibilistic constraints. Therefore,

$$\text{Poss}\,\{X = u\} = \mu_A(u) \tag{12.2}$$

where u is a generic value of X, μ_A is the membership function of A and A is a set of values in X.

A fuzzy number $A = (l, m, u)$ with $l \leq m \leq u$ is a triangular fuzzy number, if the membership function can be written as (Zadeh, 1965):

$$\mu_A(x) = \begin{cases} 0 & \text{for} \quad x \leq 1 \\ \frac{x-l}{m-l} & \text{for } l < x \leq m \\ \frac{u-x}{u-m} & \text{for } m < x \leq u \\ 0 & \text{for} \quad u < x \end{cases} \tag{12.3}$$

Generally triangular fuzzy numbers are described by a triple of real numbers as $\tilde{A} = (l, m, u)$ where $l \leq m \leq u$ represent the lower, modal and upper value of the fuzzy number. The modal value has a membership value of $\mu_A(m) = 1$. Here the highest value of the membership function ($\mu_A(u) = 1$) of hydrological load scenarios for inflows into a single reservoir was attributed to events where the bivariate copula probabilities between flood peak and volume are nearly the same as the probability of the flood peak. Such flood events seem to be most representative for a certain return period with regard to the agreement of the different statistical characteristics of the flood. If e.g. the return period, which was estimated from the joint probability of peak and volume, is greater than the return period of the peak, the event is less probable than expected from the return period of the peak alone. If this concordance between return periods is not given, e.g. if the joint volume-peak-probability indicates a more frequent event, the return period of the flood peak seems to be less plausible. To consider these differences a characteristic "plausibility" P_{Pl} is introduced, which can be derived from the differences in probabilities in the following way, using the logical "or" of the Copula statistics as described in Chapter 8:

$$P_{Pl} = \begin{cases} \text{Min}\left(\frac{T^{\vee}_{P,V}}{T_P}; \frac{2T_P - T^{\vee}_{P,V}}{T^{\vee}_{P,V}}\right), \forall T^{\vee}_{P,V} \in [0; 2T_P] \\ 0, \forall T^{\vee}_{P,V} \notin [0; 2T_P] \end{cases} \tag{12.4}$$

T_P and $T^{\vee}_{P,V}$ are return periods estimated from flood peak statistics respective from copula statistics of flood peak and volume.

12.3 Impact Assessments of Flood Control Measures

Often different objectives have to be considered simultaneously in flood control planning:

- Economic damages and the number of affected people are often the most important criteria. But also the spatial characteristics of these potential damages and endangered persons have to be considered.
- The effectiveness of flood protection depends strongly on the location. The fraction of the controlled catchment is decreasing with the watercourse downstream

of flood control systems. Thus the impacts of flood control on downstream reaches have to be considered in a spatial context.
- Costs of flood protection have to be differentiated between direct costs (e.g. construction costs) and indirect costs (e.g. limitations for land use in polders).

If flood scenarios are characterised by probabilities and imprecise probabilities, the simulated impacts of floods can be characterised by the same probabilistic characteristics. The plausibility criterion, which was introduced before, can be applied to differentiate between impacts.

12.4 Multi-Criteria Decision Making

In general it is difficult to consider probabilistic characteristics in Multi-Criteria Decision Making approaches (MCDM). Often only expected values are used. This is obviously not an appropriated way for flood management planning where several stochastic characteristics have to be considered. Here two different methodologies are presented which consider the probabilistic measures in different ways: a distance based method (TOPSIS) and a fuzzyfied version of the well-known AHP-method.

12.4.1 A Distance Based MCDM Tool – the TOPSIS Approach

Many MCDM-tools are based on a comparison of Euclidean distances between the realizations of criteria of alternatives and a (hypothetic) idealized point which is defined by the optimal values of these criteria. The TOPSIS-method (Technique for Order Preference by Similarity to Ideal Solution), which was developed by Hwang and Yoon (1981) belongs to this group of methodologies. The starting point is a decision matrix of M criteria ($F_j(j = 1 \ \ldots \ M)$), which were estimated for N alternatives $A_i(i = 1 \ \ldots \ N)$. The value of criterion j for alternative i is marked as f_{ij}:

$$
\begin{array}{c|cccc}
 & F_1 & F_2 & \cdots & F_M \\
\hline
A_1 & f_{11} & f_{12} & \cdots & f_{1M} \\
A_2 & f_{21} & f_{22} & \cdots & f_{2M} \\
\vdots & \vdots & \vdots & \cdots & \vdots \\
A_N & f_{M1} & f_{M2} & & f_{NM}
\end{array}
\qquad (12.5)
$$

To avoid dimensional problems between criteria which are different in their scales, the decision matrix has to be normalized. This normalization is done for each column of the matrix in the following way:

$$r_{ij} = \frac{f_{ij}}{\sqrt{\sum_{i=1}^{N} f_{ij}^2}} \quad \forall i = 1, \ldots, N \quad \forall j = 1, \ldots, M \qquad (12.6)$$

where r_{ij} is the normalized value of criterion j for alternative i.

In the next step the normalised criteria are weighted by multiplication with a weighting factor w_j

$$v_{ij} = w_j r_{ij} \quad \text{for } j = 1 \ldots M \tag{12.7}$$

The result of this procedure is a weighted matrix of normalized criteria:

$$\begin{bmatrix} v_{11} & v_{12} & \cdots & v_{1M} \\ v_{21} & v_{22} & \cdots & v_{2M} \\ \vdots & \ldots & \ldots & \vdots \\ v_{N1} & v_{N2} & \cdots & v_{NM} \end{bmatrix} \tag{12.8}$$

Now these criteria are differentiated into criteria which have to be maximized (e.g. damage reductions) and criteria which should be minimized (e.g. costs of constructions). The set of criteria which should be maximized are ordered by the index amount J, the set of criteria which have to be minimized by the index amount J'. The reachable positive ideal solution (PIS) has the coordinates v_j^+ $(j = 1 \ldots M)$. It is specified by the best results which were derived from all alternatives for each one of the M criteria. For criteria which have to be maximized it is defined by the maxima, for criteria which have to be minimized by the minima of all realizations:

$$\boldsymbol{V}^+ = \{v_1^+, \ldots, v_M^+\} = \left\{ \left(\operatorname*{Max}_i v_{ij} \,\middle|\, j \in J \right), \left(\operatorname*{Min}_i v_{ij} \,\middle|\, j \in J' \right) \right\} \tag{12.9}$$

The negative ideal solution NIS $\boldsymbol{V}^-$ is defined in a similar way. Here criteria, which have to be maximized, reach their minimum and criteria, which have to be minimized their maximum:

$$\boldsymbol{V}^- = \{v_1^-, \ldots, v_n^-\} = \left\{ \left(\operatorname*{Min}_i v_{ij} \,\middle|\, j \in J \right), \left(\operatorname*{Max}_i v_{ij} \,\middle|\, j \in J' \right) \right\} \tag{12.10}$$

For each one of the N alternatives two euclidean distances can be estimated over all criteria: the distance D_i^+ from the (fictitious) best-case alternative $\boldsymbol{V}^+$ and the distance D_i^- from the (fictitious) worst-case alternative $\boldsymbol{V}^-$.

$$D_i^+ = \sqrt{\sum_{j=1}^{M} \left(v_{ij} - v_j^+ \right)^2} \quad i = 1 \ldots N \tag{12.11}$$

$$D_i^- = \sqrt{\sum_{j=1}^{M} \left(v_{ij} - v_j^- \right)^2} \quad i = 1 \ldots N \tag{12.12}$$

The best alternative is located as close as possible to the ideal point ("utopia") and as far as possible from the worst point ("nadir"). Both distances have to be

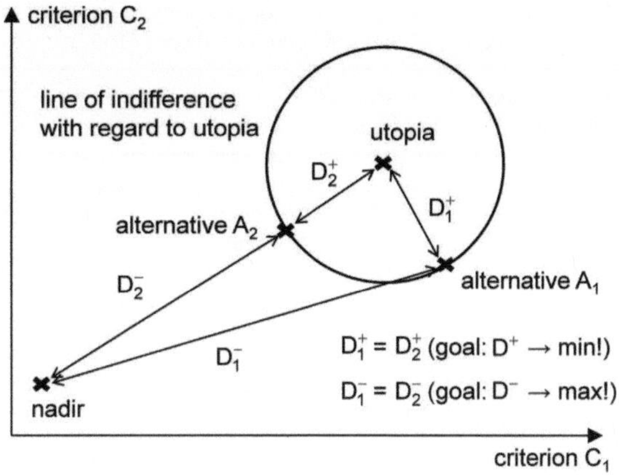

Fig. 12.1 Estimation of distances for the TOPSIS

combined to rank the alternatives. This can be done with the relative distance which is defined as follows:

$$\bar{C}_i = \frac{D_i^-}{D_i^+ + D_i^-} \tag{12.13}$$

This distance has a value between 0 and 1. The relative distance to NIS reaches its maximum value if an alternative performs better than all other alternatives. The geometrical interpretation of these distances is shown in Fig. 12.1 for the two-dimensional case. Here two alternatives have the same distance from the ideal point, but alternative A1 has a greater distance from the nadir point than alternative A2 and will therefore be preferred.

12.4.2 A Fuzzyfied Version of the Analytic Hierarchy Process Method (FAHP)

The AHP-method (Saaty, 1977, 1980) is based on a structuring of complex decision problems in a hierarchic way, where goals are subdivided into sub-goals. The realized criteria of alternatives are compared first in relationship to sub-goals. Then the results of these comparisons are combined by a weighting of sub-goals to goals. Alternatives are compared pairwise for each criterion. The results of these comparisons are summarised in a symmetric matrix specifying the pair-wise relationships of alternatives with regard to one criterion. The consistencies of these comparisons can be evaluated with eigenvectors of the resulting matrices. The ranking of alternatives is based on a weighting of criteria. The resulting weighting vector contains factors which specify the relative importance of each criterion. These factors are derived as eigenvectors from a matrix of pairwise comparisons of criteria according to the preferences of decision makers. Saaty suggested a rating scale which is shown in Table 12.1 (second column). A consistency ratio, which is derived from the eigenvector can be used to evaluate the consistency of the rating.

Table 12.1 Scale of pairwise comparisons of criteria after Saaty (1980) (modified) and in a fuzzyfied version after Srdjevic and Medeiros (2007)

Rating scale of Saaty	Definition		Fuzzyfied Saaty scale
1	Equal importance	Equal result	(1, 1, 1 + δ)
3	Somewhat more important	Somewhat better	(3 − δ, 3, 3 + δ)
5	Much more important	Much better	(5 − δ, 5, 5 + δ)
7	Very much more important	Very much better	(7 − δ, 7, 7 + δ)
9	Absolutely more important	Extremely better	(9 − δ, 9, 9)
2, 4, 6, 8	Intermediate values		(x − 1, x, x + 1), x = 2, 4, 6, 8

δ – Fuzzy-Distance $(0,5 \leq \delta \leq 2,0)$

One problem of AHP consists in the subjectivisms of evaluations. There are two groups of subjective comparisons: the criteria have to be compared in their relative importance and the outcomes of alternatives have to be compared pairwise for each criterion. The uncertainties of these comparisons can be considered by fuzzy numbers. Srdjevic und Medeiros (2007) suggested a fuzzyfied version of the AHP-rating of Saaty, where the comparisons of the relative importance of criteria and the relationships between realizations are handled as fuzzy numbers which are defined according to the column on the far rigth in Table 12.1. This so-called "Fuzzy-AHP- method" (FAHP) combines the AHP- method of Saaty (1980) with the Fuzzy-theory of Zadeh (1965). It is an extension of AHP as it gives options to consider uncertainties in ratings and weights explicitly. However, the measures of consistency which are derived in AHP from eigenvectors of the matrices are not defined here. In the following the mathematical operations of Fuzzy-AHP are described based on publications of Deng (1999) and Taslicali and Ercan (2006).

In a first step the criteria have to be ordered hierarchically by differentiating goals and sub-goals. At one hierarchic level the criteria are pairwise compared with regard to the relative importance among the two compared criteria for the superior hierarchy criterion. To give an example: the criterion "economic damages" can be subdivided into damages in domestic areas, agricultural areas and industrial areas. In AHP the rating among these damages could be based e.g. on a comparison of domestic and industrial damages, where a rating of 3 (somewhat more important) and between domestic and agricultural damages, where a rating of 5 (much more important) could be applied. In FAHP triangular fuzzy-numbers are chosen for these comparisons, based on the fuzzyfied scale of Saaty (Table 12.1). To define triangular fuzzy numbers a distance value δ is used which varies between 0.5 and 2.0. Srdjevic und Medeiros (2007) suggest to use $\delta = 2$ for uneven real numbers and $\delta = 1$ for even real numbers. In FAHP the rating among damages which was given in the example above would be (1; 3; 5) for the relationship between domestic and industrial damages, and (3; 5; 7) for relationships between domestic and agricultural damages. All further computational steps of FAHP are similar to AHP. The difference between both methods consists in the need to handle fuzzy numbers in the FAHP. The pairwise comparison of criteria is based on the fuzzyfied Saaty scale:

$$\tilde{A} = \begin{bmatrix} \tilde{a}_{11} & \tilde{a}_{12} & \cdots & \tilde{a}_{1M} \\ \tilde{a}_{21} & \tilde{a}_{22} & \cdots & \tilde{a}_{2M} \\ \vdots & \vdots & \ddots & \vdots \\ \tilde{a}_{M1} & \tilde{a}_{M2} & \cdots & \tilde{a}_{MM} \end{bmatrix} \tag{12.14}$$

where $\tilde{a}_{\mathrm{ij}}$ is a fuzzy number which expresses the preference of the decision maker for criterion i in relationship to criterion j with $\tilde{a}_{ij} = (1;\ \ 1;\ \ 1)$ for all $i = j$ $(i,j = 1,2,\ldots,M)$ and $\tilde{a}_{\mathrm{ij}} = 1/\tilde{a}_{ji}$.

The weighting of the criteria can be estimated from the normalization of this matrix as follows (Chang, 1996):

$$\tilde{\boldsymbol{w}}_i = \sum_{j=1}^{M} \tilde{a}_{ij} \otimes \left[\sum_{k=1}^{M} \sum_{l=1}^{M} \tilde{a}_{kl} \right]^{-1} \tag{12.15}$$

This results in a weighting vector **w** which consists of M fuzzy numbers, expressing the relative weight of each of the M criteria:

$$\boldsymbol{w} = [\tilde{\boldsymbol{w}}_1, \tilde{\boldsymbol{w}}_2, \ldots\ldots, \tilde{\boldsymbol{w}}_M]^T \tag{12.16}$$

For each criterion a symmetric matrix with N columns and N rows is specified to compare the N alternatives. The element a_{ij} of this matrix compares the realization of this criterion for the alternative i in relationship to alternative j. These relationships are also specified by fuzzy numbers:

$$\tilde{A}_k = \begin{bmatrix} \tilde{a}_{11} & \tilde{a}_{12} & \cdots & \tilde{a}_{1N} \\ \tilde{a}_{21} & \tilde{a}_{22} & \cdots & \tilde{a}_{2N} \\ \vdots & \vdots & \ddots & \vdots \\ \tilde{a}_{N1} & \tilde{a}_{N2} & \cdots & \tilde{a}_{NN} \end{bmatrix} \quad \text{with } k = 1 \ldots M \tag{12.17}$$

Each of these M matrices has to be normalized to specify the ranking of each of the N alternatives with regard to each criterion k for $k = 1 \ldots M$:

$$W_i^k = \sum_{j=1}^{N} \tilde{a}_{ij} \otimes \left[\sum_{l=1}^{N} \sum_{m=1}^{N} \tilde{a}_{lm} \right]^{-1} \quad \text{for } k = 1 \ldots M \text{ and } i = 1 \ldots N \tag{12.18}$$

The fuzzy-numbers W_i^k are elements of a matrix which specifies the priorities among alternatives. It consists of M columns (for each criterion) und N rows (for each alternative):

$$X = \begin{bmatrix} W_1^1 & W_1^2 & \cdots & W_1^M \\ W_2^1 & W_2^2 & \cdots & W_2^M \\ \vdots & \cdots & \cdots & \cdots \\ W_N^1 & W_N^2 & \cdots & W_N^M \end{bmatrix} \tag{12.19}$$

As result of this operation one receives a matrix of fuzzy numbers which can be multiplied with the weighting of criteria, which was estimated before as vector **w**. One receives a performance matrix which specifies the ranking of alternatives for all criteria:

$$Z = \begin{bmatrix} W_1^1 \otimes \tilde{w}_1 & W_1^2 \otimes \tilde{w}_2 & \cdots & W_1^M \otimes \tilde{w}_M \\ W_2^1 \otimes \tilde{w}_1 & W_2^2 \otimes \tilde{w}_2 & \cdots & W_2^M \otimes \tilde{w}_M \\ \vdots & \cdots & \cdots & \cdots \\ W_N^1 \otimes \tilde{w}_1 & W_N^2 \otimes \tilde{w}_3 & \cdots & W_N^M \otimes \tilde{w}_M \end{bmatrix} \tag{12.20}$$

In the last step a ranking of alternatives will be provided to decision makers by summarizing the rows of this performance matrix:

$$F_i = \sum_{j=1}^{M} W_i^j \otimes w_j \quad \text{for } i = 1 \ldots N. \tag{12.21}$$

The result of this procedure is a single fuzzy number for each of the N alternatives, which can be used to compare it with other alternatives. There are several options to compare fuzzy numbers. Here the "Total Integrated Value" method, which was suggested by Liou and Wang (1992) was applied. This de-fuzzyfication is flexible in handling the upper and lower bounds. An integrated value I_λ is estimated from a triangular fuzzy number which is specified by its lower value l, its upper value u and the modal value m as follows:

$$I_\lambda = \frac{1}{2}[\lambda u + m + (1 - \lambda)l]. \tag{12.22}$$

The parameter λ weights the lower and upper bound of the fuzzy number. For $\lambda = 1$ one receives the average of u and m, for $\lambda = 0$ the upper value is not considered and the average of m and l is estimated and for $\lambda = 0.5$ the modal value has a higher importance than the lower or upper bounds. The results will differ accordingly to the chosen λ value. With this flexible de-fuzzyfication the user gets information about the sensitivity of the ranking with regard to the range of possible outcomes.

12.5 Case Study

The methodology described above has been applied to the Unstrut river basin in the central part of Germany. This study area is the same as was presented in Chapter 8. The river basin has an area of around 6,400 km^2 and is situated in two different Federal States of Germany, upstream in the Federal State of Thuringia and downstream in the Federal State of Saxony-Anhalt (Fig. 12.2). This geographic location results in an uneven distribution of benefits and burdens of flood control. The upstream flood control system belongs to Thuringia, but a large part of flood protected areas are situated in Saxony-Anhalt. The catchment has a heterogeneous topographic structure, with lower regions in the central part, the Harz Mountains

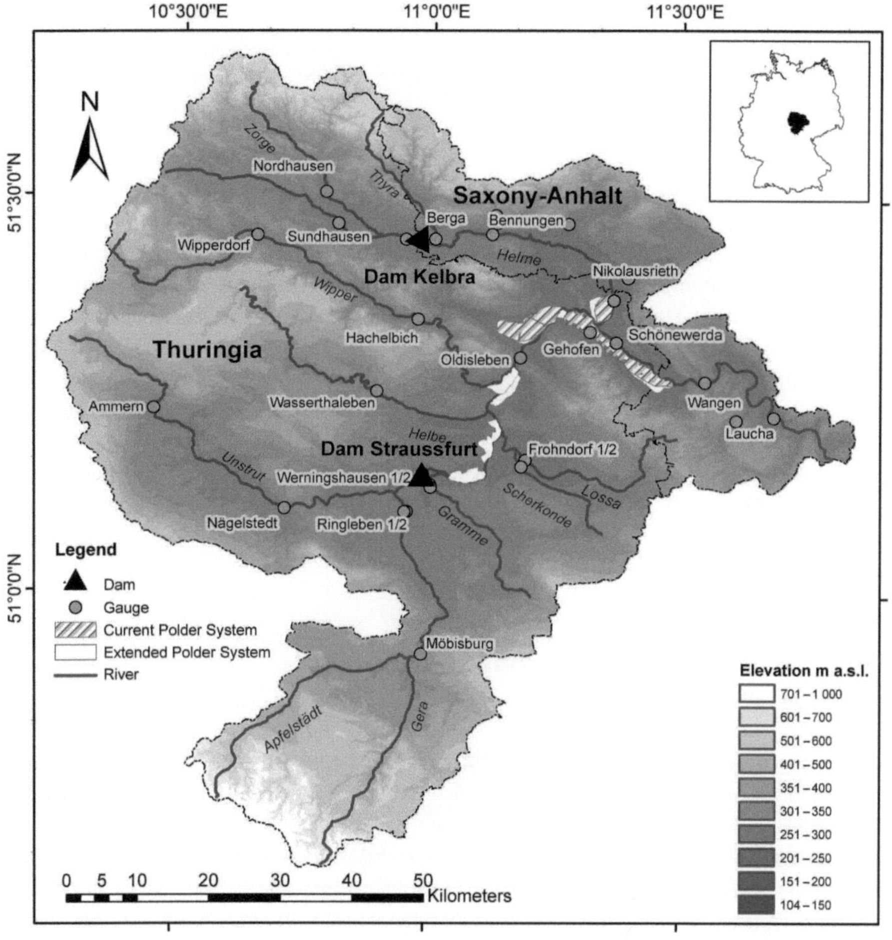

Fig. 12.2 Topographical map of the Unstrut catchment in Thuringia and Saxony-Anhalt with components of the technical flood retention system (current and extended) and important gauges within the catchment

Table 12.2 Planning states of the flood control system

State of the system	Specifications
SS1	As-is state. Only a small percentage of system retention capacity can be used due to malfunctioning technical elements
SS2	As-is state but with working inlet structures. The polder system consists of polders Oldisleben, Kannawurf, Moenchenrieth, Artern and Wiehe and is fully operational; use of check dams
SS3	Extension of the system. State 2 plus additional polders Riethgen, Waltersdorf, Soemmerda, Schluesselwiesen to increase the retention capacity
SS4	Alternative extension of the system. State 2 plus additional polders Riethgen, Waltersdorf, Wundersleben, Scherndorf; increased retention capacities in comparison with state 3 plus new polders; the polder inlet structures are not controlled
SS5	As state 4 but with controlled polder inlets
SS6	As state 5 but with widened inlet structures of new planned polders

in the North and the Thuringian Forest in the South. At the moment the technical flood retention system within this river basin consists of the reservoir Kelbra and the reservoir Straussfurt, of some other smaller reservoirs of local importance, a flood channel and a flood polder system with five polders situated between the cities of Oldisleben and Wangen (see Fig. 12.2). In total the flood retention system has a volume of $\sim 100 \times 10^6$ m^3. The reservoir Kelbra is operated by the state of Saxony-Anhalt, whereas the remaining flood retention facilities are operated by the state of Thuringia. The local planning authorities suggested a set of flood control measures, varying from the optimization of the existing polders, increase of retention time within polders by additional check dams to creation of new polders, alteration of the polder inlet structures and different types of inlet regulations (controlled and uncontrolled flooding). These measures were clustered into six states of the flood retention system. The "as-is" state is denoted as state 1 and the most complex state as state 6 (see Table 12.2).

12.5.1 Specification of Hydrological Loads

A long series of runoff (10,000 years) was simulated on a daily basis by coupling a stochastic rainfall generator and a deterministic hydrological model which was calibrated as described in Section 4.2 of Chapter 8. These data were essential for this study as the measured discharge data since the 1960s were affected by the construction of two flood reservoirs and river regulation works. Thus the long-term characteristics of the current flood regime can only be specified by simulations. The adaption of the hydrological model was based on measured rainfall-runoff data, which were available from two flood events only. A non-influenced inflow gauge situated upstream of the flood reservoir Straussfurt was used to verify the flood statistical results provided by the stochastic-deterministic simulations with the statistics of a series of 40 years measured discharge data. Based on simulated daily runoff values for 10,000 years 30 flood events were selected as described in Chapter 8. For each return period of 25, 50, 100, 200, 500 and 1,000 years five events were chosen, which cover a wide range of spatial distributions of runoff, peak-volume relationships and hydrographs. Low-volume summer events, high volume spring floods, floods with multiple peaks, floods which are influenced by antecedent conditions of high soil moisture and floods with high spatial heterogeneity in precipitation were considered to characterise the large variety of hydrological loads. The results of multivariate statistical analyses of simulated data, which are described in Chapter 8, were used to characterise the multiple probabilities of these events. Copula statistics were applied to estimate joint probabilities of the flood peaks and corresponding volumes at the inflows to the dams Straussfurt and Kelbra as well as probabilities of coincidences of flood peaks at both reservoirs. A detailed description of the model selection and the copula analyse is given by Klein et al. (2010).

In Fig. 12.3 the inflow-outflow relationships of the two reservoirs for two different floods with a return period of the peaks of 100 years are shown to demonstrate the necessity for a multivariate approach. Both events have a flood peak of about 300 $m^3\ s^{-1}$ at the inflow to the Straussfurt reservoir. This classifies both events with a statistical return period of 100 years. However, the hydrographs differ significantly. The floods in the top of Fig. 12.3 have a preliminary peak, two large peaks and a large volume and can not be buffered significantly by both reservoirs. Compared with the second group of floods (at the bottom), which is characterised by a rather regular hydrograph with a moderate volume, this event would cause damages that are 30 times higher. The analysis of the bivariate copula probability of the flood peak and volume at gauge Straussfurt reveals that the flood with the large volume has a multivariate return period of 681 years, whereas the other flood has a bivariate return period of only 134 years. In this case, it would obviously be wrong to attribute the same probability to both events.

The performance of the flood control system is significantly affected also by the spatial distribution of rainfall. If the rainfall volume is concentrated on the catchments in the Northern and Southern parts of the basin, the runoff can be controlled by the reservoirs. The effect of both reservoirs is limited if the main volume of the rainfall falls in the central part of the basin. Bivariate probabilities of synchronous flood peaks at the inflow gauges of the Kelbra and the Straussfurt reservoirs can be applied to consider such effects.

The runoff along the river course was simulated with a hydraulic model for 180 events (30 hydrological scenarios and six different states of the flood control system). The following characteristics were estimated: (i) inundation areas, (ii) maxima of water levels, (iii) maxima of flow velocity, (iv) the maxima of the products of water level and flow velocity, (v) the total duration of the flood events and (vi) the time of exceedance for certain water level thresholds. Operation schemes for reservoirs and polders were applied, which were based on analyses of the actual operation of the existing flood storage facilities. Costs and benefits of the planned measures were estimated based on these hydraulic simulations. Here costs of operation and maintenance of polders, but also costs of temporary flooding of agriculturally

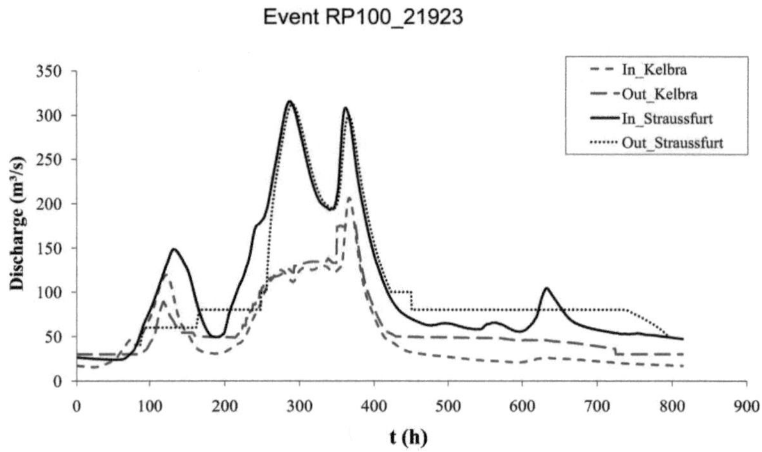

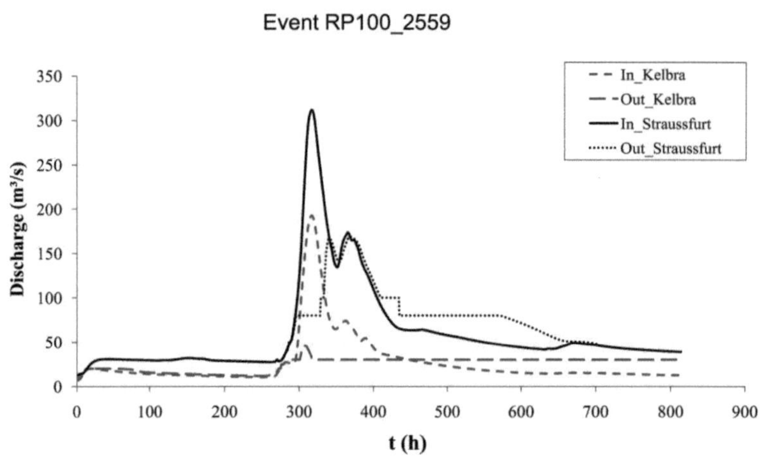

Fig. 12.3 Inflow and outflow hydrographs of two flood events with a return period of 100 years (defined by the peaks) for the reservoirs Kelbra (North) and Straussfurt (South)

used polders were considered. These costs were compared with potential reductions of damages. Since pecuniary damage varies with flood-specific parameters, the absolute values were estimated for each flood with land-use type specific damage functions. These functions specify the relative degree of damage, which can be expected from a certain hydrological load according to its water level, flow velocity and duration for a specific land use. The damage functions were combined with analyses of land use of inundation areas, which were provided by a Geographic Information System (GIS). With application of GIS it became possible to automate the calculation of geographically distributed economic risks and risks for affected persons as well as to specify vulnerable localities (e.g. schools, nursery homes, hospitals etc.) which would be affected by a flood. These characteristics were estimated and mapped for all simulated flood events.

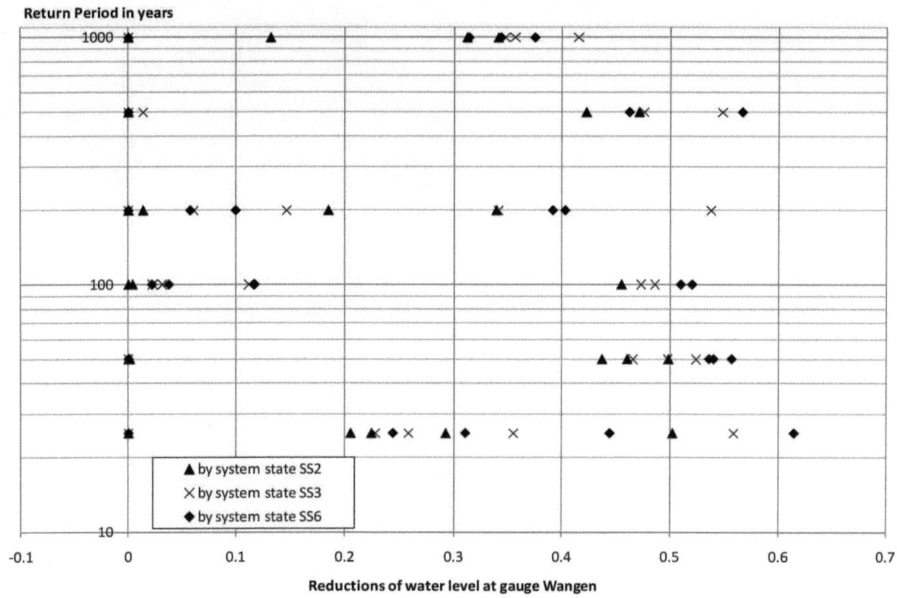

Fig. 12.4 Reduction of the flood peak at the basin outlet Wangen

In general the results of simulations show an ambivalent role of the flood control system. In Fig. 12.4 the reductions of flood peaks with system states 2 and 6 compared with the as is state of the system is shown. Under favourable conditions even peaks of very rare floods can be reduced. On the other hand the extended flood control system may have almost no impacts on floods with return periods of 50 years. Obviously a more detailed specification of floods is needed to characterise the efficiency of the flood control system.

By inspection of the hydraulic simulations it became also evident that flood damages could be increased by new polders under unfavourable conditions. One example for this effect is shown in Fig. 12.5. Here a part of the city Soemmerda is flooded if new polders are built on natural retention areas. Especially for rare flood events the hydraulic conditions may be worsened by additional dykes. Settlements may be affected if the volume-peak relationships are be unfavourable and if new polders zoned by dykes are built within natural retention areas.

The results of the bivariate statistical analyses provide indicators to characterise hydrological loads by their plausibility. There are representative flood events where peak, volume and spatial distribution of runoff have nearly the same probabilities. For other events the probabilities of these characteristics differ significantly. The plausibility of any single flood event was specified as it was discussed above. The measure of plausibility was derived from different Copula-based probabilities. For river sections which are affected by one reservoir only, the relationship between the flood peak and volume at the inflow of this reservoir was most relevant, but for river sections downstream of the conjunction of the Unstrut River with its tributary

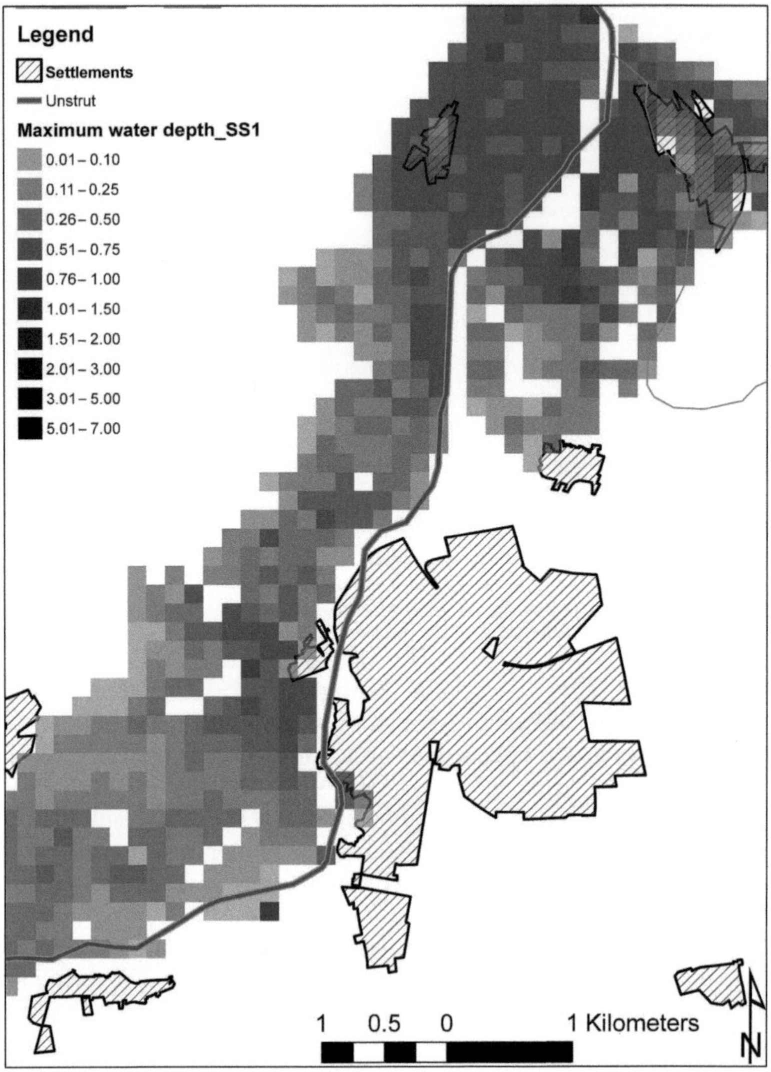

Fig. 12.5 Differences between inundation areas, here the actual state, in the next figure with additional polders (system state 6)

Helme River, the Copula probability of spatial flood coincidences is more important. Exemplary only Copula probabilities for peak and volume at the gauge Straussfurt were used here. Based on this characterisation of plausibility the most plausible ones of the set of five flood events with the same return periods of the peak were estimated. Together with the events with minima of plausibility in both directions (where Copula based return periods are lower or higher than the return period of the peak) a fuzzy number was specified, which is assumed to be representative for this return period. Thus the damage characteristics were fuzzyfied according to these triangular fuzzy numbers. An example of fuzzyfied damages is given in Fig. 12.6.

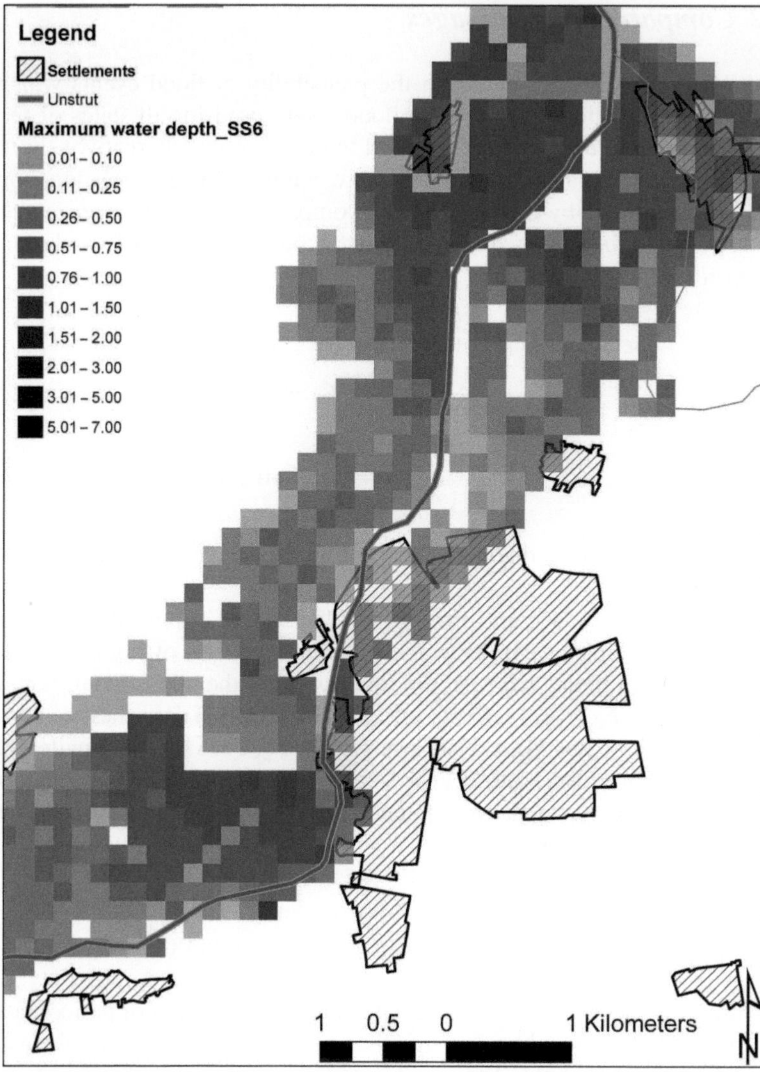

Fig. 12.5 (Contd.)

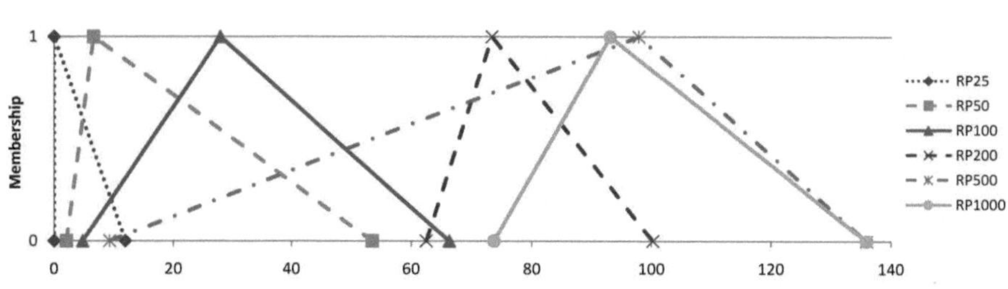

Fig. 12.6 Fuzzified damages for the reference system ("status quo") for the return periods RP 25, 50, 100, 200, 500 and 1,000 years

12.5.2 Comparison of Damages

The fuzzyfied damages are based on the plausibility of flood events which were used for their estimation. As the same floods were used for all states of the flood control system these damages can be used to compare the alternatives. In the first approach these triangular fuzzy numbers were compared using a relational operator *V* which was suggested by Chang (1996). It compares the modal values *m*, the lower (*l*) and upper (*u*) bounds of two triangular fuzzy numbers F_1 and F_2 and returns the following results $V(F_2 > F_1)$:

$$\begin{array}{ll} 1 \ \textit{if} \ m_2 \geq m_1 & \text{or} \\ 0 \ \textit{if} \ l_1 \geq u_2 & \text{and} \end{array} \tag{12.23}$$

$$\frac{l_1 - u_2}{(m_2 - u_2) - (m_1 - l_1)} \text{ in all other cases.} \tag{12.24}$$

The last value specifies the intersection of the two fuzzy numbers F_1 and F_2, as it is shown in Fig. 12.7.

Both values $V(F_1 \geq F_2)$ and $V(F_2 \geq F_1)$ have to be estimated to compare two fuzzy numbers completely. In every case one comparison of non-identical fuzzy numbers will result in the value "1". In this case only the minimum of both results $V(F_1 \geq F_2)$ and $V(F_2 \geq F_1)$ is important. This intersection describes the possibility that both fuzzy numbers have a relationship $F_1 \geq F_2$, even if m_1 is smaller than m_2. If the operator returns the result "0", this option can be excluded. To compare several triangular fuzzy numbers, pair-wise comparisons of fuzzy numbers are needed. One receives the degree of possibility that a convex fuzzy number *F* is greater than *k* other convex fuzzy numbers $F_j(i = 1, \ldots, k)$ by

$$V[(F \geq F_1) \text{ and } (F \geq F_2) \text{ and } \ldots \text{ and } (F > F_k)] = \min V(F \geq F_i)(i = 1, \ldots k) \tag{12.25}$$

The different categories of damages estimated for each return period and states of the flood control system by fuzzy numbers can be compared in this way with simulated results of other states. The result describes the degree of possibility that the

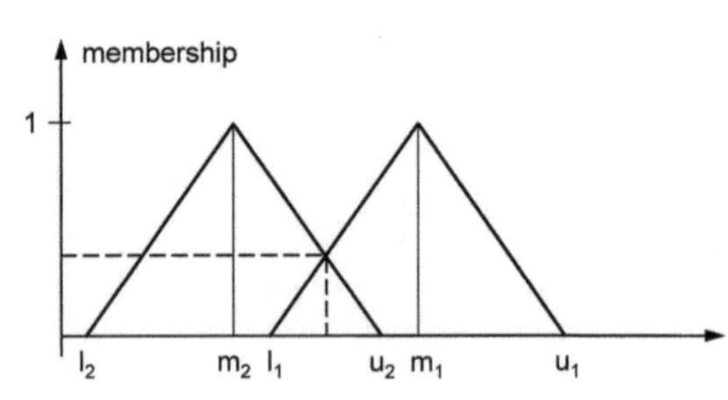

Fig. 12.7 Comparisons of two triangular fuzzy numbers with intersection

Table 12.3 Possibility that a certain state of the system (SS1–SS6) would result in higher economic damages than all other alternatives, differentiated by return periods (RP in years)

SS	RP=25	RP=50	RP=100	RP=200	RP=500	RP=1,000
1	0.167	0.167	0.196	0.095	0.144	0.085
2	0.167	0.167	0.196	0.082	0.142	0.085
3	0.167	0.167	0.151	0.116	0.152	0.106
4	0.167	0.167	0.153	0.237	0.188	0.242
5	0.167	0.167	0.152	0.234	0.187	0.239
6	0.167	0.167	0.152	0.236	0.187	0.243

alternative i will result in higher damages than all other alternatives j. An example is given for economic flood damages for each return period in Table 12.3.

These results can be interpreted as follows: For small return periods (RP = 25 and RP = 50 years) the states of the system are indifferent, all states perform in the same way. For flood scenarios with a return period of the flood peak of 100 years, the flood damages would be reduced by additional polders (SS3–SS6). For this return period the possibilities of higher damages of the system states 1 and 2 are higher than for other states of the system. For more extreme floods (with return periods of 200 and more years) the risk of additional damages increases with additional polders. It exceeds the potential benefits of additional flood storage capacities. For the system states 4–6 the risk of higher damages is more than doubled compared with states 1–3 for extreme floods with return periods of 200 and 1,000 years. This assessment demonstrates that decisions about the modification of the flood control system have to balance the options to reduce damages for low return periods with the increase of risks of additional damages caused by newly built retention facilities for events with higher return periods. In the next step the TOPSIS approach was applied to compare options and risks of the different states of the system.

12.5.3 Application of TOPSIS

The following criteria were estimated to compare the alternatives of planning:

- Economic damages differentiated between damages within and outside of settlements and (compared with the state at present) increased damages and reduced damages
- Affected people
- Reduced flood risk downstream of the outlet gauge Wangen, which is characterised here by reductions of the peak flows at Wangen.

As was shown above, damages can be reduced or increased by extensions of the flood control system according to the multiple characteristics of the flood. At different locations within the river basin damages can be decreased by improved retention or increased by side-effects of polders. Here the distance-based evaluation differentiates between risk and opportunities. The resulting DSS considers four criteria

Table 12.4 TOPSIS-distances of system states, optimal is the maximum (numbers printed in bold)

	SS1	SS2	SS3	SS4	SS5	SS6
Reduction of flood peaks at the basin outlet						
All floods	0.013	0.603	**0.899**	0.825	0.831	0.868
Frequent floods only	0.008	0.717	**0.883**	0.873	0.839	0.943
Rare floods only	0.017	0.583	**0.947**	0.821	0.812	0.822
Maximized damage reduction in the Unstrut basin upstreams Wangen						
All floods	0.023	0.507	**0.819**	0.829	0.842	0.830
Frequent floods only	0.015	0.431	**0.847**	0.862	0.872	0.865
Rare floods only	0.030	0.586	0.772	0.797	**0.811**	0.798
Minimized increase of damages in the Unstrut basin upstream of gauge Wangen						
All floods	**0.994**	0.809	0.271	0.155	0.219	0.185
Frequent floods only	**0.993**	0.742	0.185	0.144	0.252	0.204
Rare floods only	**0.994**	0.860	0.325	0.153	0.169	0.154
Combined goals: flood peak reduction, maximized damage reduction, minimized increase of damages						
All floods	0.550	0.562	**0.581**	0.517	0.539	0.516
Frequent floods only	0.509	**0.546**	0.532	0.524	0.536	0.528
Rare floods only	0.575	0.598	**0.613**	0.509	0.521	0.496

(maximized damage reductions, minimized increases of damages, affected people and reductions of flood peaks).

For each flood event one of the six states of the system gives a best result for a certain criterion and another state a worst one. For all flood scenarios within one class of return periods and for all alternatives the Euclidian distances between the results and the optimal values (minimal values for damages and maximal values for flood peak reductions) were estimated for each criterion. The distances of each alternative to the ideal point (PIS) and the nadir (NIS) were estimated for all single flood events and summarised to distance values from NIS and PIS for each class of return periods. The plausibility of flood scenarios was not considered directly in this approach. However during the aggregation of distances within each class of return periods, there is an option to consider only events above a certain threshold of plausibility. The decision maker weights the importance of the different criteria and the return periods to combine these results. In Table 12.4 some results are shown to demonstrate the flexibility of the TOPSIS results in relationships to different weightings.

The optimal distance value in each category is “1” (greatest distance from NIS). If the most important goal is flood protection downstream, measures with additional polders would be selected. One receives the same result if damage reductions within the river basins would be given priority in planning. If the risk of additional damages will be emphasised, all additional polders would be refused. If these three targets are combined with equal weights any differentiation between the alternatives becomes difficult. At least one distance from the worst point is small in all cases where the criteria are weighted equally.

12.5.4 Application of Fuzzy-AHP

In the next step Fuzzy-AHP was applied. To arrange the criteria hierarchically, goals and sub-goals have to be specified. Here five hierarchic levels, which are shown in Fig. 12.8, were applied. The flood scenarios were differentiated first of all according to their return period classes which were specified with the return period of the flood peak at the most central gauge of the river basin (gauge at the inflow into the Straussfurt reservoir). Costs are a criterion which can be differentiated into two parts, the costs of construction and the losses of agriculture (crop failures) if polder areas are flooded. The benefits of the alternatives are specified by differentiation between flood damages (here damages in settlements and damages outsides are considered separately), flooded areas, affected people and reductions of the water level at gauge Wangen.

The decision maker has to specify his priorities by a weighting of the criteria. This is done comparing two criteria with fuzzy-numbers according to Table 12.5. This weighting can be modified if the decision maker explores the decision space interactively. Other weightings which were included in the developed Decision Support System (DSS) are:

1. Weighting by return periods (fourth level): All return periods can be weighted equally, higher weightings of rare or of frequent events are possible (frequent

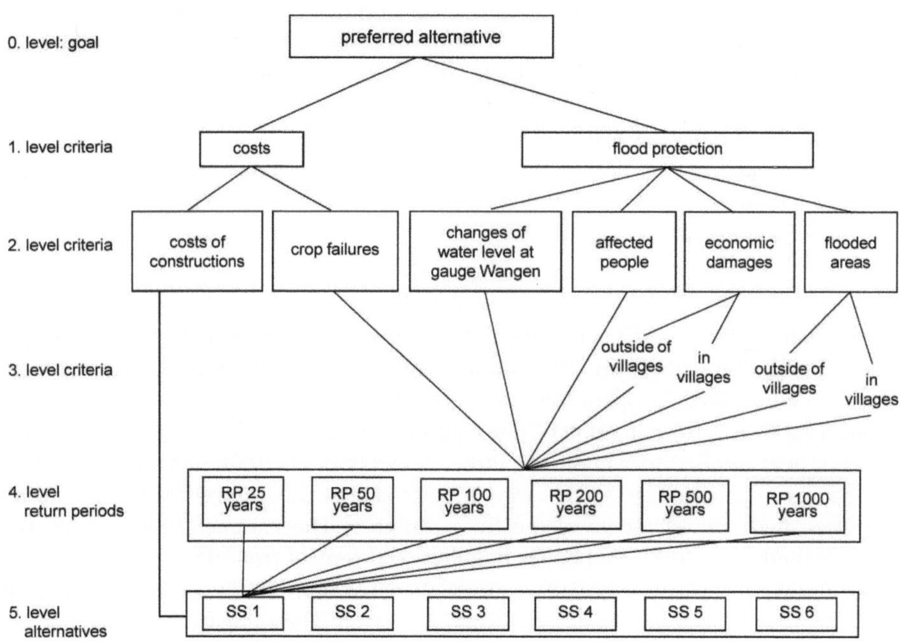

Fig. 12.8 Hierarchic structure of the FAHP-approach

Table 12.5 Impact of the parameter λ on de-fuzzyfication of the results of FAHP

λ	SS1	SS2	SS3	SS4	SS5	SS6
0	0.11	0.12	0.11	0.06	0.07	0.07
0.5	1.90	2.01	1.91	1.21	1.58	1.40
1	3.70	3.91	3.72	2.36	3.10	2.73

events are events with return periods of the peak from 25 to 100 years). Three options are offered.

2. At the third level damages can be weighted according to their locations. Three options are possible: higher weighting of damages outside of settlements or of damages within settlements, or an equal weighting of these areas (three variants).
3. At the second level the criteria of "flood protection" can be weighted in seven variants. Focus can be given on the following components:
   a. economic damages
   b. flooded areas
   c. affected people
   d. changes of water level at gauge Wangen
   e.- g. focus on changes of water level at gauge Wangen plus one of the criteria mentioned under a–c.
4. For the criterion "cost" in the second level a higher weight could be given to construction costs or to losses of agriculture. Agricultural losses were specified not by expected values, based on integration of all flood probabilities, but with losses which were derived from the applied flood scenarios. The seasonal variability of agricultural damages was considered.
5. With regard to the possible combinations three variants of weighting were offered: At the first level a higher importance could be given to costs or to flood protection efficiency. Both criteria could be weighted equally or one could be emphasized.

In total we receive 567 possible combinations (3·3·7·3·3), which can be specified within the DSS interactively. All criteria were provided as triangular fuzzy numbers for each alternative. The pair-wise comparison of the criteria of two alternatives was done after the de-fuzzyfication of these numbers. Here the "Total Integrated Value" method was applied to weight the three elements of the triangular fuzzy number according to a parameter λ. This weighting considers the differences in plausibility between the three elements of the fuzzy number. The parameter value $\lambda = 0.5$ was applied if the lower and upper bounds were derived from floods with the same plausibility. If the upper bound is more plausible than the lower one, a value of $\lambda > 0.5$ is used, in the reverse case a value below 0.5. The comparison of alternatives was based on fuzzy numbers which are specified in Table 12.1. Fuzzy numbers were applied also for the weighting of the criteria. As a result of FAHP fuzzy numbers are

Table 12.6 Results of the Fuzzy-AHP approach with focus on flood protection and equal weighting of damages at settlements and non-populated areas, De-fuzzyfication with the Total Integrated Value (λ=0.5), optimal is the maximum (numbers printed in bold)

Main goal	SS1	SS2	SS3	SS4	SS5	SS6
Reduction of flood peaks at the basin outlet						
All scenarios	1.11	1.27	**1.50**	1.15	1.25	1.19
Frequent floods only	0.78	0.95	**1.06**	0.95	1.04	0.99
Rare floods only	0.91	0.97	**1.19**	0.78	0.84	0.78
Maximized damage reductions in the Unstrut basin upstream gauge Wangen						
All scenarios	1.90	**2.01**	1.91	1.21	1.58	1.40
Frequent floods only	1.30	**1.47**	1.28	1.07	1.40	1.20
Rare floods only	**1.55**	**1.55**	**1.55**	0.72	0.94	0.85
Combined goals: flood peak reduction, maximized damage reduction, minimized increase of damages						
All scenarios	1.44	**1.57**	**1.57**	1.06	1.28	1.18
Frequent floods only	1.01	**1.18**	1.07	0.91	1.10	0.99
Rare floods only	1.16	1.20	**1.28**	0.67	0.80	0.75

estimated which have to be de-fuzzyfied to describe the relative performance of each alternative compared with all other alternatives. The decision maker has the choice to select between three versions of de-fuzzyfication with the Total Integrated Value method. He may emphasise the differences between the alternatives and criteria with a value of $\lambda = 1$ or be more indifferent with $\lambda = 0$. The effect of the choice of the parameter λ is demonstrated in Table 12.5 for the main goal "flood protection" with a focus on "economic damages", an equal weighting of "damages within and outside of settlements" and equal importance of all floods without consideration of their return periods. In this example the system state 2 would be preferred; however, the differences between the system states 1–3 and 4–6 are small if the upper bound of the resulting fuzzy numbers is not considered ($\lambda = 0$). To get a better impression about the differences among alternatives, a parameter value of $\lambda = 0.5$ was chosen. The results are presented in Table 12.6.

A comparison of the results of both methodologies (TOPSIS and FAHP) demonstrates similarities and differences:

- With the main goal of a reduction of flood peaks at the basin outlet FAHP clearly prefers System State 3. TOPSIS differentiates between rare and frequent floods. For frequent floods, additional polders with large inlet structures are preferred (System State 6) (without consideration of the different plausibilities of flood events).
- If damage reductions within the Unstrut basin were chosen as the main goal, FAHP sees no need for additional polders but prefers the version SS2 (repaired inlet structures for existing polders). Here the damage reductions and possible increase of damages were not considered separately. TOPSIS would prefer additional polders (System State 5) under consideration of possible flood damage

reductions within the basin but refuse them if the focus is directed on the possible increase of damages (also System State 2 is preferred in this case).
- For a combination of the two goals "flood peak reduction at the outlet" and "damage reduction within the basin" both methods deliver nearly the same results. Preference is given to System State 3 under consideration of all floods, with focus on frequent floods to System State 2 but with special emphasis to rare floods System State 3 would be preferred.

12.6 Conclusions

Risk oriented planning depends strongly on the information which can be used to specify hydrological risk. With regard to the limited technical capacities for flood protection the remaining risk of such systems should be identified. This specification was done here under consideration of multiple flood characteristics which are important for the functionality of technical flood retention systems. The combination of these characteristics makes the difference between system failures and effective flood control. Flood scenarios with a probabilistic characterisation by multivariate statistics can be applied to improve flood control planning with special emphasis on possible failures and remaining risks. The application of multivariate statistics demands a large data base which can be derived from simulations with a coupling of stochastic and deterministic models. However, the results will be uncertain and derived statistics should be handled as uncertain as well. This can be done with "imprecise probabilities". There are several options to evaluate flood control systems and to optimize them. It has been shown that a comparison of costs and benefits is difficult if the benefits are affected by numerous initial and boundary conditions. To consider multiple criteria (e.g. the system performance under different hydrological loads) the application of Multi-Criteria Decision Making is useful. These tools should be integrated in a DSS with a user interface which gives decision makers the opportunity to explore the decision space interactively. Two examples were shown to demonstrate the relevance of imprecise probabilities (which was considered in the Fuzzy-AHP approach) and the side-effects of flood protection measures, which were considered in the TOPSIS-approach. Both methods have advantages and disadvantages. However, it was shown that the results with common goals are similar. This demonstrates the practical value of the two proposed methodologies.

References

Blazkova S, Beven K (2002) Flood frequency estimation by continous simulation for a catchment treated as ungauged (with uncertainty). Water Resour Res 38(8): art. no.-1139.

Cameron DS, Beven KJ, Tawn J, Blazkova S, Naden P (1999) Flood frequency estimation by continuous simulation for a gauged upland catchment (with uncertainty). J Hydrol 219:169–187

Chang DY (1996) Applications of the extent analysis method on fuzzy AHP. Eur J Oper Res 95:649–655

Deng H (1999) Multicriteria analysis with fuzzy pairwise comparison. Int J Approx Reason 21:215–231

Goicoechea A, Hansen DR, Duckstein L (1982) Multiobjective decision analysis with engineering and business applications. Wiley, New York, NY

Hwang C-L, Yoon K (1981) Multiple attribute decision making: methods and applications. A state of the art survey. Springer, Berlin, 259 pp

Kachroo RK (1992) Storage required to augment low flows – a regional study. Hydrol Sci J – J des Sci Hydrologiques 37:247–261

Klein B, Pahlow M, Hundecha Y, Schumann A (2010) Probability analysis of hydrological loads for the design of flood control systems using copulas. ASCE J Hydrol Eng 15(5):360–369

Klir (1999) Uncertainty and information measures for imprecise probabilities: an overview 1st International Symposium on Imprecise Probabilities and Their Applications, Ghent, 29 June–July 2

Klir GJ, Smith RM (2001) On measuring uncertainty and uncertainty-based information: recent developments. Ann Math Artif Intell 32:5–33

Lamb R, Kay AL (2004) Confidence intervals for a spatially generalized, continuous simulation flood frequency model for Great Britain. Water Resour Res 40(7):W07501. doi:10.1029/2003WR002428

Liou TS, Wang MJ (1992) Ranking fuzzy numbers with integral value. Fuzzy Sets Syst 50:247–255

McMillan HK, Brasington J (2008) End-to-end flood risk assessment: a coupled model cascade with uncertainty estimation. Water Resour Res 44:W03419

Plate EJ, Meon G (1988) Stochastic aspects of dam safety analysis. Proceedings of the JSCE No. 393/II-9 (Hydraulic and Sanity Eng.), pp 1–8

Saaty TL (1977) A scaling method for priorities in hierarchical structures. J Math Psychol 40: 234–281

Saaty TL (1980) The analytic hierarchy process. McGraw-Hill, New York, NY

Srdjevic B, Medeiros Y (2007) Fuzzy AHP assessment of water management plans. Water Resour Manage 22(7):877–894

Taslicali AK, Ercan S (2006) The analytic hierarchy & the analytic network processes in multicriteria decision making: a comparative study. J Aeronaut Space Technol 2(4):55–65

Walley P (1991) Statistical reasoning with imprecise probabilities. Chapman and Hall, London

Zadeh LA (1965) Fuzzy sets. Inform Content 8:338–353

Zadeh LA (2005) Toward a generalised theory of uncertainty (GTU) – an outline. Inform Sci. 172:1–40

Index

A.H. Schumann (ed.), *Flood Risk Assessment and Management*,
DOI 10.1007/978-90-481-9917-4, © Springer Science+Business Media B.V. 2011

MIX
Papier aus verantwortungsvollen Quellen
Paper from responsible sources
FSC® C105338

If you have any concerns about our products,
you can contact us on
ProductSafety@springernature.com

In case Publisher is established outside the EU,
the EU authorized representative is:
Springer Nature Customer Service Center GmbH
Europaplatz 3, 69115 Heidelberg, Germany

Printed by Libri Plureos GmbH
in Hamburg, Germany